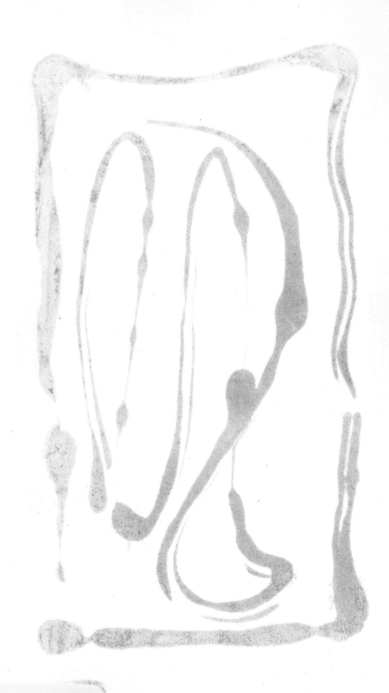

ESSENTIAL
ORGANIC CHEMISTRY

for Students of the Life Sciences

ESSENTIAL
ORGANIC CHEMISTRY
for Students of the Life Sciences

A.P. Ryles
Skelmersdale College, Northway, Lancashire

K. Smith and R.S. Ward,
Department of Chemistry, University College of Swansea, West Glamorgan

JOHN WILEY & SONS
Chichester · New York · Brisbane · Toronto

Copyright © 1980 by John Wiley & Sons, Ltd.

Library of Congress Cataloging in Publication Data:

Ryles, A. P.
  Essential organic chemistry for students of the life sciences.
  1. Chemistry, Organic. I. Smith, K., joint author.
II. Ward, R. S., joint author. III. Title.
QD251.2.R94   547   78-31504
ISBN 0 471 27582 4
ISBN 0 471 27581 6 pbk.

Typeset by Preface Ltd., Salisbury, Wilts. and printed by
Page Bros (Norwich) Ltd., Norwich.

CONTENTS

Preface xi

Abbreviations Used xiii

1 Introduction 1
  1.1 What is Organic Chemistry? 1
  1.2 Historical Perspective 1
  1.3 Bonding and Shape in Organic Compounds 3
  1.4 Molecular Orbital Theory of Bonding in Organic Chemistry . 5
  1.5 Hybridization and Hybrid Orbitals 7
  1.6 Polar Effects in Organic Compounds 10
  1.7 Functional Groups 12
  1.8 Mechanism of Organic Reactions 14
  1.9 Reactivity of Organic Compounds 16
  1.10 Classification of Organic Reactions. 17
  1.11 Physical Properties of Organic Compounds 17
  1.12 Sources of Organic Compounds 19

2 Stereochemistry 21
  2.1 Stereoisomerism 21
  2.2 Geometrical Isomerism 21
  2.3 Optical Isomerism 23
  2.4 Elements of Symmetry 24
  2.5 Optical Activity 25
  2.6 Fischer Projection Formulas 26
  2.7 Molecules Containing more than One Asymmetric Centre . . 28
  2.8 Optical Isomerism without an Asymmetric Atom 29
  2.9 Nomenclature of Stereoisomers 30
  2.10 Resolution of Enantiomers 33
  2.11 Asymmetric Synthesis 34
  2.12 Conformations of Molecules 35
  2.13 Conformations of Cyclohexane Derivatives 36

3 Techniques of Organic Chemistry 40
  3.1 Filtration and Crystallization 40
  3.2 Distillation 41
  3.3 Solvent Extraction 42
  3.4 Chromatography 43
  3.5 Electrophoresis. 46

3.6   Elemental Analysis 46
3.7   Mass Spectrometry 48
3.8   Ultra violet Spectroscopy 49
3.9   Infrared Spectroscopy 50
3.10 Nuclear Magnetic Resonance Spectroscopy 52
3.11 Diffraction Techniques 55

4 Aliphatic Hydrocarbons 59
4.1   Structures of Hydrocarbons 59
4.2   Physical Properties 60
4.3   Bonding and Reactivity 60
4.4   Reactions with Halogens 62
4.5   Reactions with Acids 66
4.6   Reactions with Borane 68
4.7   Oxidation Reactions 69
4.8   Hydrogenation of Alkenes and Alkynes 71
4.9   Polymerization of Alkenes 72
4.10 Organometallic Derivatives of Alkynes 73
4.11 The Petroleum Industry 74
4.12 Preparation of Aliphatic Hydrocarbons 75

5 Aromatic Compounds and Aromaticity 79
5.1   The Structure of Benzene 79
5.2   Aromaticity 82
5.3   Resonance Energy 84
5.4   Electrophilic Substitution 84
5.5   Substitution Effects in Electrophilic Substitution 90
5.6   Polynuclear Aromatic Hydrocarbons 95
5.7   Side Chain Reactions of Alkybenzenes 96
5.8   Preparation of Aromatic Hydrocarbons 97

6 Organic Halides, Alcohols, Phenols, Thiols, and Ethers 100
6.1   Occurrence and Uses 100
6.2   Physical Properties 102
6.3   Nucleophilic Substitution 103
6.4   Examples of Nucleophilic Substitution Reactions 105
6.5   Elimination Reactions 108
6.6   Organometallic Compounds Derived from Organic Halides . 110
6.7   Acidity of Alcohols, Phenols, and Thiols 112
6.8   Esterification of Alcohols and Phenols 112
6.9   Oxidation of Alcohols, Phenols, and Thiols 113
6.10 Electrophilic Substitution Reactions of Phenols 115
6.11 Reactions of Epoxides 116
6.12 Preparations 117

7 Organic Nitrogen Compounds 121
   7.1   General Characteristics of Amines 122
   7.2   Basicity of Amines 123
   7.3   Reactions of Amines 124
   7.4   Imines and Related Compounds 130
   7.5   Nitriles 130
   7.6   Amides 131
   7.7   Nitro Compounds 132
   7.8   Diazonium Salts 133
   7.9   Heterocyclic Nitrogen Compounds 136
   7.10 Preparation of Organic Nitrogen Compounds 138

8 Carbonyl Compounds 143
   8.1   The Carbonyl Group 144
   8.2   Physical Properties of Carbonyl Compounds 145
   8.3   Keto–Enol Tautomerism 147
   8.4   Nucleophilic Reactions of Carbonyl Compounds 148
   8.5   Acidity of Carboxylic Acids 158
   8.6   Acidity of α-Hydrogens in Carbonyl Compounds 160
   8.7   Oxidation and Reduction of Carbonyl Compounds 163
   8.8   Preparation of Carbonyl Compounds 166
   8.9   Sulphonic Acids and their Derivatives 169

9 Bifunctional Molecules 173
   9.1   Compounds Containing Cumulated Double Bonds 175
   9.2   Compounds Containing Conjugated Double Bonds 176
   9.3   Compounds Containing a Lone Pair of Electrons Conjugated
       to a Double Bond 184
   9.4   Compounds Containing Non-conjugated Double Bonds . . . 185
   9.5   Compounds Containing other Combinations of Functional
       Groups 187
   9.6   Cyclization of Bifunctional Molecules 189
   9.7   Polymers Derived from Bifunctional Molecules 191

10 Fats, Oils, Waxes, and Detergents 196
   10.1   Occurrence and Composition of Fats and Oils 197
   10.2   Analysis of Fats and Oils 199
   10.3   Hardening of Vegetable Oils; Margarine 199
   10.4   Drying Oils; Paints and Lacquers 200
   10.5   Soaps and Detergents 201
   10.6   Waxes 202

11 Carbohydrates 204
   11.1   Monosaccharides 205
   11.2   Glycosides 209

viii

11.3   Reactions of Monosaccharides 211
11.4   Disaccharides. 215
11.5   Polysaccharides 217

12 Proteins and Nucleic Acids 221
12.1   Structure of Amino Acids 221
12.2   Properties of Amino Acids 224
12.3   Reactions of Amino Acids 225
12.4   Synthesis of Amino Acids 225
12.5   Peptides 226
12.6   Amino Acid Composition of Peptides and Proteins 227
12.7   Primary Structure of Proteins 228
12.8   Gross Structure of Proteins 230
12.9   Structure and Reactivity of Enzymes 232
12.10 Synthesis of Peptides and Proteins 233
12.11 Chemical Components of Nucleic Acids 235
12.12 Primary Structure of Nucleic Acids 236
12.13 Secondary Structure of DNA 238
12.14 Replication of DNA 239
12.15 Biosynthesis of Proteins—The Role of RNA 239
12.16 Viruses 240

13 Tetrapyrrolic Compounds: Photosynthesis and Respiration . . . 242
13.1   General Characteristics of Tetrapyrroles 242
13.2   Photosynthesis 243
13.3   Respiration 248

14 Other Physiologically Important Compounds 250
14.1   The Vitamins 250
14.2   Vitamin A 250
14.3   Vitamin B Complex 250
14.4   Vitamin C 254
14.5   Vitamin D 255
14.6   Vitamin E 256
14.7   Vitamin H 256
14.8   Vitamin K 256
14.9   Hormones 257
14.10 The Thyroid and Parathyroid Hormones 257
14.11 The Pancreatic Hormones 259
14.12 The Adrenal Hormones 259
14.13 The Pituitary Hormones 260
14.14 The Sex Hormones 260
14.15 Medicinal Compounds 261

15 Metabolism and Biosynthesis 264
   15.1   Energy in Living Systems 264
   15.2   Glycolysis 266
   15.3   The Hexose Monophosphate Shunt 267
   15.4   The Citric Acid Cycle 269
   15.5   Fat Metabolism 269
   15.6   Polyketides 272
   15.7   Terpenes and Steroids 273
   15.8   Amino Acid Metabolism 276
   15.9   Flavonoids, Coumarins, Lignans, and Lignin 278
   15.10 Alkaloids 279
   15.11 Penicillins and Cephalosporins 280
   15.12 Tetrapyrrolic Compounds 281

Appendix 284

Index 293

Preface

In recent years there has been an increase in the number of students studying subjects such as biology and medicine which require a working knowledge of organic chemistry. Unfortunately such students are confronted with a formidable array of excellent, detailed textbooks written for honours chemistry students, but relatively few which fulfil their own particular needs. In writing this book it has been our aim to cater specifically for the needs of students of the life sciences.

When planning the book we have adopted the following principles:

1. The text has been kept as simple as possible consistent with producing a clear, coherent account. We have tried to avoid many peripheral topics which are of interest to organic chemists but are unlikely to help the student's understanding of biological phenomena.

2. Wherever possible we have attempted to 'demystify' biochemical reactions by showing that many apparently complex reactions which occur in living systems have parallels in the reactions which organic chemists carry out in the laboratory. We have also attempted to demonstrate the many similarities between the reactions of different classes of organic compounds by extensive cross-referencing. The early parts of the book deal with the basic principles and the reactions of functional groups, whereas the latter sections treat most of the important biological compounds from a chemical point of view.

3. In dealing with the thorny question of nomenclature we have adopted a pragmatic approach. There are two systems in common use: the systematic nomenclature recommended by IUPAC and a host of trivial names still commonly used by practising chemists. We have attempted to reconcile these two approaches by using the name which is in most common use but giving the alternative name in parenthesis when the compound is first mentioned. Afterwards only the more common name is used. For the student who requires additional information an appendix on nomenclature has been included at the end of the book.

4. Study questions have been included wherever possible in the text so that the reader can test his understanding of a topic before passing on to the next.

We should like to express our thanks to Mrs A. Franklin and Mrs L. E. Smith for typing the manuscript and to Professor H. C. Brown for the loan of a typewriter.

Another volume containing worked Examples and Problems with Solutions supplementary to this text, will be published shortly.

October 1978   A.P.R.   K.S.   R.S.W.

Abbreviations Used

Organic or other groups

R	an organic group (e.g. an alkyl or aryl group)
Ar	an organic group (e.g. an aryl group)
Me	a methyl group, CH_3—
Et	an ethyl group, C_2H_5—
Pr	a propyl group, C_3H_7—
Bu	a butyl group, C_4H_9—
Ac	an acetyl group, CH_3CO—
P	a phosphate group, HO_3P—

Important biochemical compounds

NAD^+	nicotinimide adenine dinucleotide
NADH	reduced NAD^+
$NADP^+$	nicotinimide adenine dinucleotide phosphate
NADPH	reduced $NADP^+$
ADP	adenosine diphosphate
ATP	adenosine triphosphate
FAD	flavin adenine dinucleotide
FMN	flavin mononucleotide
Chl	chlorophyll
Chl^*	excited chlorophyll

Prefixes of names

R
S
D
L
Z } indicate stereochemistry of compounds—see Chapter 2 for details
E
cis
trans

Units

m	metre
cm	centimetre = 10^{-2}m
mm	millimetre = 10^{-3}m
nm	nanometre = 10^{-9}m
pm	picometre = 10^{-12}m
Å	Ångstrom unit = 10^{-10}m
mol	mole
kJ	kilojoule = 10^3 joules
p.p.m.	parts per million

Miscellaneous Symbols

k	rate constant
[ ]	molar concentration
K_a	acid dissociation constant
pK_a	$-\log_{10}K_a$
K_b	base dissociation constant
pK_b	$-\log_{10}K_b$
$[\alpha]$	specific optical rotation
⌢	shift of two electrons
$\delta+$	small positive charge
$\delta-$	small negative charge

CHAPTER 1

Introduction

1.1 What is Organic Chemistry?

Organic chemistry is the *study of the compounds of carbon*. The name 'organic' (living) derives from earlier times when all compounds to which the name was applied were obtained from plant or animal sources. Nowadays many compounds are referred to as organic even though they have no connection with living things; indeed in many cases they could not exist in a biological system. For this reason the best definition of organic chemistry is the broad one given above.

It may seem odd to single out one element, carbon, and denote a whole branch of chemistry to its study, but *many more compounds are known that contain carbon than do not*.

The reasons that carbon forms such a large number of compounds are complex, but three features are of particular importance: (i) not only do carbon atoms form strong bonds with other atoms, but they also form strong bonds with each other, which can result in the linking together of many carbon atoms to form chains or rings; (ii) carbon is tetravalent, i.e. it forms bonds with up to four other atoms, which allows branching of the carbon chains and the formation of many complex structures; (iii) carbon has the ability to form multiple bonds (double and triple bonds, see Section 1.5) with itself and other elements.

The varied nature of organic compounds is illustrated by considering some of the materials which are important in everyday life. They range in complexity from very simple compounds such as methane (the major constituent of natural gas) and ethanol (alcohol), through compounds of intermediate molecular size and complexity such as morphine (an important analgesic) and chlorophyll (the green pigment of plants), to high molecular mass compounds such as polythene, proteins, and cellulose. In addition to their role in the maintenance of life processes, organic compounds are also involved in many other aspects of man's existence.

1.2 Historical perspective

Some organic compounds were extracted or prepared by the ancient Egyptians, Phoenicians, and Romans. Dyestuffs such as indigo and Tyrian purple, alcohol, vinegar, and common soap are all products with a long history. But not until the latter part of the eighteenth century did organic chemistry as we now understand it have its beginnings. At that time the Swedish chemist C.W. Scheele isolated a number of pure substances, such as citric acid from lemons and lactic acid from milk, and the French chemist A. Lavoisier recognized that these products from natural systems consisted mainly of different combinations of just a small group of elements (C, H, N, and O). In 1807, J.J.F. von Berzelius, a Swede, coined the term 'organic' to apply to such compounds derived from living or once-living systems.

At that time it was generally believed that organic compounds possessed a *vital force* derived from the living system from which the product was obtained. Although organic compounds might be interconverted while retaining the vital force, no organic compound could be produced from purely mineral (or inorganic) precursors. Then in 1828 Wöhler prepared urea (acknowledged as an organic compound, obtainable from urine) by heating ammonium cyanate (accepted as inorganic). A decade later, after many similar preparations, the vital force theory was finally abandoned.

By the mid-nineteenth century there were better methods of analysis, and a greater understanding of how organic compounds could be interconverted, but as yet there was little idea about *structure*. In 1858, Kekulé in Germany and Couper in Paris independently suggested that the elements C, H, N, O, and Cl exhibit characteristic *valencies* or bonding abilities (4, 1, 3, 2, and 1 respectively) in organic compounds. To obtain a structure one merely had to draw lines, representing bonds, connecting atoms in such a way as to satisfy all of their valencies. Unique structures (Figure 1.1) could then be drawn for simple compounds such as ethane (C_2H_6) and methanol (CH_4O), and thus the occurrence of two distinct compounds with the formula C_2H_6O could be understood in terms of two *isomers*, ethanol and dimethyl ether. Isomers are compounds with the same molecular formula but different structures.

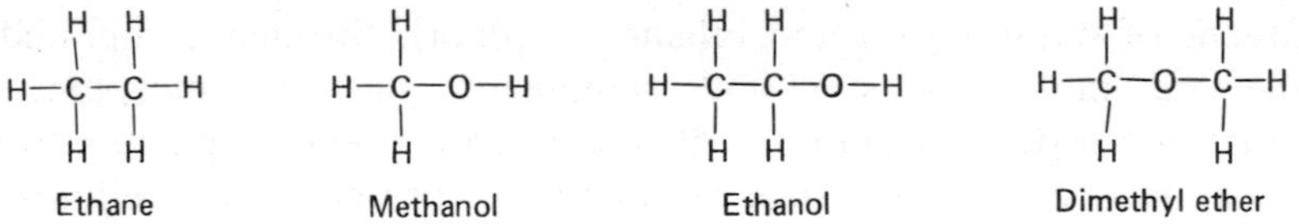

Figure 1.1 Structures for some simple organic compounds

Question 1.1   Draw the structures of (a) two isomers having the molecular formula C_3H_7Cl; (b) three isomers having the molecular formula C_5H_{12}.

In 1874 van't Hoff and Le Bel independently proposed that four 'bonds'

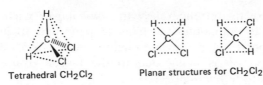

Tetrahedral CH$_2$Cl$_2$ Planar structures for CH$_2$Cl$_2$

Figure 1.2 Tetrahedral and planar structures for CH$_2$Cl$_2$

attached to a carbon atom did not all lie in the same plane, but were disposed so as to point to the corners of a tetrahedron with carbon at its centre. This could explain why only one isomer of CH$_2$Cl$_2$ was known, whereas two planar structures would be expected (Figure 1.2). (To illustrate the tetrahedral arrangement of groups around the central carbon it is sometimes convenient to draw wedge-shaped lines for bonds projecting above the plane of the paper and hatched lines for those projecting below.)

After the discovery of the electron in 1897, electronic and atomic theory was applied to organic chemistry with great success. As quantum mechanics was developed during the 1930's this too was applied to organic chemistry, and it aided an understanding of the nature of the bonds between atoms. These developments of the last century form the basis of our present understanding of organic chemistry and are dealt with in greater detail in the following sections.

1.3 Bonding and shape in organic compounds

Organic compounds are generally three-dimensional in the sense that not all of the constituent atoms lie in the same plane. This comes about because *saturated* carbon atoms (i.e. ones bonded to four other atoms) are substituted *tetrahedrally* (see Section 1.2). Although early chemists relied upon observations such as isomer numbers in order to deduce that carbon was tetrahedrally substituted, nowadays techniques such as X-ray diffraction (for crystalline solids) and electron diffraction (for gases) can be used accurately to pinpoint the atomic nuclei within a molecule (Chapter 3). These techniques confirm the basic tetrahedral arrangement about carbon atoms in a whole range of compounds.

Only symmetrical compounds such as methane and tetrachloromethane

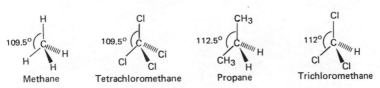

Methane   Tetrachloromethane   Propane   Trichloromethane

Figure 1.3 Bond angles in some simple tetrahedral structures

(Figure 1.3) are exactly tetrahedral, with *bond angles* each 109.5° (the tetrahedral angle). In compounds such as propane and trichloromethane (chloroform), the bond angles deviate slightly from 109.5° (Figure 1.3.). This can be thought of as being due to the bigger groups forcing an expansion of some of the bond angles.

In addition to bond angles, *bond lengths* (i.e. the distances between 'bonded' atoms) may be determined from diffraction or other data. Bond lengths depend upon the atoms which are bonded and the nature of the bond, but for the same type of bond between the same atoms they are approximately constant. For example, the lengths of simple C—C, C—H, and C—Cl bonds are 154 pm, 109 pm, and 178 pm respectively (bond lengths are sometimes expressed in Å where 1 Å$=10^{-10}$ m$=100$ pm).

With a knowledge only of bond angles and bond lengths, it is possible to build models of relatively complicated organic molecules, and several kinds of models are commercially available. These are essentially of three basic types (Figure 1.4) — ball and stick models, which are visually effective and useful for class demonstrations; more sophisticated models with accurate bond lengths and angles, which can be used to make predictions about molecular shape; and space-filling models, which indicate the space occupied by the atoms in a molecule and can thus be of use in seeing if atoms interfere or interact with each other.

Using models of these types it is easy to see (i) that most organic

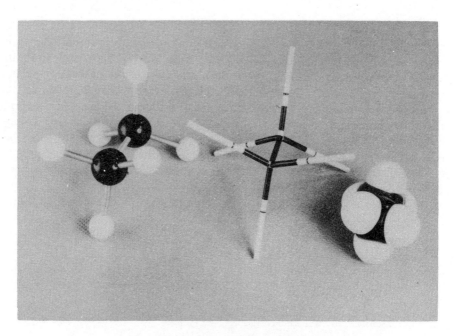

Figure 1.4 Three types of model of ethane

compounds are not flat, but occupy three dimensions; (ii) that the C—C—C chain in propane is bent, not straight as often drawn in shorthand ($CH_3CH_2CH_3$); (iii) that rings are not necessarily flat; and (iv) that some compounds are much more likely than others to be stable. (As a general rule, if a model is difficult to make without appreciably distorting bond angles or lengths the molecule is probably unstable.) More detailed treatment of the shapes or *stereochemistry* of molecules is given in Chapter 2, but it is important first to have a better understanding of the nature of chemical bonds.

1.4 Molecular orbital theory of bonding in organic chemistry

There are several ways in which the structures of simple molecules such as methane can be rationalized, but it is generally accepted that the theory which offers most help in understanding the structures of the widest range of chemical compounds is the *molecular orbital* (MO) theory. To understand this theory we must briefly consider the electronic structures of atoms.

The theory of *quantum mechanics* enables the motion of an electron to be described in terms of its energy, albeit that the mathematical solutions are only approximate. The region of space in which an electron is most likely to be found is called an *orbital*, and one can draw a contour line within which there is, say, a 95% probability of finding the electron at any time. Alternatively, an orbital can be looked upon as the region of space within which the *electron density* is greatest. In this way orbitals may be said to have both size and shape.

In atoms there are a variety of *atomic orbitals* designated 1s, 2s, $2p_x$, $2p_y$, $2p_z$, 3s, etc. Details of these designations need not concern us, but it may be helpful to consider the initial number as roughly indicative of energy, the following letter as indicative of shape, and the suffix as indicative or orientation.

The shapes of a 1s orbital and the three 2p orbitals are shown in Figure 1.5. Notice that the three 2p orbitals are orientated in such a way as to

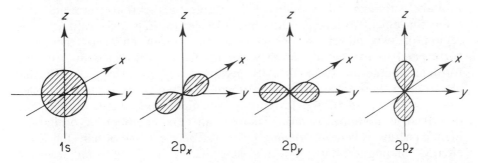

Figure 1.5 Shapes of atomic orbitals

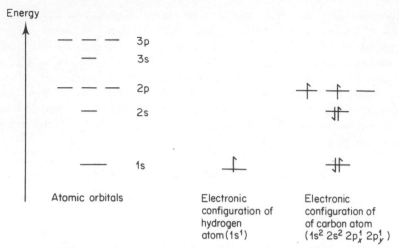

Figure 1.6 Relative energies of various atomic orbitals

point mutually at right angles. The 2s and 3s orbitals have the same shape as the 1s orbital, though they are bigger. Similarly 3p is the same shape as 2p.

Figure 1.6 indicates the relative energies of the various atomic orbitals. Only two electrons are permitted to occupy any one orbital, and to do so they must have opposite spins (i.e. their spin quantum numbers are of opposite sign). The electrons of an atom occupy the orbitals of lowest energy. If several orbitals are degenerate (i.e. have the same energy), and if there are insufficient electrons to fill them, then the electrons preferentially occupy different orbitals and have the same spin (Hund's rule). The electronic configuration of any atom can then be determined, and this is illustrated in Figure 1.6 for hydrogen (atomic number 1) and carbon (atomic number 6).

In molecules the various atomic orbitals of the constituent atoms interact with each other to form a new set of orbitals called *molecular orbitals*. The total number of molecular orbitals is equal to the total number of atomic orbitals. Although it is not strictly justifiable on theoretical grounds, it is often convenient to imagine that the molecular orbitals are localized in the region of two nuclei. When they are filled with electrons they then correspond to the 'bonds' as they are normally represented in structural formulas. Some of these orbitals occupy a region of space such that electrons in them experience the attraction of two or more atomic nuclei. This is an energetically favourable situation and these orbitals are described as *bonding orbitals*. There is an equal number of orbitals of higher energy, called *antibonding orbitals*. The electrons occupy the lower energy bonding orbitals, thus releasing energy in going from atoms to molecules. The bonding orbital for the hydrogen molecule and the atomic orbitals for the two hydrogen atoms are illustrated in Figure 1.7.

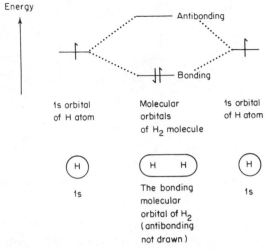

Figure 1.7 Atomic and molecular orbitals of hydrogen

Bonds formed in this way—by sharing of two electrons in a single bonding orbital—are called *covalent bonds*, as distinct from ionic bonds such as are found in many inorganic solids, e.g. NaCl. Bonds in which the electron density is highest along the line directly joining the two nuclei and spherically symmetrical with respect to this line are described as *sigma* (σ) bonds. The H—H bond in the hydrogen molecule is a covalent σ bond, and σ bonds are also found in all organic compounds, though sometimes accompanied by other types of bond, as we shall see later.

It is easy to deduce the shape of the H_2 bonding molecular orbital from a coalescence of the shapes of the two interacting atomic orbitals, but for more complex molecules many of the molecular orbitals are much harder to envisage because they are spread over the whole molecule. Localized molecular orbitals (between pairs of atoms) are much easier to envisage, and for this purpose it is convenient to invent an imaginary set of atomic orbitals called *hybrid orbitals*.

1.5 Hybridization and hybrid orbitals

The electronic configuration of the carbon atom is $1s^2 2s^2 2p_x^1 2p_y^1$. Interaction of these atomic orbitals with hydrogen 1s might be expected to give CH_2 with an H—C—H bond angle of 90° since only the $2p_x$ and $2p_y$ atomic orbitals are singly occupied and these are perpendicular. However, CH_4 is much more stable than CH_2, which can be rationalized by the transfer of an electron from the 2s to the $2p_z$ orbital. This requires the input of a little energy (see Figure 1.6), but it is more than compensated for by the energy *released* on forming two extra bonds.

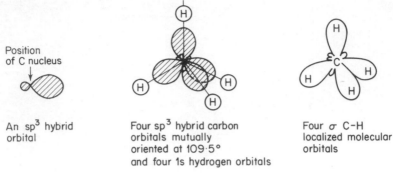

Position of C nucleus

An sp³ hybrid orbital

Four sp³ hybrid carbon orbitals mutually oriented at 109·5° and four 1s hydrogen orbitals

Four σ C–H localized molecular orbitals

Figure 1.8 Prediction of the shape of methane by the use of hybrid orbitals

However, simple interaction of the four atomic orbitals (2s, $2p_x$, $2p_y$, $2p_z$) with four hydrogen 1s orbitals would be expected to give three mutually perpendicular C—H bonds with the fourth oriented arbitrarily. However, if the four atomic orbitals of the carbon valence shell are mixed (*hybridized*) mathematically, four identical orbitals mutually oriented at 109.5° (Figure 1.8) can be generated. These orbitals are referred to as *sp³ hybrid orbitals* because they arise from mixing *one* 2s and *three* 2p orbitals. These hybrid orbitals can then form localized bonding and antibonding orbitals by interaction with atomic orbitals from other atoms, as described for the hydrogen molecule. Figure 1.8 shows how hybrid orbitals can be used for predicting the shape of a molecule like methane. Bond angles are correctly predicted, and the generated molecular bonding orbitals are localized between pairs of bonded atoms.

Four sp³ hybrid orbitals can be similarly generated for nitrogen and oxygen atoms, but in these cases only three or two covalent bonds are formed with hydrogen, giving ammonia and water respectively. On the basis of the hybridized orbital concept the angles H—N—H in NH₃ and H—O—H in water should be 109.5°. The actual angles are 107° and 104.5°. The differences can be explained by assuming that electrons in orbitals which are not associated with a second nucleus (i.e. *non-bonding* or *lone pair* electrons) occupy more space, thus compressing the bonding orbitals (Figure 1.9).

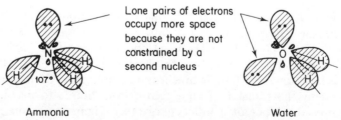

Lone pairs of electrons occupy more space because they are not constrained by a second nucleus

Ammonia

Water

Figure 1.9 Space occupation of lone pairs of electrons

There are other possible ways of 'hybridizing' carbon atomic orbitals. For example, the 2s and two of the 2p orbitals can be mixed to produce sp² *hybrid orbitals* which all lie in one plane at 120° to each other. The remaining 2p orbital is unhybridized and perpendicular to the plane of the sp² orbitals. The shapes of the derived orbitals are indicated in Figure 1.10, which also shows how two such hybridized carbon atoms combine with each other and with four hydrogen atoms to form a σ - bonded C_2H_4 framework. The remaining two 2p orbitals can be oriented in such a way as to allow lateral overlap between them. This overlap produces a bonding orbital which has two lobes, one above and one below the plane of the six atoms. The maximum electron density in this orbital is outside the line joining the two carbon nuclei so it is not a σ orbital, but is called a pi (π) orbital (Figure 1.10). The structure thus derived for the molecule C_2H_4 is very similar to the true structure of ethene (ethylene; Figure 1.10), the slight deviations from 120° in bond angles being attributable to the CH_2 groups being more bulky than the H atoms and so compressing the H—C—H angle somewhat.

The postulation of π bonds helps in understanding the chemistry of alkenes (Chapter 4), arenes (Chapter 5), and carbonyl compounds (Chapter 8).

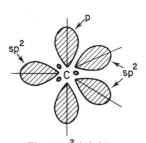

Three sp² hybrid
orbitals and one
p orbital on carbon

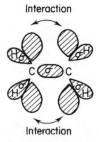

σ framework of C_2H_4
showing two unhybridized
2p orbitals, one on each carbon

C_2H_4 σ framework
represented by lines and
a π bond (two lobes)
formed by overlap of
p orbitals

Structure of ethene; the double
bond represents one σ and
one π bond

Figure 1.10 Bonding in ethene

Question 1.2 Predict the approximate C—C—O bond angles in the following molecules: (a) CH$_3$CH$_2$OH (ethanol);

(b) CH$_3$—C$\overset{O}{\underset{H}{\diagdown}}$ (ethanal);

(c) CH$_2$=C$\overset{OCH_3}{\underset{H}{\diagup}}$ (methoxyethene).

Yet a third way of hybridizing carbon orbitals is to mix the 2s and one of the 2p orbitals to produce two sp hybrid orbitals and leave two 2p orbitals unhybridized (Figure 1.11). The two sp hybrid orbitals are oriented at 180° to each other and at right angles to the unhybridized p orbitals. In C$_2$H$_2$ (ethyne) the sp hybridized orbitals of the two carbon atoms and the 1s orbitals of two hydrogen atoms combine to form the σ bonds, and the remaining p orbitals overlap to form two π bonds (Figure 1.11). This helps in understanding the chemistry of alkynes (Chapter 4) and nitriles (Chapter 7).

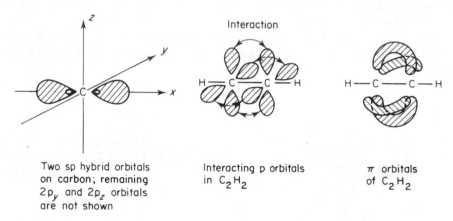

Two sp hybrid orbitals on carbon; remaining 2p$_y$ and 2p$_z$ orbitals are not shown

Interacting p orbitals in C$_2$H$_2$

π orbitals of C$_2$H$_2$

Figure 1.11 Bonding in ethyne

1.6 Polar effects in organic compounds

The elements which occur in organic compounds do not all have the same *electronegativity* (attractiveness to electrons). Those elements at the right hand side of the periodic table are more electronegative than those at the left. Although carbon and hydrogen have about the same electronegativities (*c.* 2.5 and 2.2 respectively), oxygen (3.5), nitrogen (3.0), and the halogens (F, 4.0; Cl, 3.0) are all significantly more electronegative than carbon. Thus, the electrons in a C—F, C—O, C—N, or C—Cl bond are not equally attracted by both of the bonded atoms, but are attracted more by the more electronegative element. This means that the more electronegative element, having attracted rather more than its share of the available electron density, will be partially negatively charged, and the carbon atom partially positively charged. The bond is said to be

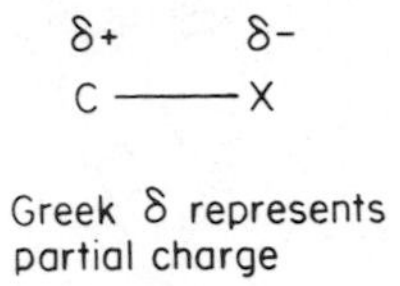

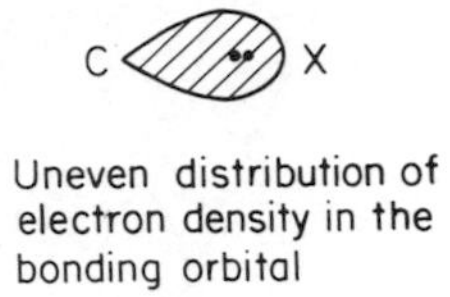

Figure 1.12 Bond polarization

polarized (i.e. there is charge separation, producing a *dipole*), and this is often represented as indicated in Figure 1.12. The effect that an atom or group of atoms has on the electron density at a neighbouring site by virtue of its electronegativity is called the *inductive effect*. Inductive effects are transmitted through the σ bonds of a molecule.

π bonds are more readily polarized than σ bonds because the electrons are less tightly held by the nuclei. Thus, more pronounced electronic effects are observed in molecules containing π bonds. In systems containing alternating single and double bonds (i.e. *conjugated* systems) these polar effects can be transmitted through the entire system. The transmission of polar effects through π bonds is called the *mesomeric effect*.

In order to be able to represent the electronic distributions in molecules containing π bonds it is often convenient to draw two or more extreme structures (*canonical forms*), each of which shows some of the features of the molecule. The real structure is said to be a *resonance hybrid* and is best thought of as being a blend of the canonical forms. The use of this concept to represent a carbonyl group and a conjugated carbonyl compound is illustrated in Figure 1.13 (double headed arrows are used to indicate the relationship between canonical forms).

In molecular orbital terms, the mesomeric effect is a reflection of the fact that the p atomic orbitals which go to make up the π bond orbitals can overlap all together to form a lower energy orbital (Figure 1.14). When an atom (represented by 'X' in Figure 1.14(b)) bearing a lone pair of electrons or an unoccupied orbital is directly attached to a π bond, then it too can adopt sp^2 hybridization and give rise to an extended π system.

The inductive and mesomeric effects of a substituent group are not necessarily in the same direction. For example, if an electronegative atom possessing a lone pair of electrons (e.g. O, N, Cl) is attached to a π system then the extended overlap results in a net flow of electron density from the electronegative atom into the π bond. This can be represented by a

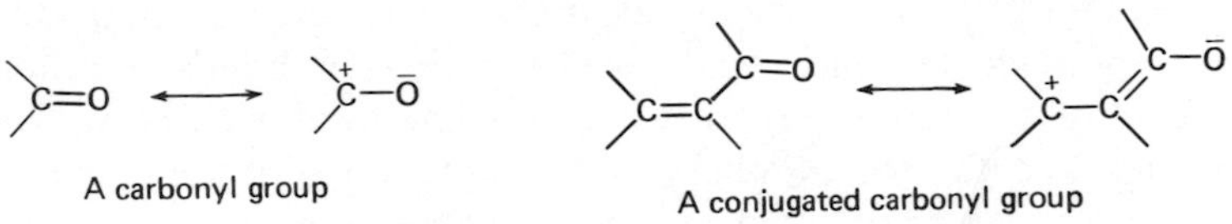

A carbonyl group

A conjugated carbonyl group

Figure 1.13 Resonance between canonical forms

12

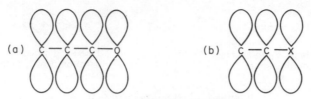

Figure 1.14 Extended overlap of p orbitals in a conjugated π system. (X is an atom with a lone pair of electrons or an unoccupied orbital.)

Figure 1.15 Resonance hybrid for an alkoxyalkene (enol ether)

resonance hybrid such as that depicted in Figure 1.15 (two dots are used to represent a lone pair of electrons). The inductive effect leads to polarization of the C—O σ bond in the opposite direction, but in such cases the mesomeric effect usually overrides the inductive effect.

The presence of an extended π system which may be represented by two or more canonical forms is often a stabilizing influence, especially for charged intermediates (Section 1.8). The charge, instead of being localized on a particular atom, can be spread out over several atoms, thus reducing the charge at any one centre, which provides the extra stability. Thus, the allyl cation, or indeed the anion, is more stable than the propyl cation, or anion, respectively. This is rationalized for the allyl cation as shown in Figure 1.16.

$$CH_2{=}CH{-}\overset{+}{C}H_2 \longleftrightarrow \overset{+}{C}H_2{-}CH{=}CH_2 \qquad CH_3{-}CH_2{-}\overset{+}{C}H_2$$

Allyl cation   Propyl cation

Figure 1.16 Increase of cation stability because of resonance

Question 1.3 Draw canonical forms for each of the following species: (a) allyl anion ($CH_2{=}CH{-}\overline{C}H_2$); (b) chloroethene ($CH_2{=}CHCl$).

1.7 Functional groups

It is convenient to group organic compounds in such a way that compounds with similar chemical properties are treated together. Carbon–carbon single bonds and carbon–hydrogen bonds are non-polar and have only a relatively small effect on chemical reactivity. They are therefore ignored for the purposes of classification of organic compounds. However, introduction of other atoms or groups of atoms into a molecule has a signficant effect on the chemical properties of the molecule. It is possible to attribute characteristic chemical properties to the presence of a particular group of atoms, and such a group is therefore termed a *functional group*. Some of the more common functional groups are listed in Table 1.1. The chemistry of the individual groups is discussed in Chapters 4–8, and subsequent chapters deal with compounds containing several functional groups, as found in many natural products.

Table 1.1   Common functional groups in organic compounds

Group	Group name	Compound classes	Chapter
$\mathrm{C{=}C}$	Carbon–carbon double bond	Alkene	4
$-\mathrm{C{\equiv}C}-$	Carbon–carbon triple bond	Alkyne	4
$\mathrm{C-Cl}$ (Br, I)	Chloro (halogeno)	Alkyl and aryl halides	6
$\mathrm{C-OH}$	Hydroxyl	Alcohol, phenol	6
$\mathrm{C-O-C}$	Ether	Ether	6
$\mathrm{C-SH}$	Thiol	Thiol	6
$\mathrm{C-NH_2}$	Amino*	Amino	7
$-\mathrm{C{\equiv}N}$	Cyano	Nitrile	7
$\mathrm{C-NO_2}$	Nitro	Nitro compounds	7
$\mathrm{C{=}O}$	Carbonyl	Aldehyde, ketone	8
$-\mathrm{C}{<}^{O}_{OH}$	Carboxyl	Carboxylic acid	8
$-\mathrm{C}{<}^{O}_{O-C}$	Alkoxycarbonyl	Ester	8
$-\mathrm{C}{<}^{O}_{NH_2}$	Carboxamide*	Amide	7, 8

*Note that compounds containing organic groups instead of hydrogen attached to nitrogen are also common

14

1.8 Mechanism of organic reactions

Although almost all organic compounds are less stable than the mixtures of CO_2 and H_2O, etc., to which they are converted by combustion, they often remain unchanged in air for very long periods under normal conditions. Energy must be put into the system in order to bring about the reaction. This phenomenon is typical of most reactions and is illustrated by a plot of energy against a parameter called the *reaction coordinate*, the path from reactants to products (Figure 1.17(a)). The energy barrier which must be overcome before the reaction can proceed to products is called the *activation energy*, and the highest point on the plot is called the *transition state*. Sometimes a reaction involves an intermediate, i.e. it proceeds in two distinct steps, in which case there is a dip in the plot, and there are two transition states (Figure 1.17(b)).

In a two-step reaction one of the steps normally has a higher activation energy than the other. It therefore occurs more slowly (i.e. at a lower *rate*) and is called the *rate-determining step* (since any sequence of events can proceed only as fast as the slowest). Often intermediates are not isolable, having only a transient existence because of a low energy barrier for their conversion into products. Three of the most important types of reactive intermediates in organic chemistry are *carbonium ions* (sometimes called carbenium ions), *carbanions*, and *free radicals* (Figure 1.18).

The actual pathway by which a reaction proceeds is called the *reaction mechanism*, and a number of factors are important in determining such mechanisms. For example, different reactive intermediates are favoured by particular conditions (e.g. weak bonds, non-polar solvents, and ultra-violet light favour free radicals; polar solvents and highly polarized bonds favour ionic intermediates).

Charged reagents are attracted to centres of the opposite charge, and most reagents may be classified into one of two categories: (a) *nucleophiles*, which are either negatively charged (e.g. HO^-, CN^-) or

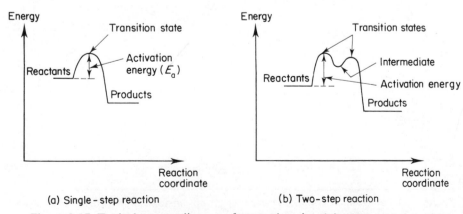

(a) Single-step reaction   (b) Two-step reaction

Figure 1.17 Typical energy diagrams for reactions involving one or two steps

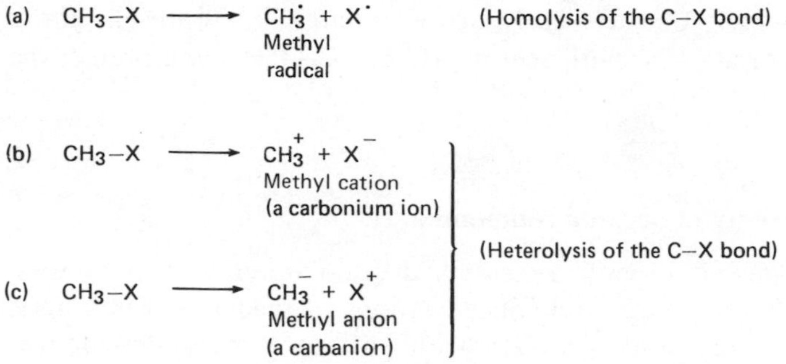

(a) $CH_3-X \longrightarrow CH_3^{\cdot} + X^{\cdot}$ (Homolysis of the C—X bond)
Methyl
radical

(b) $CH_3-X \longrightarrow CH_3^{+} + X^{-}$
Methyl cation
(a carbonium ion)

(Heterolysis of the C—X bond)

(c) $CH_3-X \longrightarrow CH_3^{-} + X^{+}$
Methyl anion
(a carbanion)

Figure 1.18 Different ways of cleaving a C—X bond

possess an electron-rich centre (e.g. the O atom in H_2O or the N in NH_3); (b) *electrophiles*, which possess a positive charge (e.g. H^+, NO_2^+) or an electron-deficient centre (e.g. the S atom in SO_3 or $SOCl_2$). Thus, nucleophiles attack electron-poor, partially positively charged centres in organic molecules (such as the carbon atom of a C—Cl bond), whereas electrophiles attack electron-rich sites (such as the π electron cloud in ethene). Much of organic chemistry can be understood in terms of this very simple principle.

It is convenient to have a shorthand notation to designate the direction of flow of electrons during an organic reaction, and a *curly arrow* is used for this purpose. The arrow represents the movement of *two* electrons. The tail of the arrow indicates the point at which the electrons reside prior to reaction, and the head of the arrow indicates their location after reaction. A simple illustration of this is provided by the reaction between a proton and hydroxide ion to give water (Figure 1.19). The electrons involved are initially a lone-pair on oxygen, formally represented by the negative charge, and these ultimately become the bonding electrons of the new O—H bond.

If an arrow head meets a carbon atom which already has an octet structure (i.e. a share in eight valency electrons), since the arrow brings two further electrons it is necessary also to displace two electrons from the atom. This is illustrated for the reaction of hydroxide ion with chloromethane (Figure 1.20)

$$HO^- \quad H^+ \longrightarrow HO-H$$

Figure 1.19 Flow of electrons in bond formation

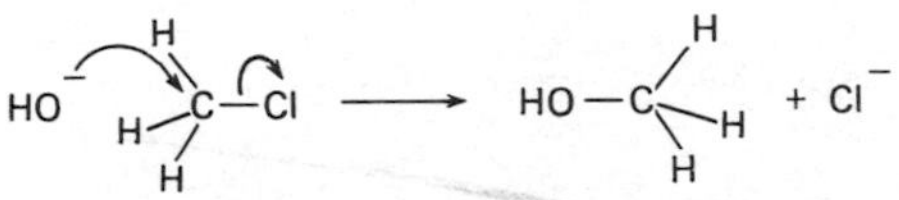

Figure 1.20 Flow of electrons in a substitution reaction

Question 1.4 Use the curly arrow notation to represent the reaction of cyanide anion ($^-$CN) with acetone (CH$_3$—C—CH$_3$) and predict the product.

$$\overset{\|}{\underset{O}{}}$$

1.9 Reactivity of organic compounds

The rates of organic reactions depend upon several factors, including solvent, temperature, and concentrations of reactants. Even under identical conditions reactions proceed with different rates depending upon the particular reagent and the structure of the *substrate* (i.e. the organic compound undergoing the reaction).

Information about reaction mechanisms can often be obtained by studying rates of reactions. Thus, under a given set of conditions, the rate of a single-step reaction of the type A + B → C is given by the expression

Rate = $k[A][B]$,

where k is a constant (dependent upon temperature and solvent) and [A] and [B] are the molar concentrations of A and B respectively. For a reaction involving more than one step, the rate equation is dependent upon the concentrations of the species involved in the rate-determining step.

For example, the reaction between chloromethane (CH$_3$Cl) and hydroxide ion (HO$^-$) could in principle proceed by two different mechanisms (see Chapter 6); (a) direct displacement of chloride ion by hydroxide ion (Figure 1.21(a)); or (b) slow dissociation of chloromethane to give methyl cation (CH$_3^+$) and chloride ion, followed by rapid reaction of the carbonium ion with hydroxide ion (Figure 1.21(b)).

These two mechanisms can be distinguished by their rate equations. For mechanism (a) the rate equation would be

Rate=$k[CH_3Cl][OH^-]$,

whereas for mechanism (b) it would be

Rate=$k[CH_3Cl]$.

In fact it is found that the rate of the reaction between chloromethane and hydroxide ion is proportional to the concentration of hydroxide ion as well

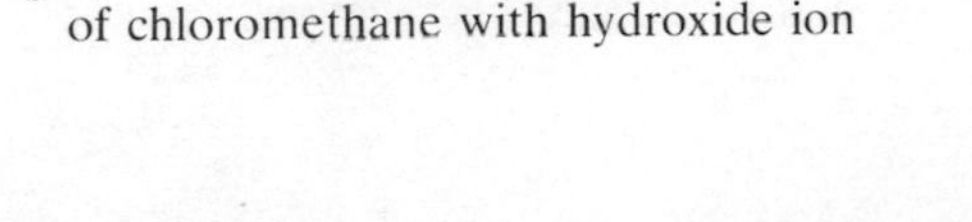

Figure 1.21 Possible mechanisms for reaction of chloromethane with hydroxide ion

as to the concentration of chloromethane. Thus, the reaction proceeds by mechanism (a).

Question 1.5 Under a standard set of conditions, the reaction between chloromethane (0.1 molar) and sodium hydroxide (0.1 molar) proceeds at a rate of 5×10^{-7} mol l^{-1} s^{-1}. Assuming mechanism (a) (Figure 1.21), what would be the rate of reaction under the same conditions if the concentration of each reactant were 0.2 molar? Assuming mechanism (b) (Figure 1.21), what would be the rate of reaction under the same conditions if the concentration of each reactant were 0.2 molar?

Even for reactions proceeding by a similar mechanism the *rate constant* (k) may vary considerably depending upon the structure of the substrate. Some understanding of the reasons underlying this variation in reactivity can be obtained by considering such things as the ease of approach to the position undergoing attack and the stability of possible intermediates. These in turn are influenced by both *electronic* (inductive and mesomeric) effects and *steric* effects (resulting from the physical size of substituent groups). For example, the direct displacement of chloride from 2-chloro-2-methylpropane (*t*-butyl chloride) by hydroxide ion (compare Figure 1.21(a)) is extremely difficult due to the presence of three relatively bulky methyl groups surrounding the carbon atom undergoing attack. On the other hand the two step replacement of chloride by hydroxide ion in *t*-butyl chloride (compare Figure 1.21(b)) is made considerably easier because the intermediate carbonium ion is stabilized by the electronic effects of the three methyl groups (for practical purposes the CH_3 group behaves as though it exerts an electron-donating inductive effect). Indeed, experimental observation indicates that the reaction between *t*-butyl chloride and hydroxide ion proceeds by the two-step mechanism (Chapter 6).

In addition to the factors described above, the existence of *strain* in some cyclic molecules (resulting from compression of bond angles) and of stereochemical features (see Chapter 2) can also affect the reactivity of organic compounds.

1.10 Classification of organic reactions

Organic reactions can be grouped into a number of categories according to the overall result. The common categories are *addition*, *elimination*, *substitution* (displacement), and *rearrangement* (isomerization) reactions, which are illustrated in Figure 1.22. The names are self-explanatory.

1.11 Physical properties of organic compounds

Organic compounds have characteristic properties such as *boiling points* (the temperature at which a liquid boils at 1 atmosphere pressure), *melting*

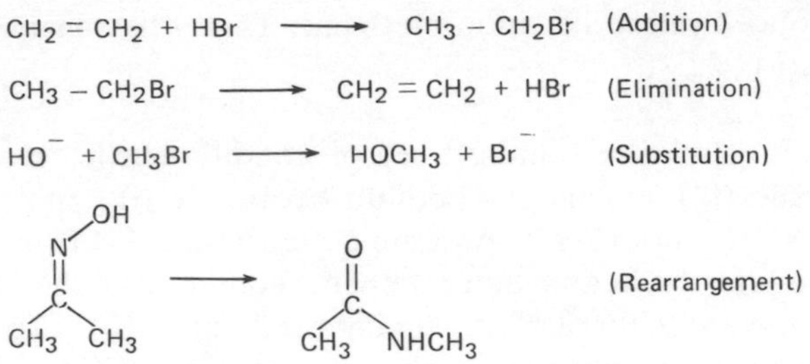

$$CH_2 = CH_2 + HBr \longrightarrow CH_3 - CH_2Br \quad \text{(Addition)}$$

$$CH_3 - CH_2Br \longrightarrow CH_2 = CH_2 + HBr \quad \text{(Elimination)}$$

$$HO^- + CH_3Br \longrightarrow HOCH_3 + Br^- \quad \text{(Substitution)}$$

(Rearrangement)

Figure 1.22 Common categories of organic reactions

points, and solubilities in particular solvents. Within a group of compounds of similar structure, the boiling and melting points rise as the molecular mass increases, but compounds with very different structures may have very different physical properties even when their molecular masses are similar. Thus, methanol (CH_3OH; boiling point 65 °C, miscible with water) has very different physical properties from ethane (CH_3CH_3; b.p. −89 °C, insoluble in water), although their molecular masses are very similar.

Boiling points depend upon the forces of attraction between molecules, *not* upon the strengths of bonds within a molecule. In non-polar compounds there are only weak gravitational forces (known as van der Waals forces) between molecules. On the other hand, polar molecules stick together by electrostatic attraction and energy is thus required to separate them and form the gas phase. Thus, the boiling points of polar molecules tend to be higher than those of non-polar molecules of similar molecular mass.

One way of expressing the overall polarity of a molecule is in terms of its *dipole moment*, which is a measure of the extent to which molecules orient themselves in an electric field. It is possible for molecules to have several highly polar bonds and yet to have little or no overall dipole moment if, as in CCl_4, individual dipoles oppose each other.

One special example of attraction between individual bond dipoles is that occurring when hydrogen is attached to an electronegative atom such as O or N. In this case the intermolecular forces of attraction are relatively large and are described as *hydrogen bonds* (Figure 1.23). Molecules such as methanol which can form hydrogen bonds have exceptionally high boiling points.

Attraction between solvent and solute molecules aids solubility of a solute in a given solvent, which is why the common observation 'like

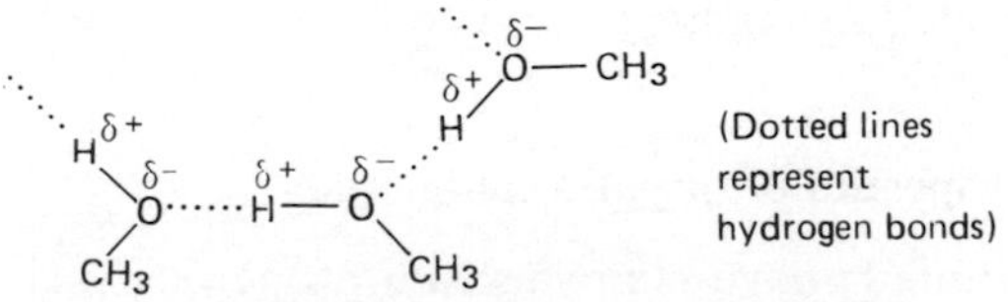

(Dotted lines represent hydrogen bonds)

Figure 1.23 Hydrogen bonding in methanol

dissolves like' (i.e. polar solvents dissolve polar compounds and vice versa) is appropriate. Organic compounds containing hydroxy or amino groups will often dissolve in water or other solvents which are capable of forming hydrogen bonds with the solute.

Melting points are more difficult to predict because they depend to a significant extent upon molecular shape, and the ease with which molecules can pack together. Nevertheless, they are highly characteristic and small amounts of impurities produce a significant depression of the melting point of an otherwise pure compound, thus providing a good means for checking purity. Indeed, if two samples have the same melting point, one simple way of establishing whether they are identical compounds is to determine the melting point of a mixture of the two. If they are different, the *mixed melting point* is significantly lower than the melting points of the pure compounds.

1.12 Sources of organic chemicals

Quite a number of organic compounds are extracted directly from plants or micro-organisms, but at the present time the main source of bulk organic material is crude oil. Coal tar had this role before crude oil was so widely available. Both coal and crude oil are old deposits of once living materials which have been trapped, compressed, and degraded. Because of the long time scale for the laying down of these deposits, supplies are being used up much more rapidly than they can be regenerated, so in the future alternative organic feedstock must be found. The production of organic chemicals from some form of biomass (e.g. by fermentation or pyrolysis) could be the answer, or, if a very cheap energy supply could be found, organic compounds could be produced by hydrogenation of carbon dioxide. This will be one of the major challenges for organic chemists during the next few decades.

Answers to questions

1.1

(a)

(b)

B

1.2 (a) 109.5°; (b) 120°; (c) 120°.

1.3 (a) $CH_2{=}CH{-}\bar{C}H_2 \longleftrightarrow \bar{C}H_2{-}CH{=}CH_2$;

   (b) $CH_2{=}CH{-}\ddot{C}l \longleftrightarrow \bar{C}H_2{-}CH{=}\overset{+}{C}l$.

1.4

1.5 (a) 2×10^{-6} mol l^{-1} s^{-1}; (b) 10^{-6} mol l^{-1} s^{-1}.

CHAPTER 2

Stereochemistry

Stereochemistry is the branch of chemistry which deals with the three-dimensional structures of molecules. A knowledge of stereochemistry is essential since many of the properties and reactions of organic compounds depend upon the spatial arrangement of atoms and groups. The first part of the chapter deals with *stereoisomerism*, which arises when molecules differ only in the arrangement of their atoms or groups in space. The second part of the chapter deals with the different flexible shapes (*conformations*) which molecules can adopt.

2.1 Stereoisomerism

Structural isomerism has already been discussed in Chapter 1. There is, however, another type of isomerism which is important in organic chemistry. This is stereoisomerism, which depends upon the existence of different arrangements of atoms or groups in space. It is convenient to divide stereoisomerism into two types, geometrical and optical, and these are dealt with in the sections which follow. A summary of the types of isomerism encountered in organic chemistry is shown in Figure 2.1.

2.2 Geometrical isomerism

A carbon–carbon double bond introduces a considerable degree of rigidity into a molecule. Thus the relatively free rotation which is possible about a single bond does not occur in the case of a double bond (Chapter 4). Rotation of one carbon atom relative to the other requires breaking and reforming the π bond which can occur only under vigorous conditions such as high temperature or irradiation with ultraviolet light.

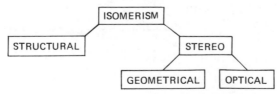

Figure 2.1 Types of isomerism encountered in organic chemistry

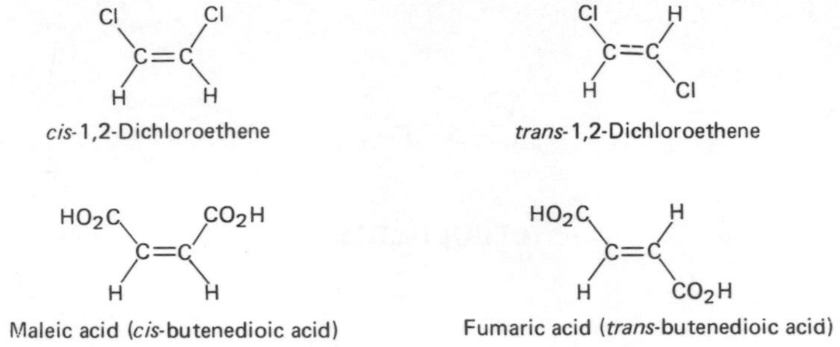

cis-1,2-Dichloroethene   trans-1,2-Dichloroethene

Maleic acid (cis-butenedioic acid)   Fumaric acid (trans-butenedioic acid)

Figure 2.2 Examples of geometrical isomerism

If the two groups attached to each carbon atom of a double bond are different, two isomers exist (Figure 2.2). The arrangement of groups about the double bond is referred to as the *configuration* of the molecule. The isomers are often called *cis* and *trans* isomers (in some cases a more general system of nomenclature must be used—Section 2.9).

Geometrical isomers usually have different physical and chemical properties. For example, maleic acid (m.p. 130 °C) readily forms a cyclic anhydride on heating (Figure 2.3), whereas fumaric acid (m.p. 287 °C) does not do so unless heated to a very high temperature, when *maleic* anhydride is formed.

Geometrical isomerism is also encountered in some cyclic compounds. In such cases the ring prevents free rotation. Two substituents can therefore have different orientations relative to each other and *cis* and *trans* isomers are possible (Figure 2.4). In order to simplify the drawing of organic structures it is often convenient to represent chains or rings of carbon atoms by stick drawings in which each apex signifies the position of a carbon atom with an appropriate number of hydrogen atoms attached.

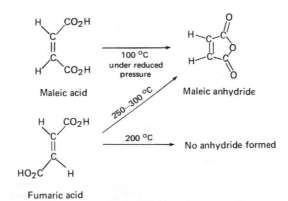

Figure 2.3 Differences in the reactions of geometrical isomers

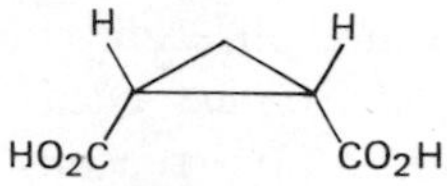

23

cis-Cyclopropane-1,2-
dicarboxylic acid

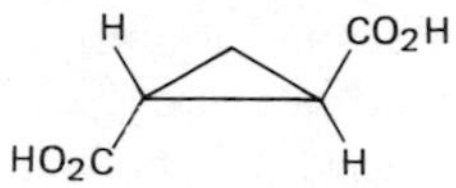

trans-Cyclopropane-1,2-
dicarboxylic acid

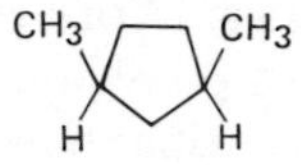

cis-1,3-Dimethyl-
cyclopentane

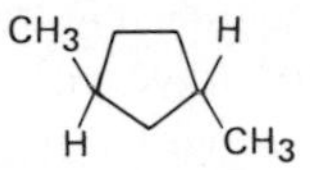

trans-1,3-Dimethyl-
cyclopentane

Figure 2.4 Examples of geometrical isomerism
in cyclic compounds

Question 2.1 (a) How many geometrical isomers of $CH_3CH=CHCH=CHCH_3$ are there? (b) Draw the structures of *cis*-1,2-dibromo-cyclopropane and *trans*-1,3-dimethylcyclobutane.

2.3 Optical isomerism

Some substances are able to rotate the plane of polarized light. This property is called *optical activity* and is described in more detail in Section 2.5. It is useful first to consider the types of molecule which exhibit optical activity.

Optical activity arises when a molecule cannot be superimposed on its mirror image (Figure 2.5(a)). Such a molecule is described as *chiral*, whereas a molecule which can be superimposed on its mirror image is described as *achiral* (Figure 2.5(b)). Only chiral molecules exhibit optical activity. (It will be useful if you make models of the compounds in Figure 2.5 to demonstrate that the two structures in (a) are different whereas those in (b) are identical.)

(a)

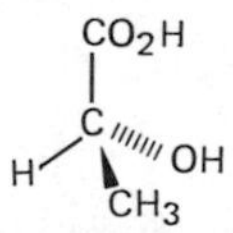

(+)-Lactic acid
((+)-2-hydroxypropanoic acid)

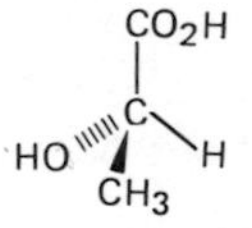

(−)-Lactic acid
((−)-2-hydroxypropanoic acid)

(b)

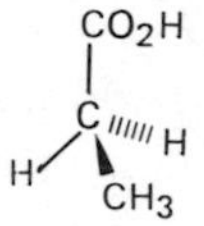

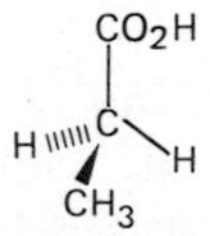

Propanoic acid (two structures identical)

Figure 2.5 Examples of chiral and achiral molecules

A chiral molecule and its mirror image are different compounds and are therefore isomers. They are called *optical isomers* because they differ in their behaviour towards polarized light. For reasons which will be explained later they are designated (+) and (−) isomers.

Optical isomers such as (+)- and (−)-lactic acid which are mirror images of one another are called *enantiomers*. In most respects enantiomers have identical physical and chemical properties. For example, (+)- and (−)-lactic acid both have the same melting point and undergo the same chemical reactions with most reagents. Enantiomers can only be distinguished by their effect upon polarized light (Section 2.5) and their behaviour towards other chiral species (Section 2.10 and 2.11).

The central carbon atom in lactic acid has four different substituents attached to it and is called an *asymmetric carbon atom*. Propanoic acid, which is achiral, does not contain an asymmetric carbon atom. The presence of an asymmetric carbon atom in a molecule is a common cause of optical isomerism.

Other atoms besides carbon atoms can also be asymmetric. For example, ammonium salts and amine oxides are tetrahedral in shape, and optical isomers exist when four different substituents are attached to the central nitrogen atom (Figure 2.6; R^1, R^2, R^3, and R^4 are alkyl or aryl groups). Compounds containing asymmetric phosphorus and sulphur atoms are also known.

Figure 2.6 Compounds containing asymmetric nitrogen atoms

Question 2.2 Which of the following compounds will exhibit optical isomerism: (a) 2-Bromopropanoic acid; (b) 2-methylpropanoic acid; (c) 2-methylbutanoic acid; (d) 3-methylbutanoic acid?

2.4 Elements of symmetry

The presence or absence of certain elements of symmetry in a molecule can serve as a useful guide to whether a molecule is chiral or achiral. As such it presents an alternative approach to drawing mirror image structures and deciding whether they are superimposable on one another.

The molecule of (+)-lactic acid does not possess a plane or centre of symmetry. It is for this reason that it cannot be superimposed on its mirror image and is chiral. Molecules which do possess a plane or centre of symmetry are achiral. For example, propanoic acid has a plane of symmetry, as does *cis*-cyclopropane-1,2-dicarboxylic acid (Figure 2.7). These are both achiral molecules.

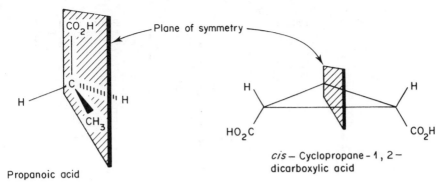

Figure 2.7 Molecules with planes of symmetry

The isomer of 2,4-dimethylcyclobutane-1,3-dicarboxylic acid shown in Figure 2.8 has no plane of symmetry but it does possess a centre of symmetry as indicated. It is therefore achiral.

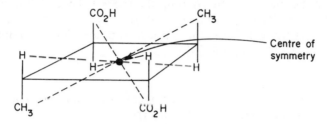

Figure 2.8 A molecule with a centre of symmetry

Question 2.3 Are the following molecules chiral or achiral:
(a)   *trans*-cyclopropane-1,2-dicarboxylic acid;
(b)   *trans*-1,3-dichlorocyclobutane?

2.5 Optical activity

Ordinary light can be thought of as a series of waves vibrating in random directions perpendicular to the direction of propagation. However if the light is passed through a sheet of Polaroid® or Nicol prism (constructed from two crystals of Iceland spar) it becomes plane polarized, i.e. it is reduced to a form in which all the waves vibrate in one plane. A second sheet of Polaroid® oriented parallel to the first allows the polarized light to pass through unimpeded (Figure 2.9(a)), but if the second sheet is oriented perpendicular to the first then the light beam is completely extinguished (Figure 2.9(b)). This effect can be readily observed using two pairs of Polaroid® sunglasses.

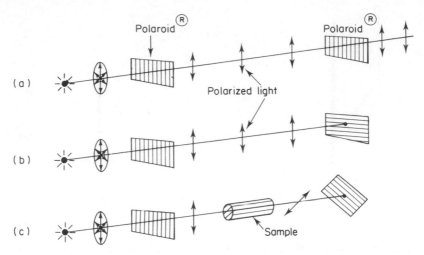

Figure 2.9 The rotation of polarized light

Optically active substances rotate the plane of polarized light. The angle of rotation can be measured by rotating the second sheet of Polaroid® until the light beam is extinguished (Figure 2.9(c)).

The characteristic value of the optical rotation for a compound is expressed as its specific rotation $[\alpha]$,

$$[\alpha] = \theta/lc,$$

where θ is the observed rotation (in degrees) of a sample of concentration c (g cm^{-3}) in a sample tube of length l (dm).

The rotation is defined as positive when it takes place in a clockwise direction and negative when anticlockwise. The specific rotation varies with temperature and also depends upon the wavelength of light used. These variables must therefore be stated. For example, the specific rotation of one enantiomer of lactic acid at 25 °C using the sodium D line is $+4°$ (i.e. $[\alpha]_D^{25} = +4°$). The specific rotation of the other enantiomer under the same conditions is $-4°$. One enantiomer is therefore called $(+)$-lactic acid and the other $(-)$-lactic acid.

A mixture containing equal amounts of two enantiomers is called a *racemic mixture*. Such a mixture is optically inactive since the effects of the two enantiomers completely cancel out.

The variation of optical rotation with wavelength (known as *optical rotatory dispersion, o.r.d.*) depends upon the configurations of the chiral centres within a molecule and can be used to elucidate such configurations.

2.6 Fischer projection formulas

Although three-dimensional formulas are very useful for representing the structures of chiral molecules, they are cumbersome to use, especially when

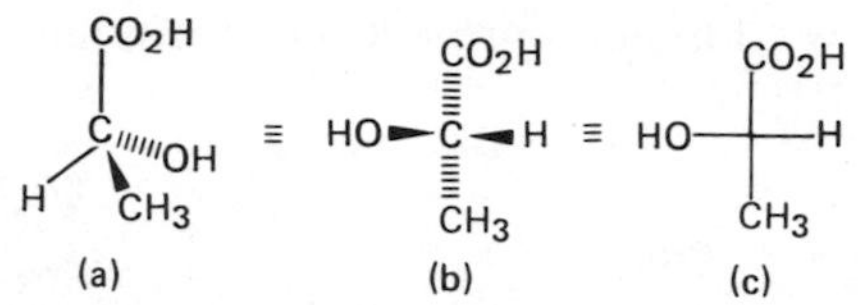

Figure 2.10 Construction of a Fischer projection formula

the molecule contains several asymmetric centres. To overcome this problem Emil Fischer devised a useful method for representing such structures in two dimensions. Consider first of all a molecule containing only one asymmetric centre (Figure 2.10(a)). In order to obtain the *Fischer projection formula* the molecule must first be turned around so that horizontal groups point towards the viewer and vertical groups point away (Figure 2.10(b)). Representation of this structure in two dimensions gives the Fischer projection formula (Figure 2.10(c)).

Similar principles apply when drawing Fischer projection formulas of molecules containing more than one asymmetric centre. Thus, in order to draw the Fischer projection formula of (−)-tartaric acid (2,3-dihydroxybutanedioic acid) (Figure 2.11(a)) one carbon atom must first be rotated through 180° so that the two carboxyl groups point downwards and the other four groups point upwards (Figure 2.11(b)). This structure (which is actually an eclipsed conformation of the molecule—Section 2.13) is then viewed from above and its representation in two dimensions gives the Fischer projection formula (Figure 2.11(c)).

The basic requirements of the Fischer convention are therefore that horizontal lines represent bonds extending above the plane of the paper, and vertical lines represent bonds extending below the plane of the paper. A certain degree of caution must be exercized in handling such formulas. For example, if the projection formula of (+)-lactic acid (Figure 2.10(c)) is rotated through 90°, or if any two of the substituents are interchanged, then the formula obtained no longer represents (+)-lactic acid but its enantiomer.

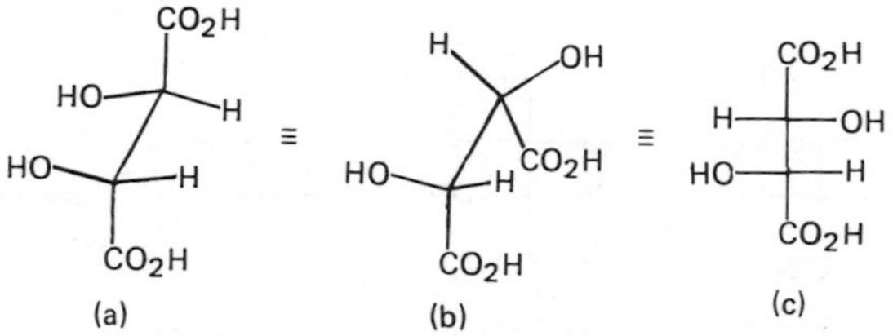

Figure 2.11 Fischer projection formula for a
molecule with two asymmetric atoms

Question 2.4 Draw Fischer projection formulas of the following molecules:

(a) 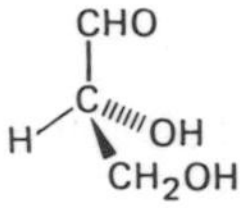 ; (b) 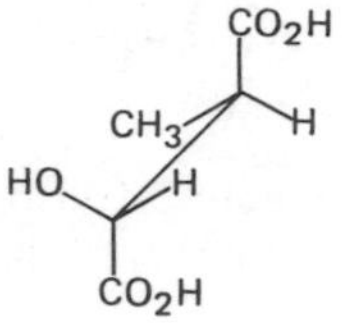

2.7 Molecules containing more than one asymmetric centre

When a molecule contains more than one asymmetric centre the number of possible stereoisomers is increased. For example, there are four stereoisomers of 2-hydroxy-3-methylbutanedioic acid (Figure 2.12). I and II are mirror images of one another and are therefore enantiomers. Similarly III and IV are enantiomers. However, although I and III are optical isomers they are not mirror images of one another. They have the same arrangement of groups (configuration) around one carbon atom but different configurations at the other. They are called *diastereoisomers*. Similarly I and IV are diastereoisomers, as are II and III, and II and IV. Since diastereoisomers are not mirror images they usually have different physical and chemical properties and can be separated without too much difficulty (Section 2.10).

Consider now the case of a molecule which contains two asymmetric centres but has the same four groups around each. In this case there are only three possible stereoisomers (Figure 2.13) because one has a plane of symmetry and is therefore superimposable on its mirror image. This isomer is called the *meso isomer* and is optically inactive.

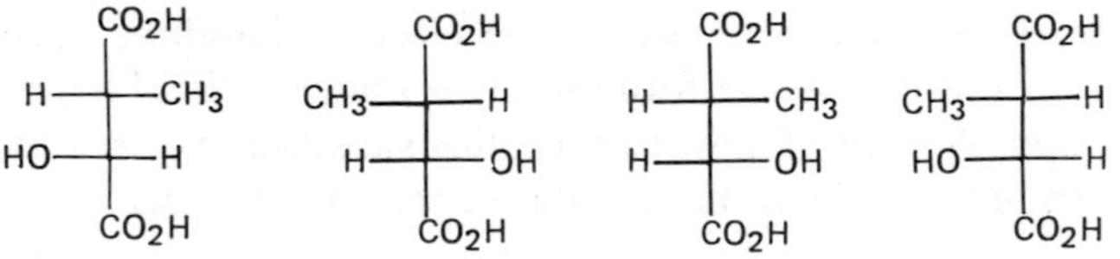

I II III IV

Figure 2.12 The stereoisomers of 2-hydroxy-3-methylbutanedioic acid

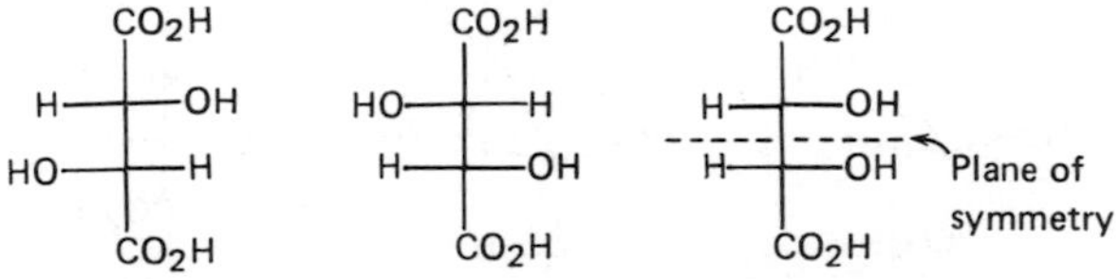

(−)-Tartaric acid (+)-Tartaric acid *meso*-Tartaric acid

Figure 2.13 The stereoisomers of tartaric acid

2.8 Optical isomerism without an asymmetric atom

In the last section we saw that it was possible for a molecule containing two asymmetric centres to be optically inactive. It is also possible for a molecule which contains no asymmetric centre to be optically active. Thus two enantiomers of penta-2,3-diene exist (Figure 2.14). The reason for this is that the molecule as a whole is asymmetric because the two ends are held in perpendicular planes. The molecule must adopt this shape because the central carbon atom is sp^1 hybridized (see Section 1.5) and the two π bonds are therefore orthogonal (i.e. at right angles to each other).

Some *ortho* substituted biphenyl derivatives are also optically active. If the groups in the *ortho* positions are large, rotation about the bond linking the two aromatic nuclei cannot occur. Consequently, when the two substituents attached to each ring are different two discrete isomers exist

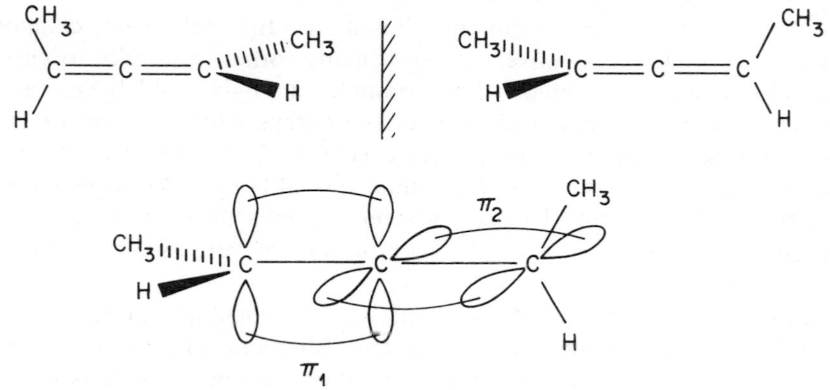

Figure 2.14 The two enantiometers of penta-2,3-diene

which cannot be superimposed (Figure 2.15).

Figure 2.15 Isomerism caused by restricted rotation

Question 2.5 Would you expect the following compounds to exhibit optical isomerism:

(a) $(CH_3)_2C{=}C{=}CHCO_2H$;   (b)

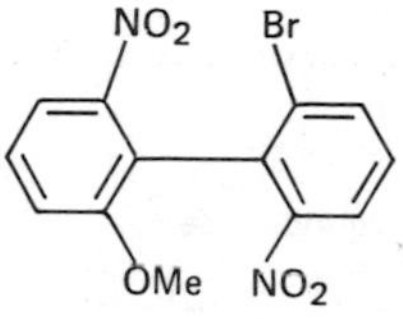

?

2.9 Nomenclature of Stereoisomers

The configurations of geometrical isomers have been denoted by the terms *cis* and *trans*. It would clearly be useful to have a similar method to denote the configurations of optical isomers. Remember that the prefixes (+) and (−) do not describe the configuration of an isomer but merely indicate the sign of its specific rotation.

For many years the configurations of optical isomers have been denoted by the symbols D and L to indicate their relationship to (+)- or (−)-glyceraldehyde (2,3-dihydroxypropanal) (Figure 2.16). Thus compounds having the same configuration as (+)-glyceraldehyde are described as having the D configuration, while those having the same configuration as (−)-glyceraldehyde are described as having the L configuration. The absolute configuration of (+)-glyceraldehyde has been verified by X-ray crystallography (Chapter 3).

Glyceraldehyde was originally chosen as the reference compound because it could be converted into many other naturally occurring, optically active compounds. For example, (+)-glyceraldehyde can be converted into (−)-lactic acid in a series of steps which do not affect the configuration of the asymmetric centre (Figure 2.17). The two molecules therefore have the same configuration (D) although the signs of their specific rotations differ. The D/L system becomes difficult to apply when the groups around the asymmetric centre are not very similar to those in glyceraldehyde.

Although the D/L system is often used, particularly in the fields of carbohydrate and amino acid chemistry (Chapters 11 and 12), a more general method for denoting the configurations of optical isomers has been devised by Cahn, Ingold, and Prelog. The four groups around an asymmetric centre are first placed in an order of priority (a,b,c,d) using a set of rules given below. The molecule is then viewed from the side opposite the group of lowest priority (Figure 2.18). If the remaining groups

$$
\begin{array}{ccc}
& CHO & \\
H & \!\!\!\!-\!\!\!\! & OH \\
& CH_2OH &
\end{array}
\qquad
\begin{array}{ccc}
& CHO & \\
HO & \!\!\!\!-\!\!\!\! & H \\
& CH_2OH &
\end{array}
$$

D-(+)-Glyceraldehyde   L-(−)-Glyceraldehyde

Figure 2.16 The optical isomers of glyceraldehyde

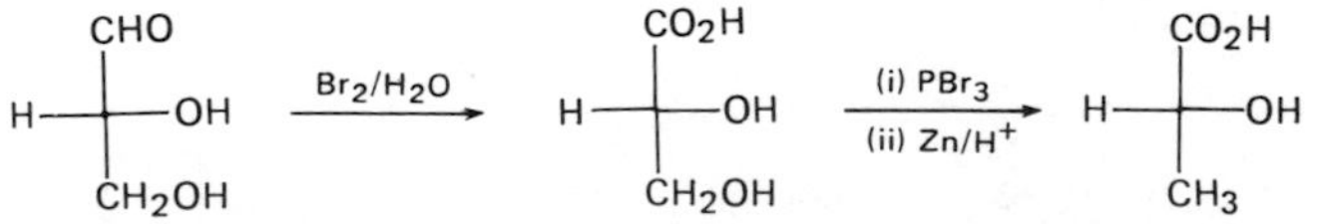

$$
\begin{array}{ccc}
& CHO & \\
H & \!\!\!\!-\!\!\!\! & OH \\
& CH_2OH &
\end{array}
\xrightarrow{Br_2/H_2O}
\begin{array}{ccc}
& CO_2H & \\
H & \!\!\!\!-\!\!\!\! & OH \\
& CH_2OH &
\end{array}
\xrightarrow[\text{(ii) Zn/H}^+]{\text{(i) PBr}_3}
\begin{array}{ccc}
& CO_2H & \\
H & \!\!\!\!-\!\!\!\! & OH \\
& CH_3 &
\end{array}
$$

D-(+)-Glyceraldehyde   D-(−)-Lactic acid

Figure 2.17 Synthesis of (−)-lactic acid from (+)-glyceraldehyde

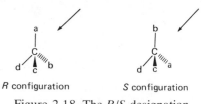

R configuration   S configuration

Figure 2.18 The R/S designation

(a,b,c) are arranged clockwise as seen from this viewpoint then the configuration is designated R (Latin, rectus = right), whereas if they are anticlockwise it is S (Latin, sinister = left).

The priority order (a,b,c,d) is decided as follows:

1. Atoms directly attached to the asymmetric centre are first arranged in order of decreasing atomic number: e.g.

   $I > Br > Cl > F$ and $OH > NH_2 > CH_3 > H$.

2. If two or more of the atoms directly attached to the asymmetric centre are alike then the priority order is determined by the atomic numbers of the second or subsequent atoms: e.g.

   $CH_2OH > CH_2NH_2 > CH_2CH_3 > CH_3$ and
   $CH_2CH_2OH > CH_2CH_2CH_3 > CH_2CH_3$.

3. If the second (or subsequent) atoms are the same but the number of such atoms is different, then the group with more substituents of higher atomic number takes priority: e.g.

   $CHCl_2 > CH_2Cl$

4. When the atom attached to the asymmetric centre is multiply bonded then the atom at the other end of the multiple bond is counted twice (for a double bond) or three times (for a triple bond): e.g.

   $CO_2H > CHO > CH_2OH$.

Question 2.6 Arrange the following groups in order of priority: NH_2, CHO, H, Br, C_2H_5, CN, OH, $COOCH_3$ C_6H_5 (phenyl), $CONH_2$, CH_2OH, CH_3, CO_2H.

On the basis of the above rules the priority order for the four groups attached to the asymmetric centre in lactic acid is $OH > CO_2H > CH_3 > H$. Consequently (+)-lactic acid has the S configuration (Figure 2.19). Similarly it can be seen that (+)-glyceraldehyde has the R configuration.

The advantage of the Cahn–Ingold–Prelog system is that it allows one to define the configuration of any asymmetric centre in a molecule from first principles and not just those which are structurally related to glyceraldehyde. Indeed, several asymmetric centres in the same molecules can be sepa-

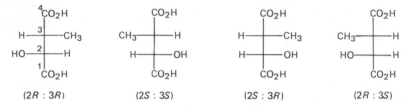

Figure 2.19 Configurations of (+)-lactic acid and
(+)-glyceraldehyde

$_4^{}CO_2H$	CO_2H	CO_2H	CO_2H
H——$_3^{}$—CH₃	CH₃——H	H——CH₃	CH₃——H
HO——$_2^{}$—H	H——OH	H——OH	HO——H
$_1^{}CO_2H$	CO_2H	CO_2H	CO_2H
(2R : 3R)	(2S : 3S)	(2S : 3R)	(2R : 3S)

Figure 2.20 The isomers of 2-hydroxy-3-methylbutanedioic acid

rately defined by this method. For example the configurations of the isomers of 2-hydroxy-3-methylbutanedioic acid (Figure 2.20) are completely
defined by adding the appropriate prefix, (2R : 3R), (2S : 3S), (2R : 3S),
or (2S : 3R), to the systematic name.

Question 2.7   Name the following compounds using the *R/S* system:

(a)   HO—|—CH₃ with CH₂CH₃ above and H below   ;   (b) CO₂H / H—OH / HO—H / CO₂H .

In most simple cases the *cis/trans* terminology can be used to define the
configuration of geometrical isomers (Section 2.2). However in some cases,
such as when four different groups are attached to the double-bonded carbon atoms of an alkene, this system is difficult to apply. The problem can
be overcome by making use of a more general system of nomenclature
based on the Cahn–Ingold–Prelog priority rules. One must first assign an
order of priority to the groups attached to each carbon atom of the double
bond. For example in the isomers of 3-bromobut-2-enoic acid (Figure
2.21), the order of priority at carbon atom 3 is Br > CH₃, and at carbon
atom 413 CO_2H > H. When the two groups of highest priority are on the same
side of the molecule it is called the Z isomer (German, zusammen = together)
whereas when they are on opposite sides of the molecule it is called the E
isomer (German, entgegen = opposite).

Question 2.8   Derive systematic *E/Z* names for the *cis* and *trans* isomers of
1,2-dichloroethene.

(Z)-3-Bromobut-2-enoic acid   (E)-3-Bromobut-2-enoic acid

Figure 2.21 The isomers of 3-bromobut-2-enoic acid

2.10 Resolution of enantiomers

Most laboratory syntheses of chiral molecules give rise to racemic mixtures. For example, in the synthesis of 2-bromopropanoic acid from propanoic acid (Figure 2.22) both of the hydrogen atoms on carbon atom 2 of the propanoic acid are equivalent and consequently equal amounts of both enantiomers of the product are obtained. It is important, therefore, to have methods for resolving such mixtures into their optically active components.

Pasteur performed the first *resolution* of an optically active compound by manually separating the crystals of (+)- and (−)-lactic acid under a microscope. However this method can only be used in a limited number of cases and is extremely tedious.

Since most physical properties of enantiomers are identical they cannot be separated by normal physical methods such as distillation or crystallization. Diastereoisomers on the other hand have different physical properties and can be separated more easily. One way of separating enantiomers therefore involves converting them into diastereoisomeric derivatives. Consider, for example the resolution of a mixture of (+)- and (−)-tartaric acid. Reacting the racemic acid with an optically active amine gives a mixture of two salts which are diastereoisomers and can be separated by fractional crystallization. The separated salts can then be reconverted into (+)- and (−)-tartaric acid by treatment with strong acid (Figure 2.23).

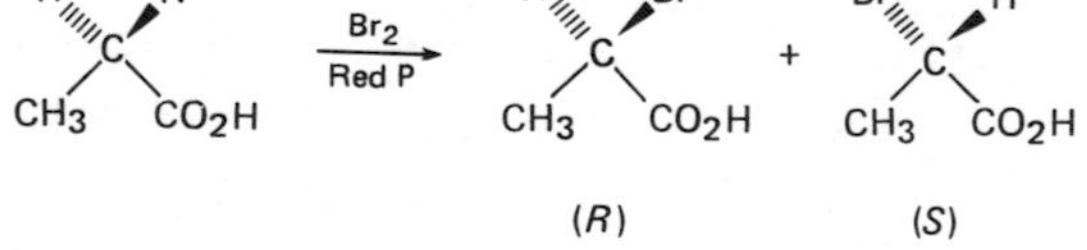

Figure 2.22 Synthesis of the two enantiomers of 2-bromopropanoic acid

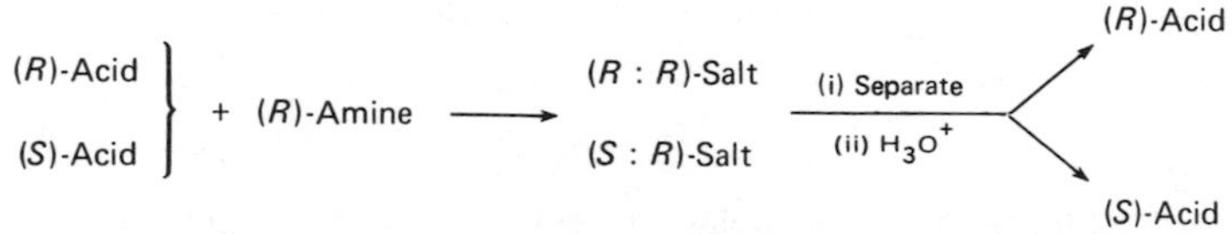

Figure 2.23 The resolution of enantiomers

Question 2.9 Suggest methods which could be used to resolve (a) a racemic amine; (b) a racemic alcohol.

Living organisms contain enzymes which are optically active and which react selectively with only one enantiomer of an optically active compound. Thus, another method which can be used to separate a racemic mixture involves feeding it to a living organism which metabolizes only one enantiomer. The unreacted enantiomer can then be recovered. For example, *Penicillium glaucum* metabolizes (+)-lactic acid but not (−)-lactic acid. It can therefore be used to give an optically pure sample of (−)-lactic acid. This technique is sometimes called biochemical resolution.

2.11 Asymmetric synthesis

Any reaction in which a chiral molecule is prepared entirely from achiral or racemic starting materials produces equal amounts of both enantiomers. For example, in the reduction of butanone with $LiAlH_4$ (Figure 2.24(a)) the nucleophilic (AlH_4^- can attack either side of the molecule with equal ease, giving equal amounts of both enantiomers. However if one of the starting materials in a reaction is already chiral then the two product molecules (diastereoisomers) are produced in unequal amounts (Figure 2.24(b)) because the two transition states leading to the products are no longer identical. Such reactions are called *asymmetric syntheses*. The chiral agent in an asymmetric synthesis may be the substrate, the reagent, or the catalyst. Since biochemical catalysts (enzymes, Chapter 12) are chiral, and many reactions occurring in nature are asymmetric syntheses, they are usually highly specific. In fact most naturally occurring compounds are present as one of their optically active forms rather than as a racemic mixture.

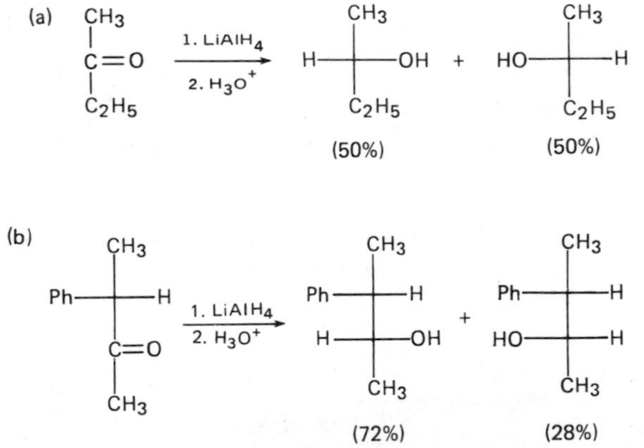

Figure 2.24 Examples of symmetric and asymmetric syntheses

2.12 Conformations of molecules

So far in this chapter we have been concerned with the configurations of molecules, i.e. the fixed arrangement of atoms or groups relative to one another. In general, different configurations can only be interconverted by breaking chemical bonds. However most molecules can adopt a number of interconvertible shapes, or *conformations*, without bond cleavage.

Consider, for example, the ethane molecule. At room temperature there is free rotation about the carbon–carbon bond and as a result the molecule can adopt an infinite number of conformations. The two extremes are called the *eclipsed* and *staggered* conformations (Figure 2.25). In the former the hydrogen atoms lie directly behind one another when viewed along the carbon–carbon bond, whereas in the latter they are staggered. The molecule prefers to adopt the staggered conformation since steric interaction between the hydrogen atoms is then at a minimum. However because there is free rotation about the carbon–carbon bond the *conformers* cannot be isolated.

In the case of the butane molecule there are two possible staggered conformations. They are called the *gauche* and *anti* conformations (Figure 2.26). These are conveniently represented by means of *Newman projection formulas* in which the molecule is viewed end-on so that the two carbon atoms are seen one behind the other.

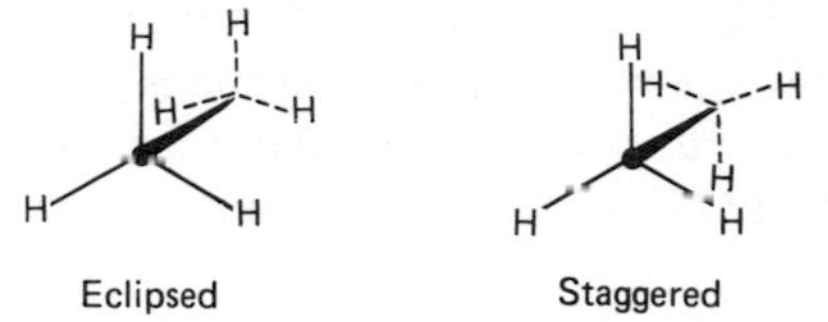

Eclipsed   Staggered

Figure 2.25 The two conformations of ethane

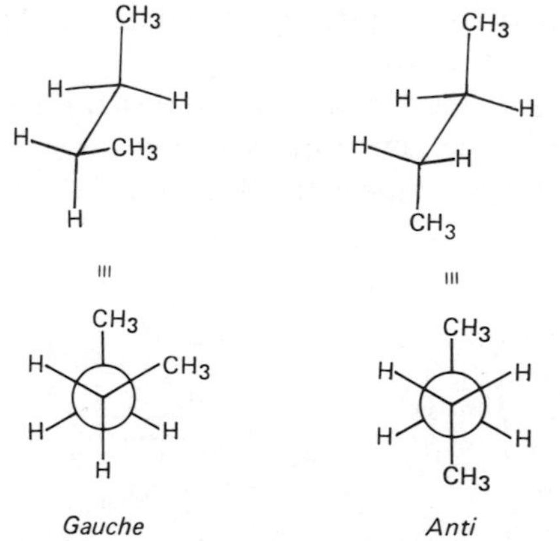

III   III

Gauche   Anti

Figure 2.26 The staggered conformations of butane

The *gauche* and *anti* conformations are the most stable conformations of the butane molecule, the *anti* conformation being slightly more stable than the *gauche* because the two methyl groups are as far apart as possible. However, although there is a significant energy barrier associated with rotation about the central carbon–carbon bond, it is not large enough to prevent rapid interconversion of the conformers at room temperature. The differing energy content of the conformations does mean, however, that at any instant 69% of the butane molecules possess the *anti* conformation and 31% the *gauche* conformation.

Question 2.10 Draw the staggered conformations of *meso*-2,3-dibromobutane and comment on their relative stability.

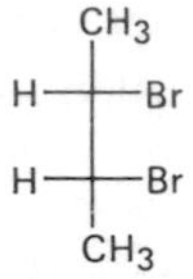

meso-2,3-Dibromobutane

2.13 Conformations of cyclohexane derivatives

Whereas open-chain molecules are very flexible, small cyclic molecules do not possess the same degree of flexibility. Cyclohexane molecules, for example, can adopt either 'chair' or 'boat' conformations which are easily interconverted (Figure 2.27). The 'boat' conformation drawn in Figure 2.27 is an idealized one. In reality it is slightly twisted and is more correctly referred to as a 'skew-boat'. (It will be useful if you make a model of the cyclohexane molecule to demonstrate that these are strain-free structures and can be interconverted.)

The chair conformation is more stable than the boat because the steric interactions between hydrogen atoms are less. In practice, therefore, nearly all cyclohexane molecules adopt chair conformations. The variation in energy content as the conformations are interconverted can be depicted by means of an energy diagram (Figure 2.28).

For cyclohexane itself there are two equivalent chair conformations. However if the cyclohexane ring carries a substituent then the two chair conformations are no longer equivalent (Figure 2.29). In one conformation

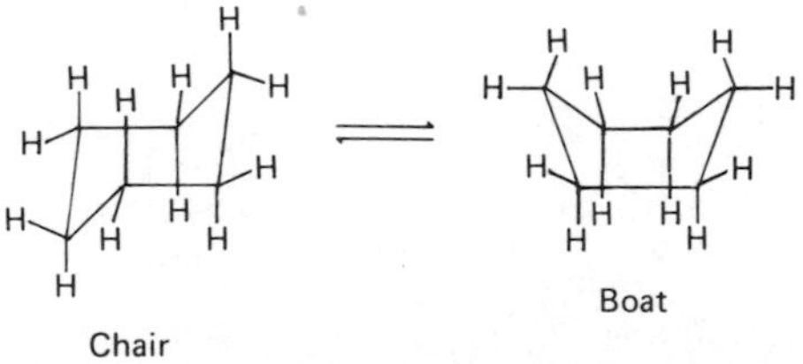

Chair   Boat

Figure 2.27 The two conformations of cyclohexane

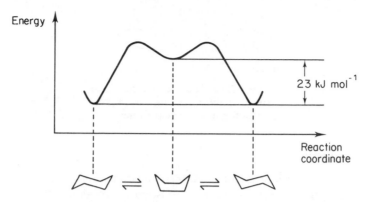

Figure 2.28 The energy diagram for chair–boat–chair interconversion

(I) the substituent occupies an *equatorial* position (i.e. approximately parallel to the 'plane' of the ring) whereas in the other (II) the substituent occupies an axial position (approximately perpendicular to the 'plane' of the ring). The conformation in which the substituent is equatorial is more stable than the one in which it is axial because the latter is destabilized by steric interactions between the substituent and axial hydrogen atoms. Most of the molecules therefore adopt conformation (I). If the substituent is particularly large (e.g. a *tert*-butyl group) nearly all the molecules adopt the conformation in which the substituent is equatorial.

In general, similar considerations apply to cyclohexanes with more than one substituent. The preferred conformation is the one in which the most bulky substituents are equatorial.

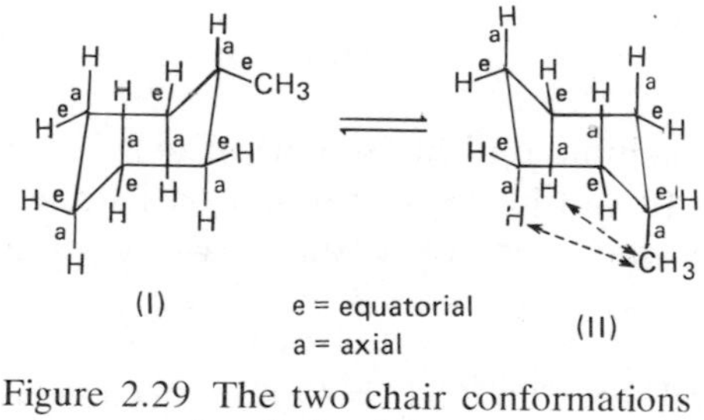

Figure 2.29 The two chair conformations
of methylcyclohexane

Question 2.11 Draw the chair conformations of (a) *cis*- and (b) *trans*-1, 2-dimethylcyclohexane. Comment on the relative stabilities of the two conformations in each case.

Rings of other sizes also adopt preferred conformations. For example, two important conformations in the cyclopentane and cyclooctane series are the 'envelope' and 'crown' respectively (Figure 2.30).

38

'Envelope'   'Crown'

Figure 2.30 The 'envelope' and 'crown' conformations of cyclo-pentane and cyclooctane

Answers to questions

2.1 (a) There are three isomers: *cis–cis*, *cis–trans*, and *trans–trans*. (In this case the *cis–trans* and *trans–cis* isomers are identical.)

(b)

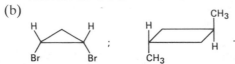

2.2 (a) and (c) both contain an asymmetric carbon atom and are therefore chiral. (b) and (d) do not contain asymmetric carbon atoms and are achiral.

2.3 (a) does not possess a plane or centre of symmetry and is chiral.
     (b) contains both a plane and a centre of symmetry and is therefore achiral.

2.4

(a) 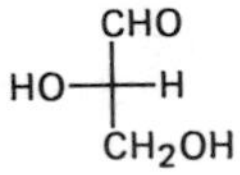 $\begin{array}{c} \text{CHO} \\ \text{HO}\!-\!\!\!-\!\!\!-\!\text{H} \\ \text{CH}_2\text{OH} \end{array}$ (b) 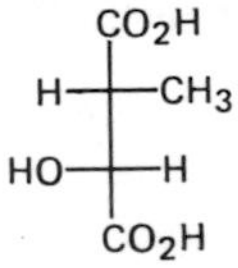 $\begin{array}{c} \text{CO}_2\text{H} \\ \text{H}\!-\!\!\!-\!\text{CH}_3 \\ \text{HO}\!-\!\!\!-\!\text{H} \\ \text{CO}_2\text{H} \end{array}$.

2.5 (a) does not exhibit optical isomerism since it contains a plane of symmetry (it can be superimposed on its mirror image); (b) does exhibit optical isomerism since two discrete enantiomers are possible.

2.6 Br > OH > NH_2 > $COOCH_3$ > COOH > $CONH_2$ > CHO
     > CH_2OH > CN > C_6H_5 > C_2H_5 > CH_3 > H.

2.7 (a) (*R*)-2-butanol; (b) (2*R*:3*R*)-2,3-dihydroxybutanedioic acid.

2.8 The *cis* and *trans* isomers are (*Z*)- and (*E*)-1,2-dichloroethene respectively since Cl has higher priority than H.

2.9 (a) A racemic amine could be resolved by reacting it with an optically active acid, e.g. (+)-tartaric acid, and proceeding as described in the text. (b) A racemic alcohol could be resolved by forming diastereoisomeric esters with an optically active acid, followed by hydrolysis of the separated esters.

2.10

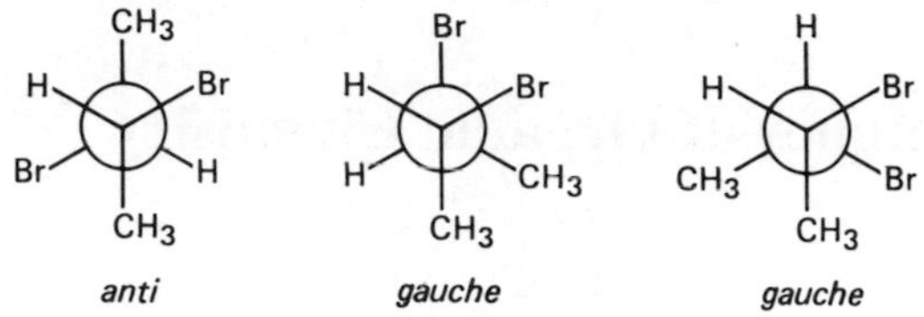

anti   gauche   gauche

The *anti* conformation is the most stable because the large bromine atoms are as far apart as possible.

2.11 (a)

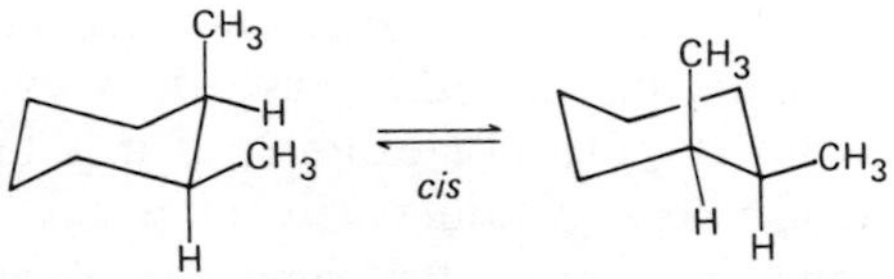

cis

In this case the conformers are of equal stability since each has one axial and one equatorial methyl group.

(b)

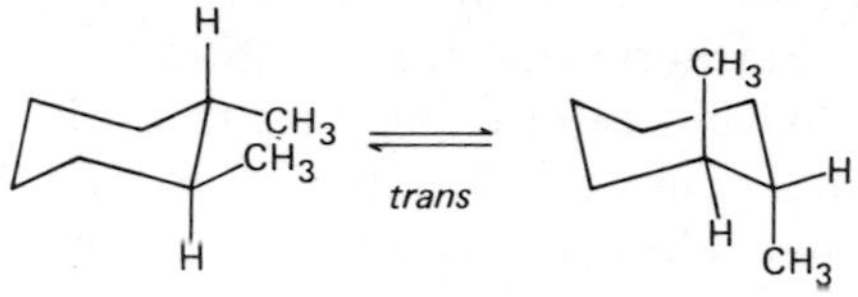

trans

In this case the diequatorial conformer (I) is more stable than the diaxial one (II).

CHAPTER 3

Techniques of Organic Chemistry

The success of organic chemistry relies upon the efficiency of the techniques available for the separation and identification of organic compounds. For some purposes fairly crude mixtures of organic compounds may be acceptable, such as in petroleum (gasoline) for use in the internal combustion engine, but in other cases it is essential that an organic compound be absolutely pure, for example if it is to be used as a food additive or pharmaceutical product. Even when a mixture is acceptable as a final product, it is usual for some partial separation from undesirable components to be carried out. (Petroleum, for example, is a very much less complicated mixture than the crude oil mixture from which it is obtained.) In any event, it is always an advantage to know the composition of a mixture. Furthermore, for identification of unknown compounds it is essential to have a pure compound to avoid being misled by information arising from impurities. A pure compound having been obtained, its spectroscopic and other properties then enable the structure to be determined.

In this chapter we study the separation and identification techniques commonly used in organic chemistry.

SEPARATION TECHNIQUES

3.1 Filtration and crystallization

Probably the simplest method of separation of two compounds is the *filtration* of a solid from a liquid. Filtration is rarely carried out directly on simple mixtures, but is more usually combined with *crystallization*. This technique relies upon the different solubilities of different compounds in a particular solvent, and the fact that most compounds are more soluble in hot solvent than in cold. Thus, when a hot, saturated solution of a mixture is allowed to cool much of the least soluble or most abundant compound separates out as crystals which can be filtered. The minor or more soluble components are concentrated in the mother liquor. In order to obtain a really pure product, *recrystallization* is often necessary.

40

3.2 Distillation

The purification of a liquid by vaporization and condensation is known as *distillation*. Non-volatile impurities are left behind in the distillation flask. Since different liquids have different boiling points, with careful regulation of temperature the components of a liquid mixture may be separated by *fractional distillation*. Some liquids form co-distilling mixtures or *azeotropes*, the compositions of which remain unchanged on distillation. For example, ethanol and water distil as a mixture containing 95% ethanol and 5% water.

In order to separate compounds with similar boiling points, it is necessary to use a *fractionating column*. This is a device which effectively increases the surface area of contact between liquid and vapour, so that vaporization and recondensation takes place many times in the column, thus enhancing the efficiency of separation. A further possible modification is to carry out distillation under reduced pressure. The boiling point of a liquid is dependent upon the pressure. The lower the pressure, the lower the boiling point. Thus, by carrying out distillation under reduced pressure, even high boiling or thermally unstable materials may be distilled at temperatures low enough to avoid decomposition. A typical laboratory apparatus for fractional distillation under reduced pressure is shown in Figure 3.1.

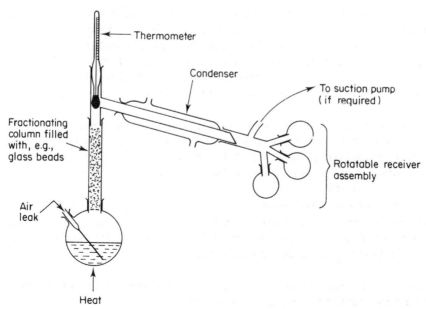

Figure 3.1 Typical laboratory apparatus for fractional distillation under reduced pressure

Probably the most important industrial application of fractional distillation is in the refining of crude oil. This is discussed in greater detail in Chapter 4.

Under appropriate conditions some solid organic compounds are conveniently purified by *sublimation*. This is a process analogous to distillation but involving direct vaporization of the solid, without an intermediate liquid state, followed by resolidification.

3.3 Solvent extraction

Different compounds have different solubilities in any particular solvent and this property can be used to separate them. An extreme example is provided by a mixture of the two dyes methylene blue and azobenzene (yellow). Methylene blue is very soluble in water, but sparingly soluble in diethyl ether, whereas azobenzene is very soluble in diethyl ether but sparingly soluble in water. Since water and ether are immiscible (i.e. do not mix), addition of water and ether to a mixture of azobenzene and methylene blue gives a yellow, ether (upper) layer and a blue, aqueous (lower) layer. Simple separation of the layers in a separating funnel, followed by a removal of the solvents by evaporation, gives the pure components. In general, however, organic mixtures are not so readily separated, and many repeated extractions are required in order to achieve a reasonable separation.

Question 3.1 The partition coefficient (concentration in ether divided by concentration in water) for a compound X is 3.0. Sixty grams of X is dissolved in 300 cm^3 water. Calculate the total amount of X extracted in each of the following experiments: (a) one extraction with 300 cm^3 of ether; (b) two successive extractions each with 150 cm^3 of ether; (c) three successive extractions each with 100 cm^3 of ether.

Solvent extraction is useful for separation of acidic compounds (soluble in aqueous base) or basic compounds (soluble in aqueous acid) from neutral organic compounds, which for the most part are insoluble in water but soluble in a water-immiscible organic solvent such as ether, chloroform, or ethyl acetate.

Countercurrent distribution is a process involving automated multiple solvent extractions. It is particularly useful for separation of closely related compounds (such as might be found in a plant extract), for which many individual extractions must be performed.

Question 3.2 Suggest ways by which the following pairs of compounds could be separated by extraction procedures:
(a) benzoic acid and naphthalene (a neutral organic compound);
(b) aniline (phenylamine) (a basic compound) and naphthalene.

3.4 Chromatography

Chromotography is a general name applied to a variety of separation techniques which all depend on the rate at which different compounds move in a *mobile phase* with respect to a fixed or *stationary phase*. In many cases the mobile phase is a liquid solvent moving over a solid stationary phase which may or may not have a liquid adsorbed on its surface. The different compounds move at different rates dependent upon their affinities for the mobile and stationary phases.

Chromatography may be used analytically, to indicate the number of components in a mixture and to allow their identification by comparison with known compounds, or preparatively, to enable separation and purification of individual components. A brief outline of each of the main types of chromatography is given below.

(a) Paper chromatography

In paper chromatography, a spot of a mixture of compounds (e.g. amino acids, see Chapter 12) is put about 2 cm from the bottom of a strip of absorbent paper, which is then suspended in a jar with the bottom of the paper just dipping into a layer of solvent (Figure 3.2). The solvent moves up the paper by capillary action, taking compounds from the mixture with it. When the solvent front has moved a fair distance, the paper is removed from the jar and the solvent is allowed to evaporate. The separated spots must be made visible, if they are not so already, by spraying the paper with some reagent which reacts with the compounds to form coloured derivatives. The technique is used mainly for analytical purposes. The ratio of the distance moved by a spot to the distance moved by the solvent is a constant for a given compound with given mobile and stationary phases. It is referred to as the R_f value for the compound in that system.

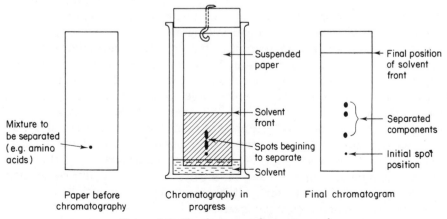

Figure 3.2 Simple paper chromatography

(b) Thin-layer chromatography

Thin-layer chromatography (t.l.c.) is very similar to paper chromatography, except that the stationary phase is an adsorbent material such as silicia or alumina spread as a thin film on the surface of a glass (or aluminium) plate. The technique is used for both analytical and preparative purposes. For preparative purposes, a thicker layer of stationary phase is used, and the mixture to be separated is spread as a narrow band near the bottom of the plate. By scraping off the stationary phase in sections after running the plate, and extracting each of them separately with a suitable solvent, the individual components of the mixture can be obtained in pure form.

(c) Column chromatography

In column chromotography an adsorbent material (e.g. silica or alumina) is packed evenly into a vertical glass column. The solvent is passed *down* the column under the influence of gravity or backed by a small positive pressure. The *eluate* (liquid emerging from the bottom of the column) is collected in small aliquots which are then evaporated to yield the separated compounds.

Sometimes the stationary phase material used in column chromatography is a charged ion-exchange resin (i.e. a polymer with ionic peripheral groups such as $-\overset{+}{N}Me_3\ Cl^-$ or $-SO_3^-\ Na^+$) or a permeable gel such as Sephadex®, in which case the techniques are usually called *ion exchange chromatography* or *gel permeation chromatography* respectively. While the operations in these cases are very similar, the principles of separation are different.

Ion exchange chromatography separates compounds according to their degree of charge. For example, carboxylic acids which dissociate to give anions RCO_2^- pass very slowly through a column of anion exchange resin (e.g. possessing $-^+NMe_3\ Cl^-$ groups), whereas species which are neutral or positively charged pass through more quickly. With a cation exchange resin, neutral and negatively charged species pass through quickly, whereas cations are retained longer. The amino acid analyser (a machine designed to give automatic analyses of amino acid mixtures—see Chapter 12), operates on the basis of ion exchange chromatography.

In gel permeation chromatography the stationary phase consists of a crosslinked polysaccharide material which swells up in water (or other solvent) to form a gel. Small molecules are able to penetrate (i.e. permeate) holes in the polymer matrix, whereas larger molecules cannot. Thus, the large molecules pass quickly through the spaces between gel particles and elute first, whereas the smaller ones are held longer by the gel and elute later. Separation is therefore on the basis of physical size of the molecules.

Use of finely graded silica or other stationary phase, together with precision pumping equipment and an eluate monitor (e.g. ultraviolet light or a refractive index detector), allows rapid and reproducible analytical application of column chromatography. The technique is then usually referred to as high performance liquid chromatography (h.p.l.c.).

(d) Gas chromatography

In *gas chromatography* the mobile phase is an unreactive gas, usually nitrogen or helium. The stationary phase is either a solid adsorbent or, more usually, a thin film of a viscous liquid deposited on the surface of an inert solid or as a coating on the walls of a very fine capillary column. The column is normally coiled to enable it to be fitted within a relatively small oven, and is then heated to an appropriate temperature to give adequate separation within as short a time as possible. A schematic diagram of a gas chromatograph is given in Figure 3.3.

A sample of material to be analysed, usually dissolved in solvent, is injected directly into the stream of gas by means of a syringe inserted through a rubber septum at the head of the column. As they pass through the column, the components separate according to volatility and affinity for the stationary phase. As the separated components elute from the column they pass through a detector which gives an electronic signal which is printed out as a peak on a chart recorder. A common type of detector is a *flame ionization detector* (FID), which works on the principle that the ions formed when a compound is burnt increase the electrical conductivity of the gas.

For preparative applications of gas chromatography the stream of eluting gas is split, a small part going to the detector, while the bulk passes into a cooled trap where it is collected.

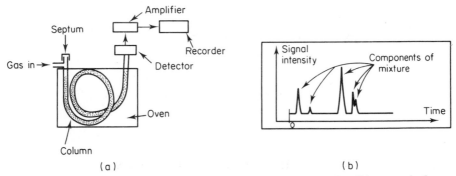

Figure 3.3 (a) Schematic diagram of a gas chromatograph; (b) a typical gas chromatogram (recorder trace)

Question 3.3 Suggest the most appropriate chromatographic techniques for each of the following applications: (a) analysis of a mixture of acetone (propanone) and methanol (both low boiling liquids); (b) separation of a large amount (10 g) of a mixture of benzamide and naphthalene (both neutral organic solids); (c) checking the purity of the separated components in (b); (d) separation of a mixture of mono-, di-, and trisaccharides (molecules containing one, two, or three sugar units).

3.5 Electrophoresis

Electrophoresis is a technique for separation of compounds which are capable of ionization, for example carboxylic acids (Chapter 8) and amines (Chapter 7). As shown in Figure 3.4, such molecules dissociate to give either negative or positive ions respectively. The extent of dissociation depends upon the pH of the solution and differs from compound to compound depending upon their K_a or K_b values.

In electrophoresis a mixture of ionizable compounds is placed at the centre of a piece of filter paper, or on a plate of agar jelly or similar material, and the paper or plate is soaked with a buffered electrolyte solution. When an electric field is applied, the anions of those molecules with high K_a values move rapidly towards the positive electrode, and those molecules with high K_b values move rapidly towards the negative electrode. Those molecules which are essentially neutral at the pH of the buffer solution move little. Thus, a separation is effected, and that part of the paper or jelly which contains a particular compound can be removed and extracted. This technique is particularly applicable to natural substances which contain both acidic and basic functions, such as amino acids and proteins (Chapter 12).

$$RCO_2H + H_2O \rightleftharpoons RCO_2^- + H_3O^+$$

$$K_a = \frac{[RCO_2^-][H_3O^+]}{[RCO_2H]}$$

$$RNH_2 + H_2O \rightleftharpoons RNH_3^+ + OH^-$$

$$K_b = \frac{[RNH_3^+][OH^-]}{[RNH_2]}$$

Figure 3.4 Dissociation of organic acids and bases

IDENTIFICATION TECHNIQUES

3.6 Elemental analysis

The first step in elucidating the structure of a compound is to determine its molecular formula. Once a compound has been thoroughly purified using one or more of the above techniques, this can be determined by combustion analysis or mass spectrometry (Section 3.7).

Complete combustion or an organic compound leads to the conversion of all the constituent carbon into carbon dioxide and all of the hydrogen into water. Measurement of the quantity of these products obtained from a known weight of compound leads to a knowledge of the proportions of carbon and hydrogen in the compound. In practice, combustion analyses are nowadays carried out on small samples (3–4 mg) and the amounts of CO_2 and H_2O are measured automatically. A catalyst is used to ensure complete combustion. Fritz Pregl was awarded a Nobel Prize for development of the first apparatus to enable such *microanalyses* to be performed.

For example, combustion of 10.00 mg of a compound X, containing only C, H, and (possibly) O atoms, was found to give 19.14 mg of CO_2 and 11.79 mg H_2O. Since 44 g of CO_2 (1 mol) is produced from 12 g of C, and 18 g of H_2O is produced from 2 g of H, the amounts of C and H in 10.00 mg of compound can be calculated as follows:

$$\text{Amount of C in 10.00 mg of compound} = 19.14 \times \frac{12}{44} = 5.22 \text{ mg};$$

$$\text{Amount of H in 10.00 mg of compound} = 11.79 \times \frac{2}{18} = 1.31 \text{ mg}.$$

Therefore

$$\text{Percentage carbon} = \frac{5.22}{10.00} \times 100 = 52.2\%;$$

$$\text{Percentage hydrogen} = \frac{1.13}{10.00} \times 100 = 13.1\%.$$

Since the only other element which could be present in the compound is oxygen,

$$\text{Percentage oxygen} = 100\text{-}52.2\text{-}13.1 = 34.7\%.$$

In order to calculate the *empirical formula*, proportions by mass must be converted into ratios of atoms by dividing each percentage by the atomic mass of the element concerned. Thus C:H:O atomic ratio is $\frac{52.2}{12} : \frac{13.1}{1} : \frac{34.7}{16}$, i.e. 4.35 : 13.1 : 2.17, i.e. 2:6:1. The formula thus obtained, i.e. C_2H_6O, is called the *empirical formula*. To determine the *molecular formula*, it is necessary to know the molecular mass of the compound. This can be determined by mass spectrometry or by measurement of any property which is dependent upon molecular mass, such as vapour density, osmotic pressure, or depression of the freezing point of a solvent. (These techniques are covered in most texts on physical chemistry.)

Mass spectrometry shows that the molecular mass of compound X is 46, and therefore the molecular formula is C_2H_6O.

Question 3.4 Combustion of 10.0 mg of a compound X, of molecular mass 28, gives 31.4 mg CO_2 and 12.9 mg H_2O. What is the molecular formula of X?

For compounds containing other elements, such as N,Cl, and S, methods are also available for determining the proportions of each of these present in the compound.

48

3.7 Mass spectrometry

A *mass spectrometer* is an instrument for separating charged particles according to their mass-to-charge ratio. In organic chemistry it is used primarily for determination of molecular mass and molecular formula, and as an aid in elucidation of structure. A schematic representation of a mass spectrometer is given in Figure 3.5. The whole instrument is maintained at a very low pressure.

The sample is introduced into the ion source where it is ionized (usually to give positive ions; i.e. $M \longrightarrow M^+ + e^-$). There are several methods for ionization. Some, such as field desorption (the repulsion of ions from a fine point at high voltage) or chemical ionization (ionization by bombardment with a previously ionized gas) impart little energy to the molecules, just enough to cause ionization. Thus, the only ions produced are those derived from the whole molecule (i.e. *molecular ions*). These are very useful for determination of molecular mass. The most common method of ionization, however, involves bombardment with high energy electrons (electron impact), which gives molecular ions possessing excess energy which break down to give smaller, *fragment ions*.

The ions are repelled from the ion source, and accelerated into a strong magnetic field which deflects the charge particles in a curved path. The lightest and most highly charged ions are deflected most. For practical purposes the charge on almost all ions is unit positive, so that the degree of deflection depends only upon the mass. At different magnetic field strengths, a different mass value passes through the collector slit, so a spectrum of the masses of the ions formed can be obtained by varying the field. A typical *mass spectrum* (for bromobenzene) is shown in Figure 3.6. (Notice that since bromine contains roughly a 1:1 mixture of two isotopes, ^{79}Br and ^{81}Br, two major molecular ions are observed.) All other ions are formed initially from the molecular ions by various sequential fragmentation processes. These processes occur in well-defined ways and organic chemists frequently use such *fragmentation patterns* to gain clues

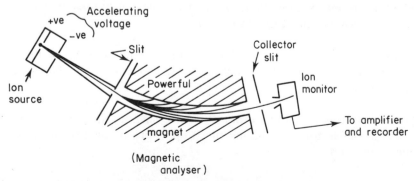

Figure 3.5 Schematic representation of a mass spectrometer

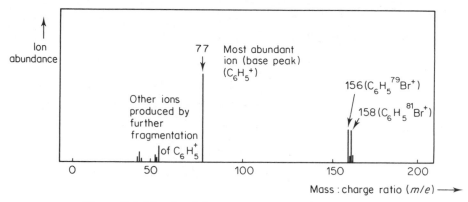

Figure 3.6 Sketch of the mass spectrum of bromobenzene

about the structures of molecules. A fragmentation pattern can also be used as a fingerprint for a particular compound.

If additional focusing of the ion beam is carried out using an electric field (electrostatic analyser), the masses of ions can be measured with very high resolution. Masses may then be measured to within a few parts per million. With this degree of accuracy it is possible to distinguish between ions of the same nominal mass, but different elemental compositions. For example, the accurate masses of the molecular ions derived from $^{12}C_2$ $^1H_4$ and $^{12}C^{16}O$ (both nominally 28) are significantly and measurably different (28.0313 and 27.9949 respectively). Thus, accurate mass measurement provides a method for direct determination of molecular formulas.

3.8 Ultraviolet spectroscopy

Organic compounds absorb electromagnetic radiation at certain wavelengths, and the wavelengths and intensities of the absorptions can give valuable structural information. Absorption in the infrared (i.r.) region (Section 3.9) gives information about the functional groups present in a molecule, whereas absorption in the ultraviolet (u.v.) and visible region gives information about the extent of conjugation.

The energy of electromagnetic radiation is inversely proportional to its wavelength. Wavelengths in the u.v. and visible region correspond to energy differences between electronic levels of molecules. Molecules with only σ electrons can undergo electronic transitions only from bonding (σ) to antibonding (σ^*) orbitals, which requires very high energy, outside the range of most u.v. spectrometers. Molecules with non-bonding (i.e. lone pair) electrons (n) or π electrons, on the other hand, undergo lower energy transitions (n $\longrightarrow \pi^*$ or $\pi \longrightarrow \pi^*$; e.g. Figure 3.7), which are observable.

In conjugated molecules, the separation between the highest π orbital and lowest π^* orbital is reduced, with the result that such molecules absorb

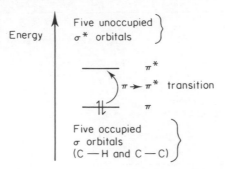

Figure 3.7 MO diagram for ethene

radiation of lower energy (i.e. longer wavelength). Thus, whereas the absorption maximum of ethene is about 190 nm, that for all-*trans*-octa-2,4,6-triene, with three conjugated double bonds, is 275 nm (Figure 3.8). Compounds containing many conjugated double bonds often absorb in the visible region, and are therefore coloured (e.g. carotene, see Section 9.2(a), and chlorophyll, Chapter 13).

Ultraviolet spectra are normally measured on dilute solutions in solvents with little or no absorption above 200 nm. The absorption maxima (λ_{max}) and molar absorbance values are characteristic physical properties useful for structure identification and for monitoring concentrations.

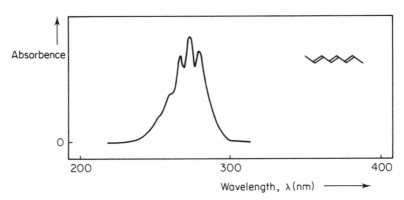

Figure 3.8 Sketch of the u.v. spectrum of all-*trans*-octa-2,4,6-triene

3.9 Infrared spectroscopy

Light with wavelengths in the infrared region has insufficient energy to promote electrons from one electronic energy level to another, but it has sufficient energy to convert a molecule from a ground state vibrational level to a higher vibrational state. These vibrations tend to be associated with particular bonds or groups of bonds in a molecule, each of which

Table 3.1   Typical infrared absorption ranges of organic groups

Group	Frequency of absorption (cm^{-1})
O−H (strong)	3600–3200 (sometimes lower)
N−H (strong)	3500–3000
C−H (strong)	3300–2800
C≡N (weak)	2250–2200
C≡C (weak)	2250–2150
C=O (very strong)	1850–1650
C=C (weak)	1680–1600
C−O (strong)	1300–1000
C−Cl (strong)	800–700

absorbs at a fairly specific wavelength which is therefore characteristic of that bond or group of bonds. Thus, i.r. spectroscopy is very useful for the identification of the functional groups present in an unknown compound, or for providing a fingerprint for characterization of a known compound by reference to a spectrum of authentic material.

Table 3.1 lists the characteristic frequencies for a number of common organic groupings, expressed in cm^{-1} (reciprocal of wavelength), the usual units in which i.r. spectra are measured. In addition, many absorptions which are characteristic of the molecule as a whole occur in the region 1500–700 cm^{-1}.

Within the basic range for a particular group (e.g. the >C=O group) it is often possible to deduce further structural information from the precise absorption frequency. For example, the C=O group frequency is different for an amide $(RCONR_2^1)$, a ketone $(RCOR^1)$, and an ester $(RCOOR^1)$, and even within the ketone category the C=O group frequency depends upon whether it is conjugated or forms part of a ring. Thus, whereas acetone (propanone) absorbs at 1710 cm^{-1}, cyclopropanone (a cyclic ketone with the same number of carbon atoms) absorbs at 1815 cm^{-1}.

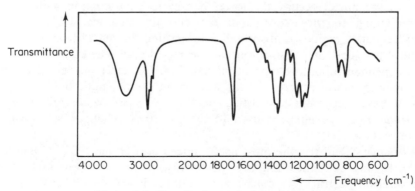

Figure 3.9 Sketch of a typical i.r. spectrum

c

Infrared spectra can be obtained on solutions in appropriate solvents, using cells made of sodium chloride (NaCl does not absorb in the i.r. region). Alternatively, liquids can be run as thin films between NaCl plates, whereas solids are often run as a mull (ground up with a viscous liquid) or as a KBr disc (compressed disc of a ground up mixture of the compound and KBr).

A typical i.r. spectrum is shown in Figure 3.9. Since transmittance rather than absorption is plotted on the vertical axis, absorption peaks point downwards.

Question 3.5 Deduce which of the following compounds is responsible for the i.r. spectrum shown in Figure 3.9:

(a) $CH_3CH_2\overset{\underset{\displaystyle |}{OH}}{C}HCH_3$;

(b) $(CH_3)_2\overset{\underset{\displaystyle |}{OH}}{C}-CH_2\overset{\underset{\displaystyle ||}{O}}{C}CH_3$

(c) Cl_3CCN

(d)

$C_6H_5\overset{\underset{\displaystyle ||}{O}}{C}-OCH_2CH_3$.

3.10 Nuclear magnetic resonance spectroscopy

Just as electrons possess spin, which is defined by the spin quantum number and which dictates that only two electrons with opposite (i.e. 'paired') spins can occupy a given molecular orbital, so also nuclear particles, protons and neutrons, possess the property of spin. In any given nucleus some of the spins may be paired, but there may be a residium of unpaired spins. Clearly, this must always be the case for nuclei with an odd mass number (odd number of protons plus neutrons). A spinning, charged body can be regarded as a small magnet, and, when placed in a magnetic field, it can adopt two different orientations, aligned with or opposed to the applied field, which have different energies. Under normal conditions more of the nuclei occupy the lower energy level. Radiation with energy corresponding to the energy gap between the two levels (i.e. in the radio-frequency region) is absorbed in promoting the nuclei from one level to the other, and this absorption can be monitored. The exact frequency (v) is dependent upon the type of nucleus (^1H, ^{13}C, etc.) and the electronic environment in which it resides, as well as on the magnetic field strength. A schematic representation of a *nuclear magnetic resonance (n.m.r.) spectrometer,* the equipment used to monitor these changes, is shown in Figure 3.10.

In organic chemistry, the most useful nuclei are ^1H and ^{13}C (^{12}C has zero spin), because of the ubiquity of hydrogen and carbon in organic compounds. Spectra are recorded on solutions in solvents such as $CDCl_3$ (deuteriochloroform) which do not contain ^1H atoms. Sophisticated

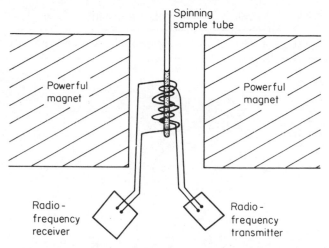

Figure 3.10 Schematic representation of an n.m.r. spectrometer

instruments are necessary in order to record ^{13}C n.m.r. spectra due to the low abundance of ^{13}C (1.1% of natural carbon) and its lower signal response relative to ^1H.

N.m.r. absorptions are expressed as *chemical shifts* (δ), measured in parts per million (p.p.m.) from the absorption position of tetramethylsilane (TMS), where

$$\delta = \frac{v - v_{TMS}}{v_{TMS}} \times 10^6 \text{ p.p.m.}$$

Sometimes ^1H n.m.r. chemical shifts are expressed as τ values, where $\tau = 10 - \delta$.

Chemical shifts depend upon the electronic environment of a nucleus. Any effect which reduces the δ value of a nucleus is referred to as a *shielding* effect, whereas an effect which increases the δ value is called a *deshielding* effect. The actual chemical shifts are highly characteristic of particular locations of nuclei. Some typical shifts for proton (^1H) spectra are given in Table 3.2. Variations within the ranges given are caused by differences in the neighbouring substituent groups.

A similar table of chemical shift values for ^{13}C nuclei in different environments can also be drawn.

The ^1H n.m.r. spectrum of methyl formate (methyl methanoate) is shown in Figure 3.11. In addition to the chemical shift values, *integration* of the peak areas gives an indication of the number of protons giving rise to each peak (integration of ^{13}C spectra is less reliable).

Very few ^1H n.m.r. spectra are as simple as that shown in Figure 3.11. Signals may appear as *singlets* (as in Figure 3.10), but often appear as *doublets, triplets*, or more complex *multiplets* due to spin–spin *coupling* between nuclei. In most circumstances, the multiplicity is determined by

Table 3.2 Typical ^{1}H n.m.r. chemical shifts

Location of ^{1}H nucleus	Range of values of δ (p.p.m.)
H—C—C—	0.8–1.4
H—C—C— (C=O)	2.0–2.3
H—C—O—C	3.3–3.7
H—C—O—C— (C=O)	3.8–4.2
H—C=C—	4.5–6.0
H— (benzene ring)	6.5–8.0
H—C—O— (C=O)	7.9–8.1
H—C—C— (C=O)	9.7–9.9
H—O—C—	Very variable
H—O—C— (C=O)	>10.0

the number of hydrogen atoms attached to adjacent carbon atoms. The presence of one adjacent hydrogen atom gives rise to a doublet, two to a triplet, and so on (i.e. $n + 1$ lines for n adjacent hydrogens). In ^{13}C n.m.r. spectra ^{13}C–^{13}C coupling is not observed due to the low probability of finding two ^{13}C nuclei in the same molecule, but ^{13}C–^{1}H coupling is observed. ^{13}C n.m.r. spectra, therefore, could be very complex, but in practice they are usually simplified by *decoupling* the ^{1}H nuclei (by irradiation with a second radio-frequency supply). Figure 3.12 records the ^{1}H n.m.r. spectrum of butanone. Note that there are three signals, which integrate in the ratio 2:3:3, and that one signal is a singlet ($\underline{CH_3}C$—, with C=O), one is a quartet (—C—$\underline{CH_2}CH_3$, with C=O) and one a triplet (—CH_2—$\underline{CH_3}$).

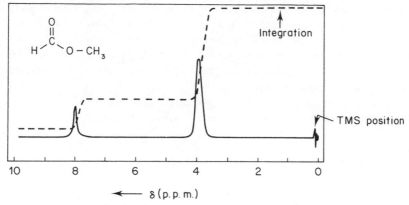

Figure 3.11 ^{1}H n.m.r. spectrum of methyl formate

Question 3.6 Predict the approximate appearance of the ^{1}H n.m.r. spectrum of methyl 4,4-dimethylpent-2-enoate:

3.11 Diffraction techniques

The regular arrays of atoms in a crystal diffract a beam of incident X-rays. The diffraction pattern produced can be recorded on a photographic film and analysed by a computer to give very precise information about the location of atoms within the crystal. In this way a

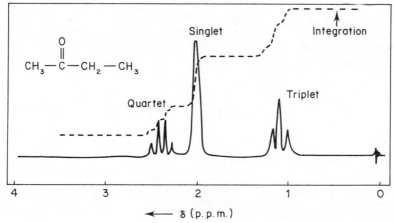

Figure 3.12 ^{1}H n.m.r. spectrum of butanone

detailed picture of a molecule can be obtained, not only enabling its gross structure to be determined, but also providing accurate bond lengths and bond angles. The technique is known as *X-ray crystallography* or *X-ray diffraction*. The main drawbacks of the technique are the difficulty in obtaining suitable crystals and its great expense in terms of computer time, especially for complex molecules.

A similar technique suitable for gaseous molecules involves diffraction of a beam of electrons, and is therefore called *electron diffraction*.

Answers to questions

3.1 $\dfrac{\text{Concentration in ether}}{\text{concentration in water}} = 3.$

(a) Let mass of X in ether be x g and hence mass of X in water is $60\text{-}x$ g. Thus

$$\frac{x/300}{(60\text{-}x)/300} = 3,$$

$$x = 3(60\text{-}x),$$
$$4x = 180,$$
$$x = 45 \text{ g},$$

i.e. amount extracted = 45 g.

(b) For the first extraction:

$$\frac{x/150}{(60\text{-}x)/300} = 3,$$

and so $x = 36$ g.

Therefore amount remaining in water = 24 g.

For the second extraction:

$$\frac{x/150}{(24\text{-}x)/300} = 3$$

and so $x = 14.4$ g.

Therefore total amount of X extracted = $36 + 14.4$ g
$$= 50.4 \text{ g}.$$

(c) For the first extraction:

$$\frac{x/100}{(60\text{-}x)/300} = 3$$

and so $x = 30$ g.

Therefore amount remaining in water $= 60 - 30$ g
$= 30$ g.

For the second extraction:

$$\frac{x/100}{(30-x)/300} = 3$$

and so $x = 15$ g.

Therefore amount remaining in water $= 30 - 15$ g
$= 15$ g.

For the third extraction:

$$\frac{x/100}{(15-x)/300} = 3$$

and so $x = 7.5$ g.

Thus total extraction $= 30 + 15 + 7.5$ g
$= 52.5$ g.

3.2 (a) Dissolve mixture in ether and extract with aqueous base (NaOH); naphthalene left in ether layer; benzoic acid dissolves as its salt (sodium benzoate) in the aqueous layer, and can be recovered by acidification. (b) Dissolve mixture in ether and extract with aqueous acid (HCl); naphthalene left in ether layer; aniline dissolves as its salt (anilinium chloride) in the aqueous layer, and can be recovered by basification.

3.3 (a) Gas chromatography; because of the volatility of the compounds; (b) column chromatography; because of the relatively large amount of material involved; (c) thin layer chromatography; each component should give only one spot; (d) gel permeation chromatography; because of the different molecular sizes.

3.4 Weight of carbon in 10.0 mg of compound Y $= \dfrac{31.4 \times 12}{44}$

$= 8.58$ mg,

Weight of hydrogen in 10.0 mg of compound Y $= \dfrac{12.9 \times 2}{18}$
$= 1.43$ mg

58

Total weight of C and H = 10.01 mg,
i.e. no other elements present.

Percentage carbon in Y = $\dfrac{8.58}{10.0} \times 100 = 85.8\%$

Percentage hydrogen in Y = $\dfrac{1.43}{10.0} \times 100 = 14.3\%,$

Atomic ratios, C:H = $\dfrac{85.8}{12} : \dfrac{14.3}{1} = 7.15 : 14.3 = 1:2.$

Therefore

Empirical formula is CH_2. Since molecular mass = 28, Molecular formula is C_2H_4.

3.5 Spectrum indicates both O—H and C=O groups, and is therefore due to compound (b).

3.6 Singlet in the range $\delta = 0.8–1.4$ p.p.m. due to nine H atoms; singlet in the range $\delta = 3.3–3.7$ p.p.m. due to three H atoms; two doublets in the range $\delta = 4.5–6.0$ p.p.m., each due to one H atom.

CHAPTER 4

Aliphatic Hydrocarbons

Hydrocarbons (containing only carbon and hydrogen) comprise a wide range of natural and synthetic compounds. The simplest hydrocarbon is methane (CH_4) which is the end product of the anaerobic decomposition of living organisms and is the major constituent of natural gas. At the other end of the molecular mass scale is polythene which is obtained by polymerizing up to a thousand or more molecules of ethene (ethylene, C_2H_4). The importance of hydrocarbons lies in their use as fuels and as raw materials for the organic chemicals industry.

Hydrocarbons can be conveniently classified as *aliphatic* or *aromatic*. Aromatic hydrocarbons are dealt with in Chapter 5. The term aliphatic is applied to compounds which do not possess aromatic properties. Aliphatic hydrocarbons are of three main types, *alkanes, alkenes, and alkynes,* and these are the subject of this chapter.

4.1 Structures of hydrocarbons

Acyclic (i.e. non-cyclic) alkanes have the general formula C_nH_{2n+2}. Therefore their structures differ by units of CH_2 and thus they constitute a homologous series. They are described as *saturated* since they contain no multiple bonds. Alkenes possess a carbon–carbon double bond whereas alkynes possess a triple bond. Such hydrocarbons are described as *unsaturated,* and acyclic examples have the general formulas C_nH_{2n} and C_nH_{2n-2} respectively. Cyclic alkanes, alkenes, and alkynes possess two fewer hydrogen atoms per ring than their acyclic counterparts. Some examples of aliphatic hydrocarbons are shown in Figure 4.1.

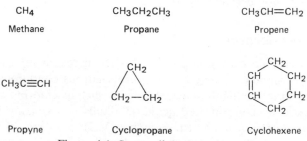

Figure 4.1 Some aliphatic hydrocarbons

59

Although there is only one compound of formula CH_4, one of formula C_2H_6, and one of formula C_3H_8, there are two of formula C_4H_{10}, and as the number of carbon atoms increases the number of possible structural isomers increases. Furthermore, propene and cyclopropane are isomers, since they both have the formula C_3H_6. In the alkene series, as the number of carbon atoms increases the number of isomers increases even more rapidly than in the alkane series due to the different possible locations of the double bond and the possibility of geometrical (*cis–trans*) isomerism (Section 2.1). In cyclic compounds the possibilities of different ring sizes and *cis–trans* isomerism again lead to many possible isomers.

Question 4.1   Draw all possible isomers of formula C_5H_{10}.

4.2 Physical properties

Carbon–carbon and carbon–hydrogen bonds are essentially non-polar in character and consequently the only forces attracting hydrocarbon molecules to each other are the very weak van der Waals forces (Section 1.11). As a result hydrocarbons have low melting and boiling points compared with many other compounds of similar molecular mass. Table 4.1 shows the melting and boiling points of a few representative examples of the different types of hydrocarbons.

It can be seen from Table 4.1 that within a homologous series (e.g. methane, ethane, propane, etc.) the boiling points progressively increase, but that branched-chain hydrocarbons have lower boiling points than their straight-chain isomers. With increased branching the molecular shape becomes more compact so that the surface area decreases, and with it the strength of the van der Waals forces.

Hydrocarbons follow the general rule of thumb that 'like dissolves like' (Section 1.11) and are soluble in non-polar solvents such as light petroleum, diethyl ether (ethoxyethane), benzene, and carbon tetrachloride (tetrachloromethane), but are virtually insoluble in polar solvents such as water.

Alkenes and alkynes exhibit characteristic bands due to the C=C and C≡C groups in their i.r. spectra, and all hydrocarbons exhibit bands due to C—H and C—C bonds (Chapter 3).

4.3 Bonding and reactivity

Since alkanes contain only non-polar σ bonds, they are relatively unreactive. Their only significant reactions require vigorous conditions (e.g. combustion or cracking) or highly reactive species such as chlorine atoms. They are not attacked by concentrated acids or alkalis or oxidizing agents such as potassium permanganate (manganate VII) solution. Cycloalkanes behave in general in the same way as acyclic alkanes, except when the ring

Table 4.1 Melting and boiling points of hydrocarbons

Hydrocarbon	Structure	m.p. (°C)	b.p. (°C)
Alkanes			
Methane	CH_4	-183	-162
Ethane	CH_3CH_3	-172	-88.5
Propane	$CH_3CH_2CH_3$	-187	-42
Butane	$CH_3CH_2CH_2CH_3$	-138	0
Pentane	$CH_3(CH_2)_3CH_3$	-130	36
2-Methylpropane	$(CH_3)_2CHCH_3$	-159	-12
2-Methylbutane	$(CH_3)_2CHCH_2CH_3$	-160	28
2,2-Dimethylpropane	$(CH_3)_4C$	-17	9.5
Alkenes			
Ethene	$CH_2{=}CH_2$	-169	-102
Propene	$CH_3CH{=}CH_2$	-185	-48
But-1-ene	$CH_3CH_2CH{=}CH_2$	-185	-6.5
cis-But-2-ene		-139	4
trans-But-2-ene		-106	1
2-Methylpropene		-141	-7
Alkynes			
Ethyne	$HC{\equiv}CH$	-82	-75
Propyne	$CH_3C{\equiv}CH$	-101.5	-23
But-1-yne	$CH_3CH_2C{\equiv}CH$	-122	9
But-2-yne	$CH_3C{\equiv}CCH_3$	-24	27
Cyclic hydrocarbons			
Cyclobutane		-80	13
Cyclopentane		-94	49
Cyclopentene		-93	46

is strained, as in cyclopropane and to a lesser extent cyclobutane. These compounds are much more reactive, and undergo some addition reactions (e.g. Figure 4.2). In this respect they resemble alkenes (see below).

Alkenes and alkynes contain π bonds (Section 1.5), which are weaker than σ bonds because their electrons are less tightly held. As a result of

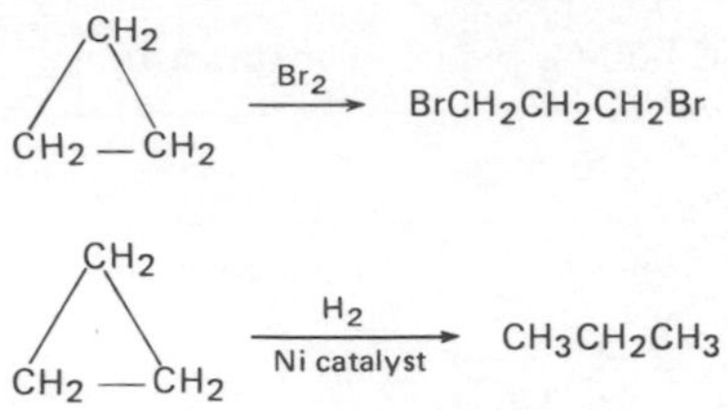

$$CH_2 \overset{\displaystyle CH_2}{\underset{\displaystyle CH_2}{\bigtriangleup}} \xrightarrow{\text{Br}_2} BrCH_2CH_2CH_2Br$$

$$\overset{\displaystyle CH_2}{\underset{\displaystyle CH_2 - CH_2}{\bigtriangleup}} \xrightarrow[\text{Ni catalyst}]{\text{H}_2} CH_3CH_2CH_3$$

Figure 4.2 Addition reactions of cyclopropane

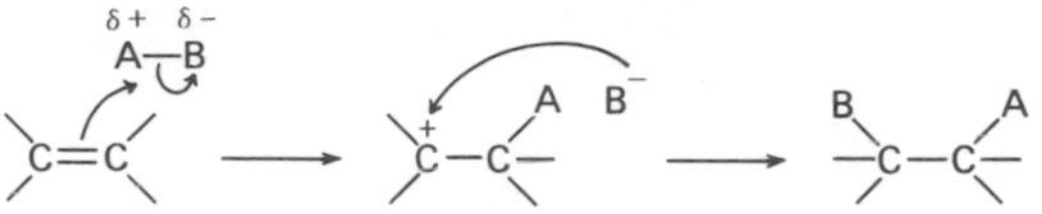

Figure 4.3 The mechanism of electrophilic addition reactions

$$CH_3C\equiv CH + Na^+ \overset{-}{N}H_2 \longrightarrow CH_3C\equiv \overset{-}{C} Na^+ + NH_3$$

Figure 4.4 Formation of the sodium salt of propyne

this these hydrocarbons are much more reactive than alkanes and undergo addition reactions. Such reactions usually proceed by attack of an electrophilic molecule on the electrons of the π bonds, and are therefore known as electrophilic addition reactions (Figure 4.3).

Since the order of electronegativities of hybridized carbon atoms is $sp^3 < sp^2 < sp$, hydrogen atoms attached to sp hybridized carbon atoms are slightly acidic. As a result alk-1-ynes have acidities somewhere between water and ammonia. Thus, alk-1-ynes form salts by reaction with strong bases (e.g. Figure 4.4).

4.4 Reactions with halogens

Alkanes react with halogens other than iodine to give mixtures of halogenoalkanes. For example, if chlorine is allowed to react with methane, the hydrogen atoms are progressively substituted by chlorine

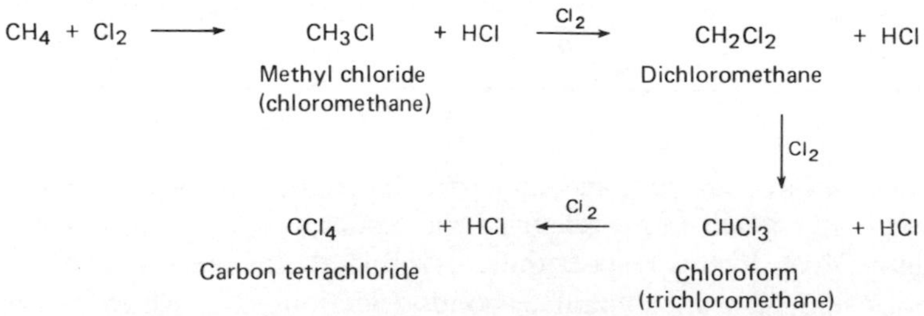

Figure 4.5 The progressive chlorination of methane

$$\text{Cl}_2 \xrightarrow[\text{or light}]{\text{Heat}} \text{Cl}^{\cdot} + \text{Cl}^{\cdot}$$

Initiation step

Propagation steps

$$\text{Cl}^{\cdot} + \text{CH}_4 \longrightarrow \text{HCl} + \text{CH}_3^{\cdot}$$

$$\text{CH}_3^{\cdot} + \text{Cl}_2 \longrightarrow \text{Cl}^{\cdot} + \text{CH}_3\text{Cl}$$

Termination steps

$$2\,\text{Cl}^{\cdot} \longrightarrow \text{Cl}_2$$

$$2\,\text{CH}_3^{\cdot} \longrightarrow \text{CH}_3\text{CH}_3$$

$$\text{CH}_3^{\cdot} + \text{Cl}^{\cdot} \longrightarrow \text{CH}_3\text{Cl}$$

Figure 4.6 The free radical chain process
for the chlorination of methane

atoms (Figure 4.5). The ratio of the products depends upon the proportions of the reactants. The higher the concentration of chlorine the greater the proportion of highly chlorinated products.

The reaction can be carried out either in the dark at 250 °C or at room temperature in the presence of light. These conditions are necessary to provide the energy required to convert the chlorine molecules into atoms. This is the initiation step of the free radical chain process (Figure 4.6).

Chlorine atoms have seven valency electrons, one of which is unpaired. They are extremely reactive because of their tendency to gain an electron to complete their octets. When a chlorine atom collides with a CH_4 molecule, it abstracts a hydrogen atom, producing a molecule of HCl and a highly reactive methyl radical, $CH_3^{\cdot}$ (Section 1.8). This reacts with a chlorine molecule in a second *propagation step,* producing CH_3Cl and a further chlorine atom which can start the cycle all over again. In principle just one photon of light is required to bring about the entire reaction, although in practice this is not achieved because *termination steps* remove free radicals from the system.

Similar reactions take place when methane reacts with fluorine or bromine. The order of reactivity is $F_2 > Cl_2 > Br_2$. Indeed the reaction with fluorine is explosive even in the dark at room temperature and is usually controlled by dilution with an inert gas and by reducing the temperature. Bromine reacts under similar conditions to chlorine though less vigorously.

Similar reactions occur when halogens react with other alkanes, though the more complex the alkane the greater the number of possible products. For example when propane and 2-methylpropane react with chlorine there are two monosubstituted products in each case (Figure 4.7).

Question 4.2 Predict how many monochloro isomers are obtained when chlorine reacts with 2-methylbutane.

Notice that rather more 2-chloropropane (Figure 4.6(a)) and 2-chloro-2-methylpropane (Figure 4.6(b)) are obtained than would be pre-

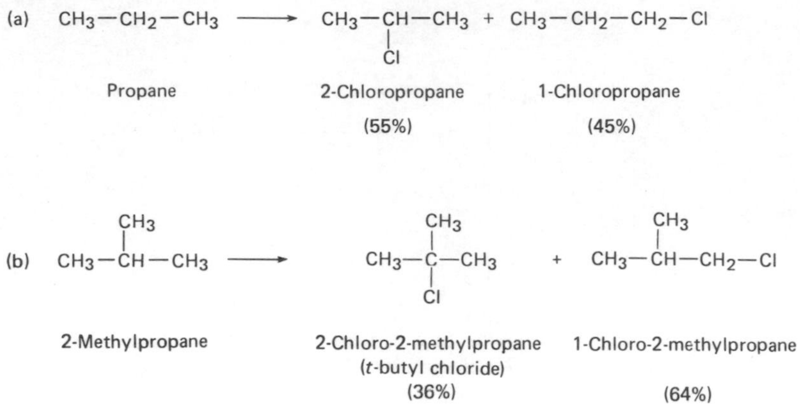

(a) $CH_3-CH_2-CH_3 \longrightarrow CH_3-\underset{\underset{Cl}{|}}{CH}-CH_3 + CH_3-CH_2-CH_2-Cl$

Propane   2-Chloropropane   1-Chloropropane

(55%)   (45%)

(b) $CH_3-\underset{\underset{CH_3}{|}}{CH}-CH_3 \longrightarrow CH_3-\underset{\underset{Cl}{|}}{\overset{\overset{CH_3}{|}}{C}}-CH_3 + CH_3-\underset{\underset{CH_3}{|}}{CH}-CH_2-Cl$

2-Methylpropane   2-Chloro-2-methylpropane   1-Chloro-2-methylpropane
(t-butyl chloride)
(36%)   (64%)

Figure 4.7 Reactions of propane and 2-methylpropane with chlorine

dicted simply on the basis of statistical replacement of hydrogen atoms. The crucial step which determines the proportions of the products formed is the abstraction of a hydrogen atom by a chlorine atom. Consider, for example, the two radicals formed by abstraction of a hydrogen atom from propane (Figure 4.8). Since the 2-propyl radical is more stable than the 1-propyl radical, more 2-chloropropane is obtained than would be predicted on a statistical basis. In general, the order of stability of free radicals is tertiary > secondary > primary > methyl, i.e. $R_3C^{\cdot} > R_2CH^{\cdot} > RCH_2^{\cdot} > CH_3^{\cdot}$ (the symbol R is generally used to represent an alkyl or aryl group).

Alkenes react with halogens even at moderate temperatures in the dark. The products are formed by simple addition of the halogen molecule across the multiple bond. Indeed, since the products are colourless the rapid decolourization of bromine can be used as a simple visual test for the presence of the alkene functional group in a compound. Alkynes react similarly but more slowly.

The mechanism involves electrophilic attack by the bromine molecule on the π electrons (Figure 4.9(a)). Although there is no permanent dipole in a bromine molecule, in the proximity of π electrons a dipole is induced and the molecule can then behave as an electrophile. An intermediate bromonium ion is formed which is attacked from the rear by bromide anion to give the dibromo product. Thus, cyclohexene reacts to give trans-1,2-dibromocyclohexane (Figure 4.9(b)).

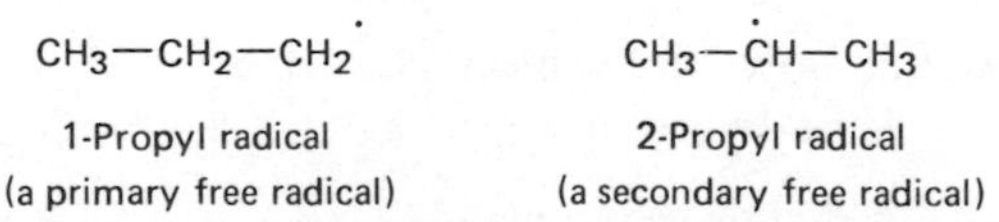

$CH_3-CH_2-\overset{\cdot}{CH_2}$ $CH_3-\overset{\cdot}{CH}-CH_3$

1-Propyl radical   2-Propyl radical
(a primary free radical)   (a secondary free radical)

Figure 4.8 The two radicals derived from propane

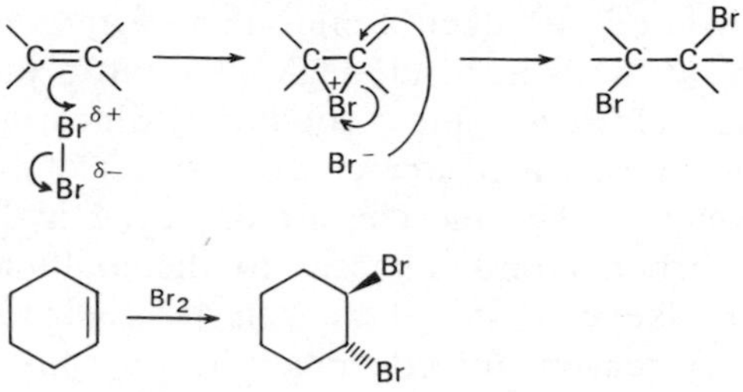

Figure 4.9 Electrophilic attack by the bromine molecule on a C=C bond

$$R^1C{\equiv}CR^2 \xrightarrow{X_2} R^1CX{=}CXR^2 \xrightarrow{X_2} R^1CX_2{-}CX_2R^2$$

Figure 4.10 Halogenation of an alkyne

Question 4.3 Give the products formed in the reaction of bromine with (a) *cis*-but-2-enedioic acid (maleic acid) and (b) *trans*-but-2-enedioic acid (fumaric acid).

Addition of one mole of a halogen to one mole of an alkyne gives a dihalogenoalkene, which can react with a further mole of halogen to give a tetrahalogenoalkane (Figure 4.10).

In the presence of water, the reaction of an alkene with bromine gives a bromohydrin (Figure 4.11). The mechanism involves reaction of the intermediate bromonium ion with water rather than with bromide ion. Chlorohydrins are similarly formed from the reactions of alkenes with chlorine water.

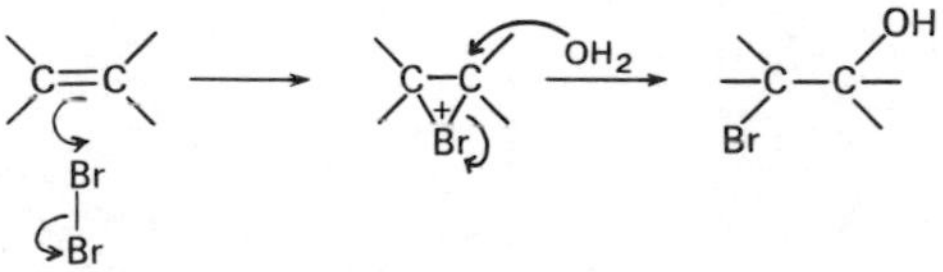

Figure 4.11 The mechanism of production of a bromohydrin from an alkene

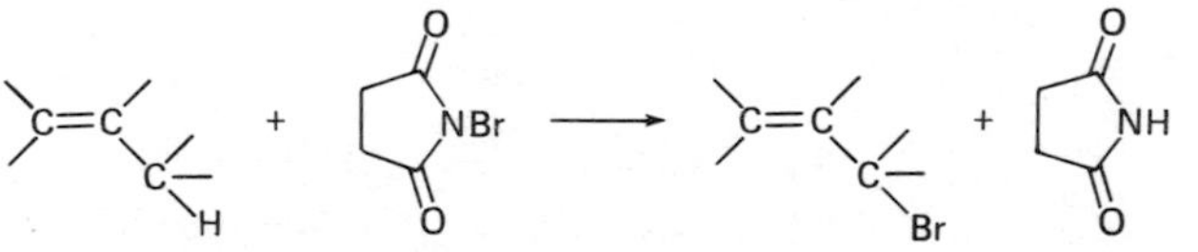

N-Bromosuccinimide   Succinimide

Figure 4.12 Allylic bromination with N-bromosuccinimide

Question 4.4 Predict the products formed when a solution of bromine and sodium nitrate react with ethene.

Carbon–hydrogen bonds in alkenes and alkynes are susceptible to substitution by halogen as in alkanes, although addition to the multiple bonds usually predominates. However, under conditions which favour free radical reactions and in the presence of low concentrations of free halogen substitution can be achieved. The most easily displaced hydrogen atoms are those attached to carbon atoms adjacent to the multiple bond (i.e. the allylic positions in alkenes), and these can be replaced selectively. A convenient laboratory reagent for effecting *allylic bromination* is *N*-bromo succinimide (Figure 4.12).

4.5 Reaction with acids

Alkanes other than highly strained cycloalkanes, such as cyclopropane, do not react with acids. Alkenes and alkynes, on the other hand, readily undergo addition reactions with hydrogen halides and sulphuric acid. The reaction involves initial electrophilic attack by the electron-deficient hydrogen atom of HX to give a carbonium ion (Section 1.8), which then reacts with the anion derived from the acid to give the addition product (e.g. Figure 4.13).

Unsymmetrical alkenes usually give predominantly one of the two possible products. For example, addition of HBr to propene gives almost entirely 2-bromopropane. This can be understood by considering the relative stabilities of the two possible intermediate carbonium ions (Figure 4.14). Addition of a proton to carbon atom 1 of the alkene gives a secondary carbonium ion, which is more stable than the isomeric primary carbonium ion formed by the alternative mode of addition. This is because

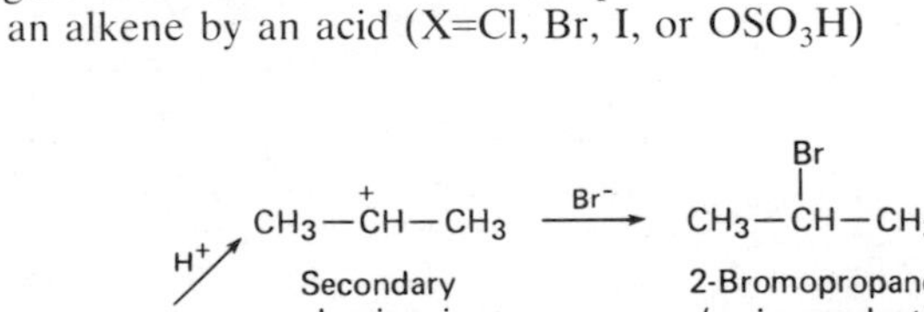

$(X = Cl, Br, I, OSO_3H)$

Figure 4.13 Mechanism of electrophilic addition to
an alkene by an acid ($X=Cl$, Br, I, or OSO_3H)

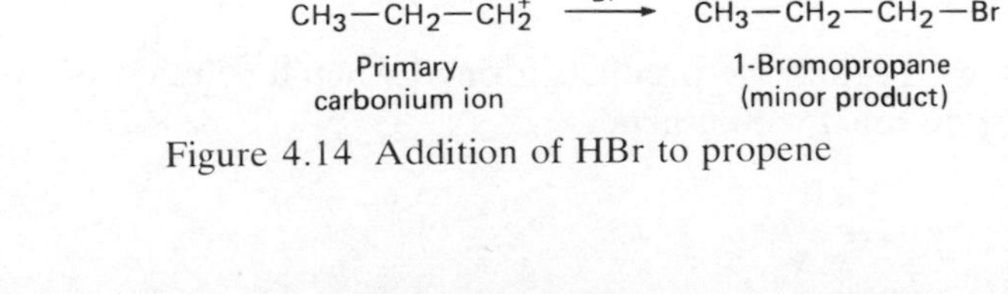

Figure 4.14 Addition of HBr to propene

secondary carbonium ions have two alkyl groups capable of delocalizing the positive charge, whereas primary carbonium ions have only one such group. Indeed, the order of stability of carbonium ions is tertiary > secondary > primary > methyl, i.e. $R_3C^+ > R_2CH^+ > RCH_2^+ > CH_3^+$.

Question 4.5 Predict the major product formed in each case when HCl reacts with (a) 2-methylpropene; (b) 2-methylbut-2-ene.

The addition of HBr to alkenes can also be carried out under free radical conditions by addition of a catalyst such as a peroxide. In this case the orientation of the addition is reversed due to the different reaction mechanism.

The initial product in the reaction of alkenes with sulphuric acid is an alkyl hydrogen sulphate, but on warming with water this is hydrolysed to give an alcohol. This is an important industrial procedure for production of alcohols (Figure 4.15(a)). A more convenient laboratory method for addition of the elements of water to a double bond involves reaction with mercury(II) acetate (ethanoate), $(CH_3CO_2)_2Hg$, followed by reduction with sodium borohydride, $NaBH_4$ (Figure 4.15(b)).

Hydrogen halides add to alkynes in a stepwise fashion giving first halogenoalkenes and then dihalogenoalkanes (Figure 4.16). The orientation of the addition of the second molecule of HX can again be understood by considering the relative stabilities of the two possible intermediate carbonium ions. The one leading to the product is stabilized by interaction of

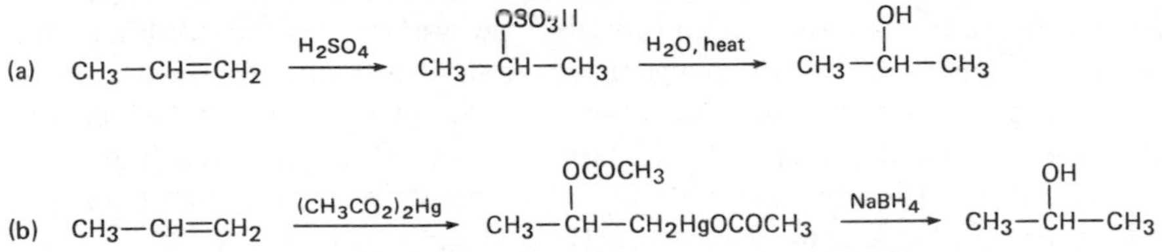

(a) $CH_3-CH=CH_2$ $\xrightarrow{H_2SO_4}$ $CH_3-\overset{\overset{\displaystyle OSO_3H}{|}}{CH}-CH_3$ $\xrightarrow{H_2O, \text{ heat}}$ $CH_3-\overset{\overset{\displaystyle OH}{|}}{CH}-CH_3$

(b) $CH_3-CH=CH_2$ $\xrightarrow{(CH_3CO_2)_2Hg}$ $CH_3-\overset{\overset{\displaystyle OCOCH_3}{|}}{CH}-CH_2HgOCOCH_3$ $\xrightarrow{NaBH_4}$ $CH_3-\overset{\overset{\displaystyle OH}{|}}{CH}-CH_3$

Figure 4.15 Addition of the elements of water to propene

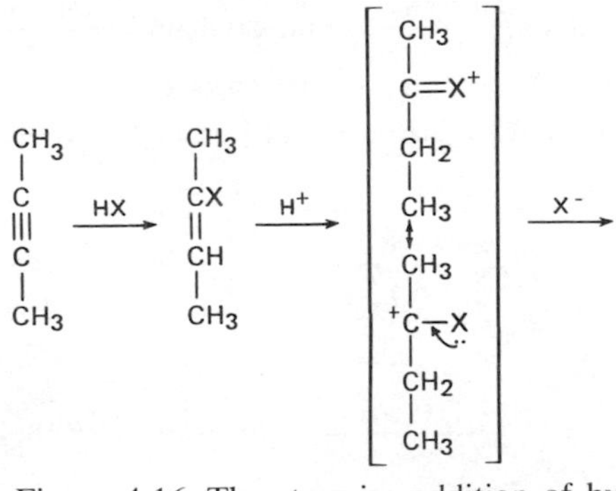

Figure 4.16 The stepwise addition of hydrogen halides to propyne

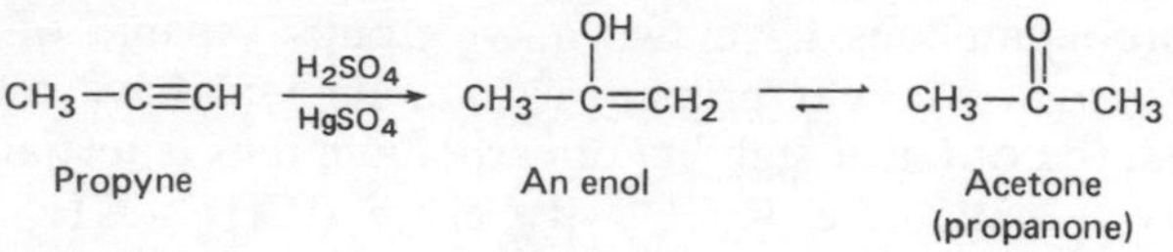

$$CH_3-C\equiv CH \xrightarrow[\text{HgSO}_4]{\text{H}_2\text{SO}_4} CH_3-\overset{\overset{\displaystyle OH}{|}}{C}=CH_2 \rightleftharpoons CH_3-\overset{\overset{\displaystyle O}{\|}}{C}-CH_3$$

Propyne   An enol   Acetone
(propanone)

Figure 4.17 Addition of the elements of water to propyne

the lone pairs of electrons on the halogen atom already present with the positive centre (Figure 4.16).

Addition of the elements of water to an alkyne can be carried out using sulphuric acid and a catalyst of mercury(II) sulphate, $HgSO_4$. The product is a ketone (or ethanal in the case of ethyne itself) formed by isomerization of the enol formed initially (Figure 4.17).

4.6 Reactions with borane

Under normal conditions alkanes do not react with solutions of borane (BH_3) in ether solvents. On the other hand alkenes and alkynes undergo rapid addition reactions (hydroboration) to give organoboranes. Hydroboration of simple mono- or di-substituted double bonds gives trialkylboranes (Figure 4.18). For more hindered alkenes the reaction may stop at the mono- or dialkyl borane stage.

Since the polarization of the H—B bond in borane is such that hydrogen bears a partial negative charge (in contrast to $H^{\delta+}$—$Br^{\delta-}$) the orientation of addition of H—BH_2 to an unsymmetrical alkene is opposite to that of H—Br. Furthermore, the reaction involves *cis* addition to the double bond.

Organoboranes are useful synthetic reagents. For example, on treatment with alkaline hydrogen peroxide they react to give alcohols in which the OH group takes the place of the boron atom. Thus, this reaction gives alcohols which are isomeric with those obtained by treatment of alkenes with either sulphuric acid or mercury(II) acetate (e.g. Figure 4.19).

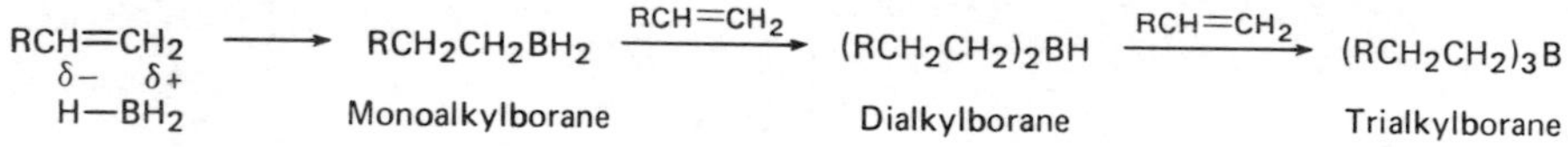

$$\underset{\underset{\text{H}-\text{BH}_2}{\overset{\delta-\quad\delta+}{}}}{RCH=CH_2} \longrightarrow \underset{\text{Monoalkylborane}}{RCH_2CH_2BH_2} \xrightarrow{RCH=CH_2} \underset{\text{Dialkylborane}}{(RCH_2CH_2)_2BH} \xrightarrow{RCH=CH_2} \underset{\text{Trialkylborane}}{(RCH_2CH_2)_3B}$$

Figure 4.18 The stepwise alkylation of borane by an alkene

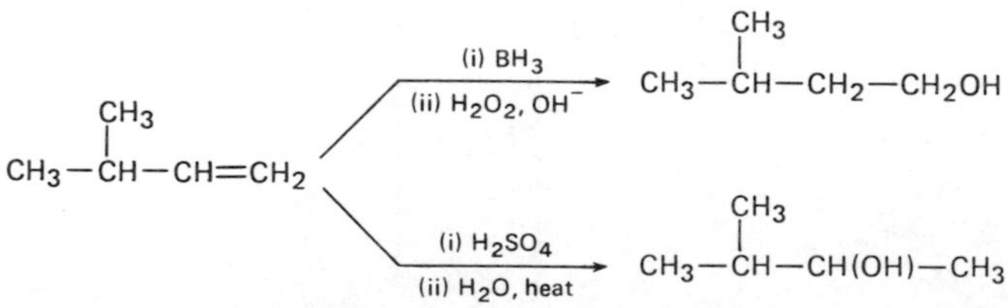

$$CH_3-\overset{\overset{\displaystyle CH_3}{|}}{C}H-CH=CH_2$$

(i) BH_3
(ii) H_2O_2, OH^- → $CH_3-\overset{\overset{\displaystyle CH_3}{|}}{C}H-CH_2-CH_2OH$

(i) H_2SO_4
(ii) H_2O, heat → $CH_3-\overset{\overset{\displaystyle CH_3}{|}}{C}H-CH(OH)-CH_3$

Figure 4.19 Synthetic methods for the preparation of isomeric
alcohols from 3-methylbut-1-ene

Question 4.6 Draw out the structure of the monoalkylborane formed during hydroboration of 1-methylcyclohexene. What would be the alcohol formed on treatment of this organoborane with alkaline hydrogen peroxide?

Alkynes react with borane to give products of mono- or di- addition to the triple bond.

4.7 Oxidation reactions

Hydrocarbons burn in air or oxygen to give water and carbon dioxide with the liberation of a considerable amount of energy (Figure 4.20). This is the basis of their use as fuels, such as natural gas and petroleum (gasoline).

The reaction involves a free radical chain mechanism. When the supply of oxygen is restricted, combustion may produce carbon and carbon monoxide as well as carbon dioxide.

$$CH_4 + 2\,O_2 \longrightarrow CO_2 + 2\,H_2O \qquad \Delta H = -882 \text{ kJ mol}^{-1}$$

$$C_2H_2 + 2\tfrac{1}{2}\,O_2 \longrightarrow 2\,CO_2 + H_2O \qquad \Delta H = -1305 \text{ kJ mol}^{-1}$$

Figure 4.20 The combustion of methane and ethyne

Alkanes are inert to most laboratory oxidizing agents, but alkenes (and alkynes) are readily oxidized by reagents such as potassium permanganate, osmium tetroxide, peracids, and ozone.

(a) Reactions of alkenes with potassium permanganate and osmium tetroxide

Alkenes react with an alkaline solution of potassium permanganate, $KMnO_4$, to give 1,2-diols (glycols). The reaction is thought to involve the formation and hydrolysis of a cyclic intermediate (Figure 4.21(a)), which

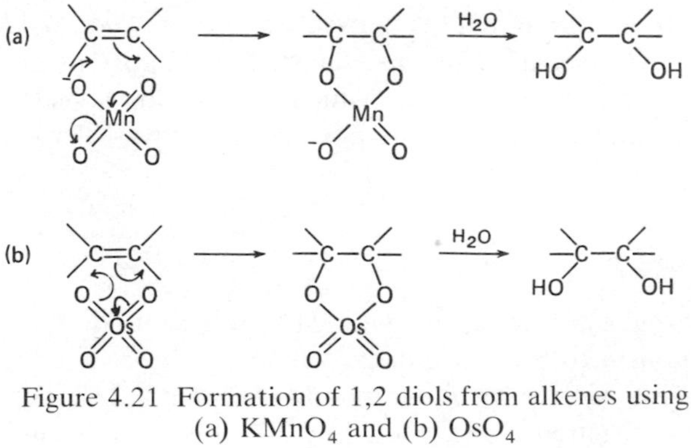

Figure 4.21 Formation of 1,2 diols from alkenes using
(a) $KMnO_4$ and (b) OsO_4

explains the observed *cis* addition of the two OH groups. A similar oxidation reaction occurs with osmium tetroxide, OsO_4 (Figure 4.21(b)).

The reaction with permanganate provides a useful test for the presence of carbon–carbon double bonds, since the deep purple colour of the MnO_4^- ion is discharged in the course of the reaction.

(b) Reactions of alkenes with peracids

Treatment of alkenes with a peracid such as perbenzoic acid, $PhCO_3H$, proceeds by electrophilic attack and results in the formation of an epoxide (Figure 4.22(a)). Epoxides (also called oxiranes) can be hydrolysed under acidic or alkaline conditions to give 1,2-diols (Figure 4.22(b)). Since overall this sequence of reactions introduces the two OH groups *trans* to each other, the glycols are usually isomers of those obtained by oxidation of alkenes with $KMnO_4$ or OsO_4.

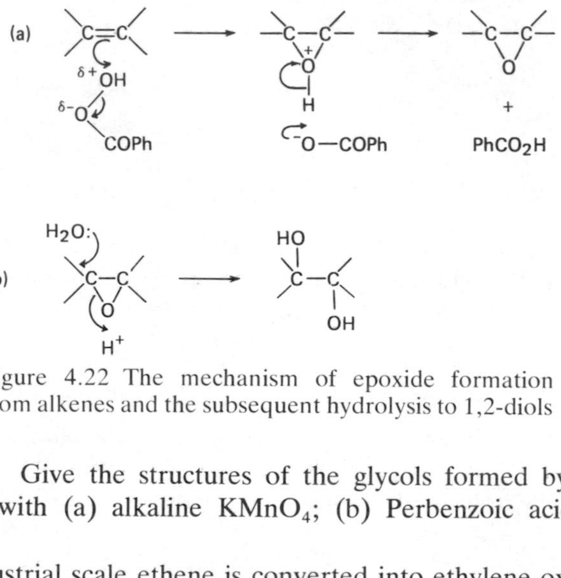

Figure 4.22 The mechanism of epoxide formation from alkenes and the subsequent hydrolysis to 1,2-diols

Question 4.7 Give the structures of the glycols formed by treatment of cyclohexene with (a) alkaline $KMnO_4$; (b) Perbenzoic acid followed by aqueous acid.

On an industrial scale ethene is converted into ethylene oxide by passing it with oxygen over a heated silver catalyst. It is important in the synthesis of ethylene glycol (ethan-1,2-diol), which is a major constituent of 'anti-freeze' mixtures and a precursor of polyesters such as Terylene® (Section 9.7).

(c) Reactions of alkenes with ozone

Ozone (trioxygen), O_3, reacts with carbon-carbon double bonds by a complex mechanism to form ozonides, which can be cleaved with zinc and acetic acid to give two carbonyl compounds (Figure 4.23). The main use of the ozonolysis reaction is in locating double bonds in unsaturated

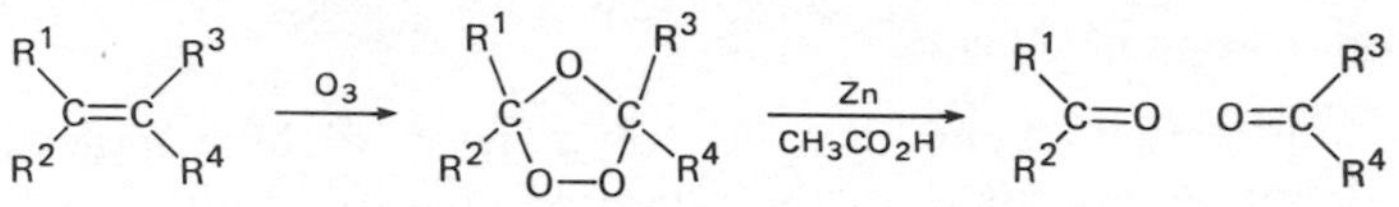

Figure 4.23 The ozonolysis reaction

molecules. Cleavage in this way usually gives smaller molecules which can be more readily identified (Section 8.4).

Question 4.8 Ozonolysis of an alkene of formula C_7H_{12} gives a single product, shown below:

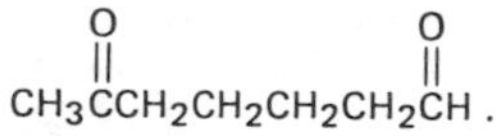

$$CH_3\overset{\overset{\displaystyle O}{\|}}{C}CH_2CH_2CH_2CH_2\overset{\overset{\displaystyle O}{\|}}{C}H .$$

Deduce the structure of the alkene.

4.8 Hydrogenation of alkenes and alkynes

Alkenes and alkynes react with hydrogen in the presence of a finely divided metal catalyst (e.g. Pt, Pd, or Ni) to give alkanes (Figure 4.24(a)). In the case of alkynes it is possible to stop the reaction at the alkene stage by use of a partially poisoned (less active) Pd catalyst called Lindlar's catalyst. The *cis* isomer of the alkene is obtained. An alternative method for reducing alkynes to alkenes involves treatment with sodium in liquid ammonia, in which case the *trans* alkene is obtained (Figure 4.24(b)).

Catalytic hydrogenation of carbon–carbon double bonds in vegetable oils causes them to harden and become fats. This process forms the basis of the margarine industry (Section 10.3).

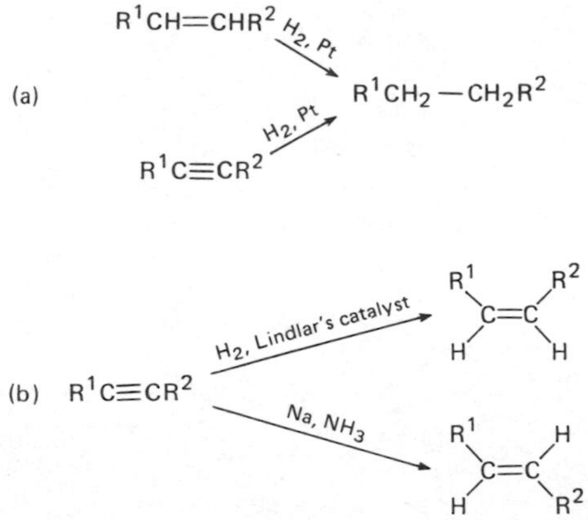

Figure 4.24 Hydrogenation of alkenes and alkynes

4.9 Polymerization of alkenes

Ethene and other alkenes can be polymerized to high molecular mass compounds by heating them in the presence of a free radical catalyst such as a peroxide (Figure 4.25).

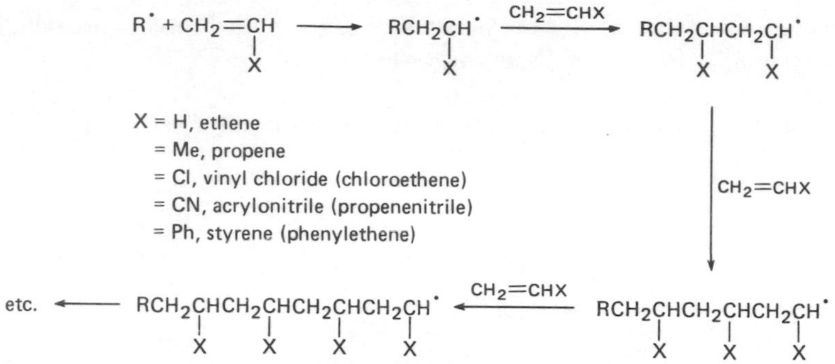

$$R^{\cdot} + CH_2{=}CH{\underset{X}{|}} \longrightarrow RCH_2CH^{\cdot}{\underset{X}{|}} \xrightarrow{CH_2=CHX} RCH_2CHCH_2CH^{\cdot}{\underset{X \quad X}{| \quad |}}$$

X = H, ethene
= Me, propene
= Cl, vinyl chloride (chloroethene)
= CN, acrylonitrile (propenenitrile)
= Ph, styrene (phenylethene)

$CH_2{=}CHX$

etc. $\longleftarrow$ $RCH_2CHCH_2CHCH_2CHCH_2CH^{\cdot}$ $\xleftarrow{CH_2=CHX}$ $RCH_2CHCH_2CHCH_2CH^{\cdot}$
$\quad\quad\quad\quad\quad\quad\;\; X \quad X \quad X \quad X \quad\quad\quad\quad\quad\quad\quad\quad X \quad X \quad X$

Figure 4.25 Polymerization of alkenes

Polymerization is a chain reaction and is important industrially for the preparation of polythene (polyethene), polypropene (polypropylene), polyvinyl chloride (PVC), polyacrylonitrile (Orlon®), polystyrene, poly(methylmethacrylate) (perspex) and polytetrafluoroethane (PTFE, Teflon®). In addition to the free radical catalysis, polymerization can also be brought about in some cases by catalysts such as acids, bases, or organometallic reagents. Indeed, polymerization of ethene and propene is nowadays usually carried out using a so-called Ziegler-Natta catalyst composed of Et$_3$Al and TiCl$_4$. The reaction in this case occurs by insertion of alkene molecules into metal-alkyl bonds, and gives polymers with very

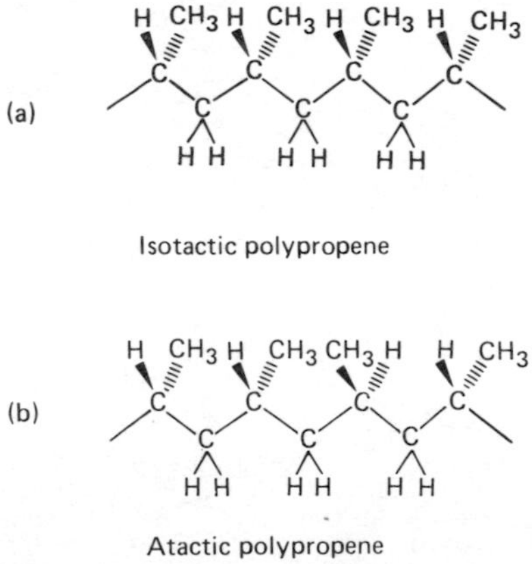

(a)

Isotactic polypropene

(b)

Atactic polypropene

Figure 4.26 The structures of isotactic and atactic polypropene

regular structures and greater strength and higher softening temperature. For example, polypropene obtained in this way has molecules in which the methyl groups are uniformly orientated (isotactic, Figure 4.26(a)), whereas free radical polymerization gives polypropene with randomly orientated (atactic) methyl groups (Figure 4.26(b)).

4.10 Organometallic derivatives of alkynes

As noted in Section 4.3 the slightly acidic hydrogens of terminal alkynes can be removed by strong bases such as sodamide (Section 4.3), Grignard reagents (RMgX, Chapter 6), or alkyl lithiums (RLi, Chapter 6). The products are organometallic derivatives of the alkyne (Figure 4.27).

Organometallic derivatives of alkynes are of considerable importance in organic synthesis. They are powerful nucleophiles and react with electron-deficient centres in organic molecules such as alkyl halides (Chapter 6) and carbonyl compounds (Chapter 8) giving rise to new carbon–carbon bonds (e.g. Figure 4.28).

Heavy metal derivatives of alk-1-ynes are explosive when dry. Copper acetylide (HC≡CCu) is sometimes formed around the valves of acetylene cylinders having brass fittings, so care must be taken when removing the cylinder valves. Copper derivatives of alkynes are also intermediates in the oxidative coupling of alk-1-ynes to conjugated diynes (Figure 4.29).

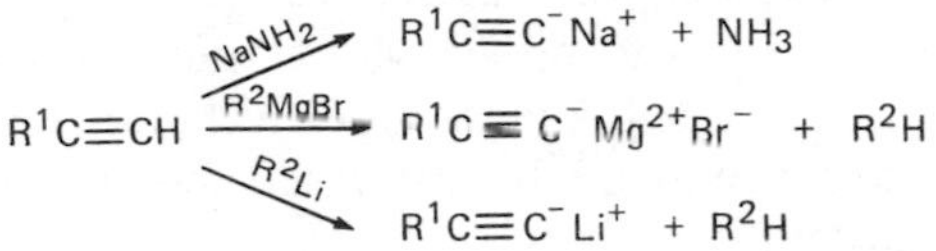

$$R^1C \equiv CH \xrightarrow[R^2Li]{\overset{NaNH_2}{\underset{R^2MgBr}{\longrightarrow}}} \begin{array}{l} R^1C \equiv C^- Na^+ \ + \ NH_3 \\ R^1C \equiv C^- Mg^{2+} Br^- \ + \ R^2H \\ R^1C \equiv C^- Li^+ \ + \ R^2H \end{array}$$

Figure 4.27 The preparation of some organometallic derivatives of alkynes

$$CH_3C \equiv C^- Na^+ + CH_3CH_2I \longrightarrow CH_3C \equiv CCH_2CH_3 + NaI$$

$$CH_3CH_2C \equiv C^- Li^+ + CH_3COCH_3 \longrightarrow CH_3CH_2C \equiv CC(CH_3)_2$$
$$\underset{O^- Li^+}{|}$$

$$\Big\downarrow H_3O^+$$

$$CH_3CH_2C \equiv CC(CH_3)_2$$
$$\underset{OH}{|}$$

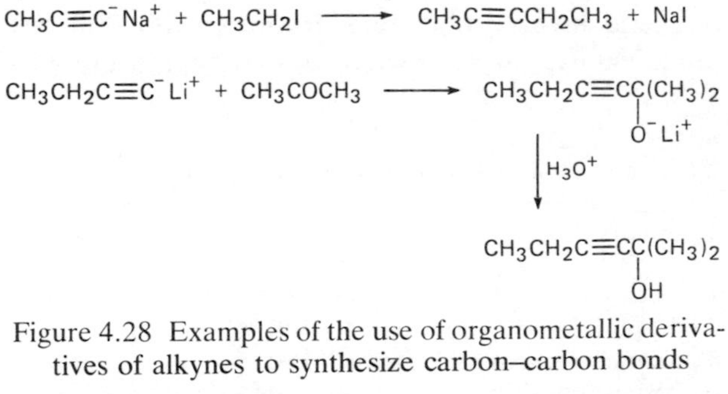

Figure 4.28 Examples of the use of organometallic derivatives of alkynes to synthesize carbon–carbon bonds

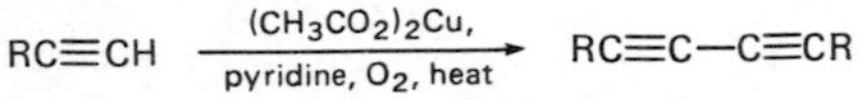

$$RC \equiv CH \xrightarrow[\text{pyridine, } O_2, \text{ heat}]{(CH_3CO_2)_2Cu,} RC \equiv C-C \equiv CR$$

Figure 4.29 Oxidative coupling of alk-1-ynes to form conjugated diynes

4.11 The petroleum industry

Anaerobic decomposition of animal and vegetable matter over many millennia has led to the accumulation of underground deposits of liquid and gaseous hydrocarbons known as crude oil and natural gas respectively.

The refining of crude oil involves separation into fractions with different boiling ranges (Table 4.2). The gas obtained consists mainly of methane, though there are appreciable quantities of ethane, propane, and butane. Propane and butane are easily liquified and are sold as bottled gases. Light petroleum is used as a solvent and as a feedstock for the organic chemicals industry. Petroleum and kerosine are used as fuels in motor car and jet engines respectively while kerosine and fuel oil are used for heating purposes.

The petroleum fraction consists mainly of straight-chain hydrocarbons containing from about six to ten carbon atoms. However, this distillate is unsuitable for use in modern car engines because of its tendency to 'knock' (i.e. ignite prematurely during compression of the petrol–air mixture causing damage to the engine and lowering efficiency). Branched-chain hydrocarbons are much less of a problem in this respect. The tendency of a particular motor fuel to pre-ignite is expressed arbitrarily as its *octane rating*, which is a scale on which heptane is assigned a value of zero and '*iso*-octane' (2,2,4-trimethylpentane) a value of 100.

Octane ratings can be improved by adding lead tetraethyl, $Pb(Et)_4$, which limits premature explosion, although causing environmental pollution. Nowadays, the tendency is to reduce the amount of lead in petroleum and to improve octane ratings by blending with branched-chain alkanes and aromatics which are produced by catalysed isomerization and dehydrogenation of the petroleum distillate.

Since the petroleum distillation fraction by itself is insufficient to meet the demand for motor fuel, processes have been devised for converting higher boiling fractions into petroleum. These are known as *thermal* and *catalytic cracking* and invlove high temperatures (500–600 °C) or the use of a silica–alumina catalyst at somewhat lower temperatures. Cracking pro-

Table 4.2   Major fractions from crude oil

Fraction	Approximate boiling range
Gas	Up to 20 °C
Light petroleum	20–100 °C
Petroleum (gasoline)	80–180 °C
Kerosine (paraffin)	160–260 °C
Fuel oil	240–340 °C
Lubricating oil ⎱	Separated by distillation
Wax ⎰	under reduced pressure
Asphalt (bitumen)	Residue

$$CH_3(CH_2)_{14}CH_3 \longrightarrow \begin{cases} CH_3(CH_2)_6CH_3 + CH_3(CH_2)_5CH\!=\!CH_2 \\ CH_3(CH_2)_7CH_3 + CH_3CH\!=\!CH_2 + 2\,CH_2\!=\!CH_2 \\ etc. \end{cases}$$

Figure 4.30 A simplified example of the petroleum cracking process

cesses are extremely complex and give a range of products including substantial quantities of alkenes. A simplified example of such a process is shown in Figure 4.30.

4.12 Preparation of aliphatic hydrocarbons

(a) Alkanes

On an industrial scale virtually all alkanes are obtained either directly or indirectly from crude oil. If it is necessary to synthesize an alkane in the laboratory, the following methods can be used: hydrogenation of an alkene or alkyne (Section 4.8); reduction of halogenoalkanes (Chapter 6); hydrolysis of a Grignard reagent (Chapter 6).

(b) Alkenes

The lower molecular mass alkenes are obtained by cracking of petroleum and other crude oil fractions (Section 4.11). Production of ethene by this method is especially important. In the laboratory alkenes are prepared by controlled hydrogenation of alkynes (Section 4.8); elimination of hydrogen halides from halogenoalkanes (Chapter 6); dehydration of alcohols (Chapter 6).

(c) Alkynes

Acetylene (ethyne) can be manufactured by addition of water to calcium carbide (CaC_2), which is itself obtained by heating limestone and coke in an electric furnace (Figure 4.31). Alternatively, acetylene can be produced by partial oxidation of methane at 1500 °C.

In the laboratory alkynes can be synthesized by elimination of hydrogen halides from dihalogenoalkanes (Figure 4.32) or by reactions of organometallic derivatives of alkynes with alkyl halides (Section 4.10).

$$CaO \xrightarrow[\text{heat}]{C} CaC_2 \xrightarrow{H_2O} HC\!\equiv\!CH$$
$$+CO \qquad + Ca(OH)_2$$

Figure 4.31 The synthesis of acetylene from limestone (CaO) and coke

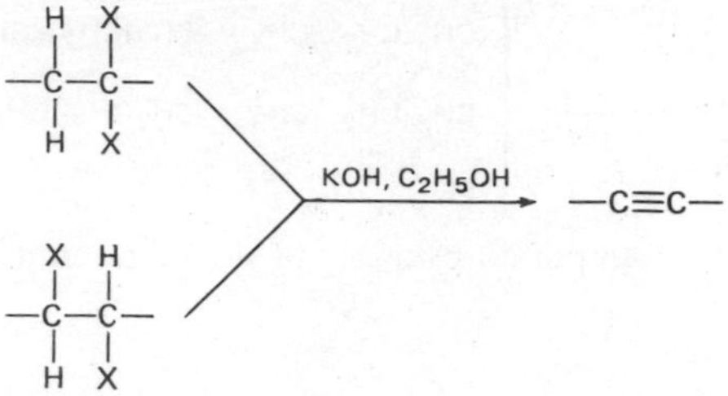

Figure 4.32 The synthesis of alkynes by elimination of hydrogen halides from dihalogenoalkanes

Answers to questions

4.1

CH₃CH₂CH₂CH=CH₂

Pent-1-ene

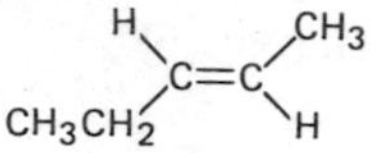

trans-Pent-2-ene

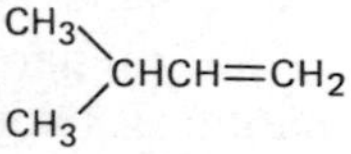

3-Methylbut-1-ene

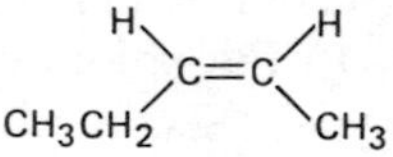

cis-Pent-2-ene

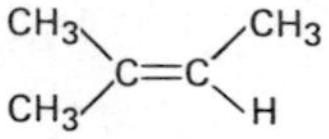

2-Methylbut-2-ene

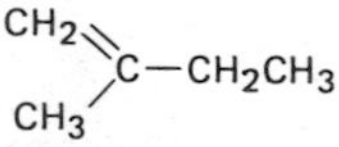

2-Methylbut-1-ene

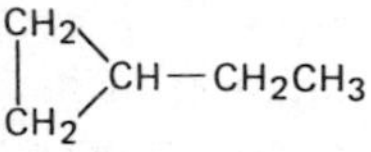

Ethylcyclopropane

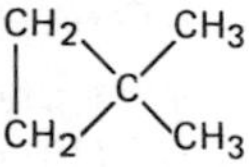

1,1-Dimethylcyclopropane

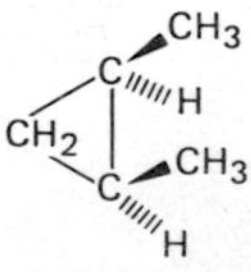

cis-1,2-Dimethylcyclopropane

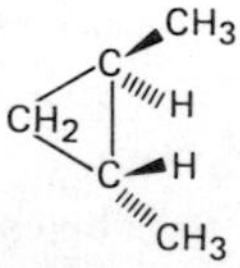

trans-1,2-Dimethylcyclopropane

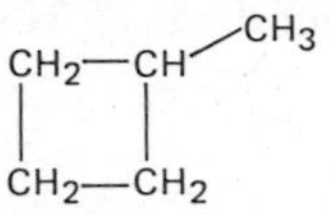

CH_2—CH—CH_3 / CH_2—CH_2

Methylcyclobutane

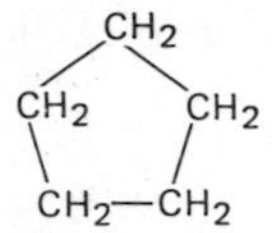

CH_2 / CH_2 CH_2 / CH_2—CH_2

Cyclopentane

4.2 There are four monochloro products:

$(CH_3)_2CHCH_2CH_2Cl$, $(CH_3)_2CHCHClCH_3$,

$(CH_3)_2CClCH_2CH_3$, $ClCH_2\overset{\underset{|}{CH_3}}{C}HCH_2CH_3$.

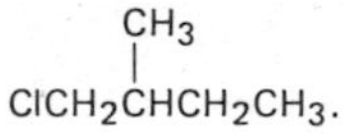

4.3

(a)

Br
H—CO₂H
Br—CO₂H
H

+

Br
HO₂C—H
HO₂C—Br
H

(b)

Br
H—CO₂H
H—CO₂H
Br

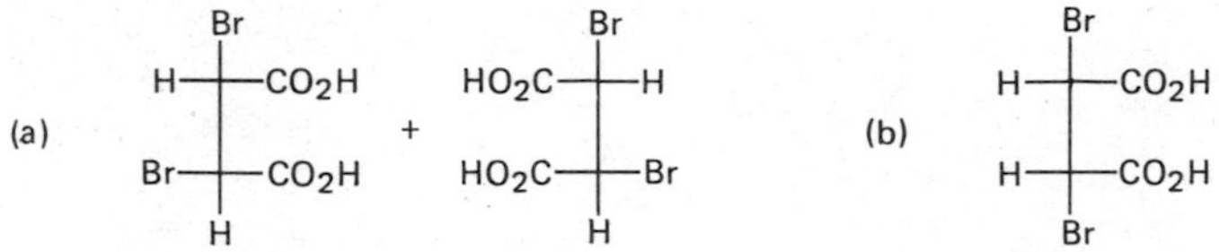

4.4

CH_2BrCH_2Br, $CH_2BrCH_2ONO_2$, and CH_2BrCH_2OH

4.5

(a) $(CH_3)_2\overset{\underset{|}{Cl}}{C}CH_3$ (b) $(CH_3)_2\overset{\underset{|}{Cl}}{C}CH_2CH_3$

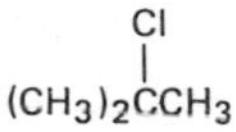

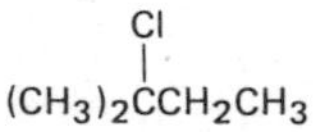

4.6

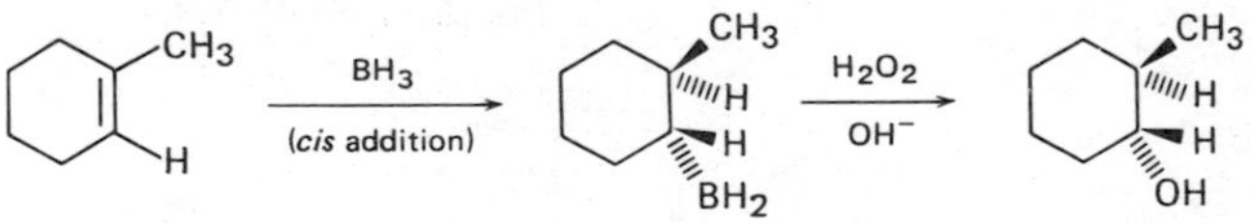

trans-2-Methylcyclohexanol

4.7

(a)   (b)

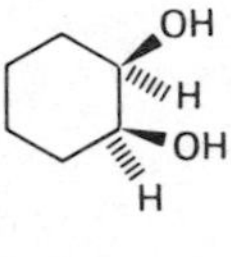

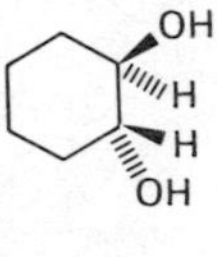

78

Since only one product, a dicarbonyl compound, is formed, the alkene must be cyclic:

CHAPTER 5

Aromatic Compounds and Aromaticity

Benzene, C_6H_6, is the parent member of a series of compounds described as *aromatic*. It is usually represented by a Kekulé structure (Figure 5.1), although, as we shall see, this is not entirely satisfactory (Section 5.1). The term 'aromatic' was originally applied to derivatives of benzene (e.g. benzaldehyde and vanillin) because of their characteristic odours. Nowadays, however, the term is used to describe compounds which resemble benzene in terms of their electronic structure and reactivity (Section 5.2).

The importance of aromatic compounds for human welfare is illustrated by their contribution to the relief of hunger, through the use of insecticides such as DDT, and to the control of disease, by the use of drugs such as Paludrin®. They are also important constituents of our diet, e.g. vitamin E, and have helped to improve our standard of living, through the use of materials such as polystyrene (Figure 5.2). In this chapter we deal mainly with the chemistry of aromatic hydrocarbons. The chemistry of functional groups attached to aromatic rings is dealt with in subsequent chapters

5.1 The structure of benzene

In 1865 Kekulé proposed that benzene was a monocyclic molecule containing three double bonds. However this structure does not satisfactorily explain the symmetry or the characteristic reactivity of benzene. Thus X-ray and electron diffraction measurements show that benzene is a planar regular hexagon (with all carbon–carbon bonds of length 139 pm) whereas Kekulé's structure would have alternating single and double bonds (having bond lengths of 154 and 134 pm respectively).

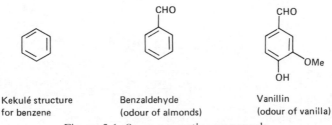

Kekulé structure
for benzene

Benzaldehyde
(odour of almonds)

Vanillin
(odour of vanilla)

Figure 5.1 Some aromatic compounds

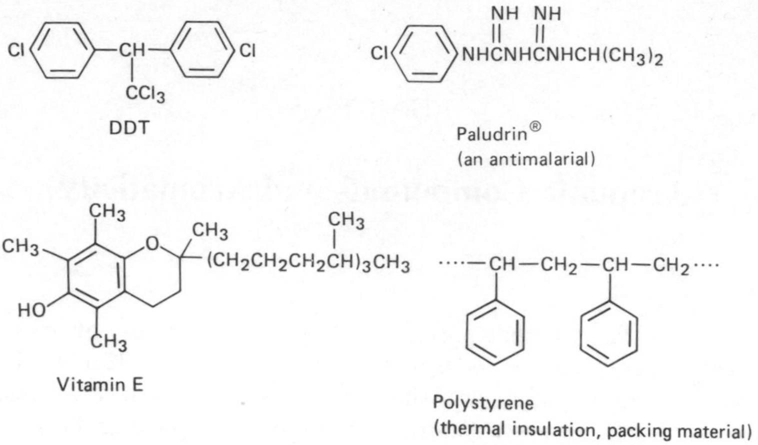
Cl
CH
Cl
CCl3
DDT
NH NH
Cl
NHCNHCNHCH(CH3)2
Paludrin®
(an antimalarial)
CH3
CH3
CH3
O
CH3
(CH2CH2CH2CH)3CH3
HO
CH3
Vitamin E
CH—CH2—CH—CH2
Polystyrene
(thermal insulation, packing material)

Figure 5.2 Some useful aromatic compounds

Furthermore, benzene does not behave like a typical alkene, as can be seen by comparing some of its reactions with those of alkenes:

1. Benzene reacts with bromine only in the presence of a catalyst such as $AlBr_3$ and undergoes *substitution*, whereas alkenes undergo *addition* reactions with bromine even in the absence of a catalyst (Figure 5.3(a)).
2. Benzene is not easily oxidized by dilute aqueous potassium permanganate solution whereas alkenes are rapidly oxidized.
3. Benzene undergoes hydrogenation only with difficulty whereas alkenes are hydrogenated more easily (Figure 5.3(b)).

Thus benzene is relatively unreactive—it reacts only slowly with reagents which react rapidly with alkenes. Furthermore the typical reactions of benzene are substitution reactions (Section 5.4) whereas alkenes typically undergo addition (Chapter 4).

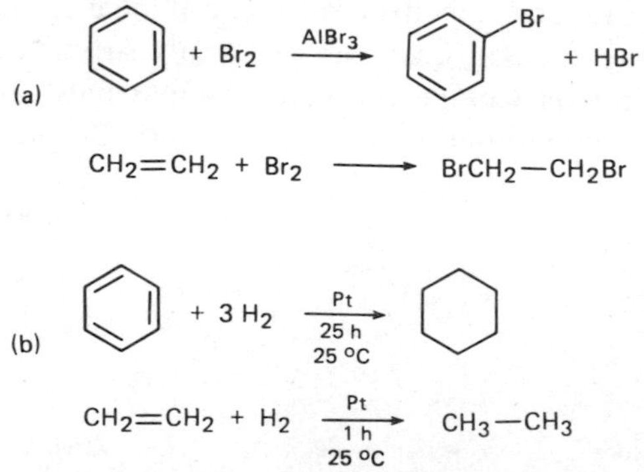
(a)
+ Br2
AlBr3
Br
+ HBr
CH2=CH2 + Br2
BrCH2—CH2Br
(b)
+ 3 H2
Pt
25 h
25 °C
CH2=CH2 + H2
Pt
1 h
25 °C
CH3—CH3

Figure 5.3 A comparison of some reactions of benzene and ethene

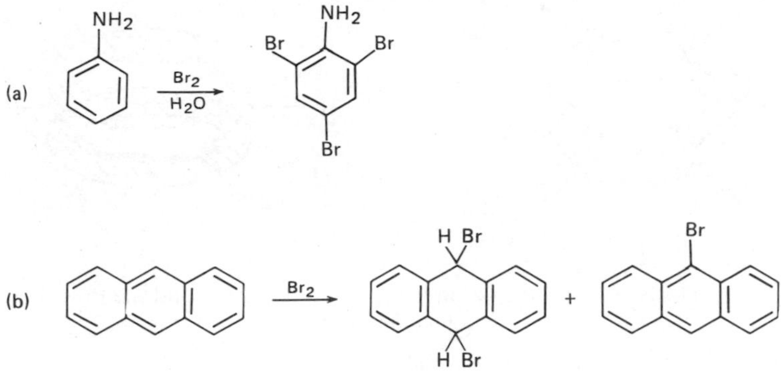

Figure 5.4 The bromination of aniline and anthracene

It is worth noting, nevertheless, that not all aromatic compounds are unreactive and not all aromatic compounds undergo substitution in preference to addition. For example, aniline (benzenamine) undergoes bromination even in the absence of a catalyst (Figure 5.4(a)), and anthracene undergoes addition with bromine as well as substitution (Figure 5.4(b)).

A better understanding of the properties of benzene is provided by two theories of organic chemistry, of which only very brief outlines are given here. *Resonance theory* (see Chapter 1) considers the benzene molecule to be a hybrid of two major canonical forms (Figure 5.5). The actual structure of the molecule is not adequately represented by either canonical form alone, and is *not* an equilibrium mixture of the two, but should be regarded as a blend of the two forms. Calculations show that this resonance hybrid has a lower energy content than either of the two contributing structures.

According to *molecular orbital theory* (see Chapter 1) sp^2 hybridized carbon atoms form the σ bond framework of the benzene molecule, and the six unhybridized 2p atomic orbitals overlap to form π bonds (Figure 5.6(a)). However, in contrast to the situation in alkenes, where two adjacent p orbitals overlap to form a single π bond, in benzene each p orbital is flanked by two exactly equivalent p orbitals with which interaction can occur. As a result, in the π molecular orbitals which are formed the electrons are not confined to the region between any two carbon atoms but spread over all six. For example, the orbital of lowest energy has two unbroken rings of electron density above and below the

Figure 5.5 The two major canonical
forms of benzene

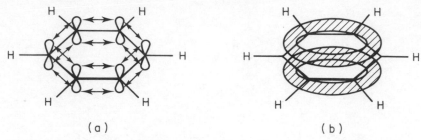

Figure 5.6 The overlapping 2p atomic orbitals of benzene and the lowest energy
molecular orbital

plane of the molecule (Figure 5.6(b)). These electrons are therefore said to be 'delocalized', and this delocalization gives the molecule enhanced stability.

In view of the above considerations a more satisfactory representation of the benzene molecule might be a circle inscribed in a hexagon (Figure 5.7). However Kekulé structures are still widely used because of their convenience for depicting reaction mechanisms, but it should be remembered that they are only symbols and not precise representations of structure.

Figure 5.7 Symbolic representation
of benzene molecule

5.2 Aromaticity

According to the molecular orbital theory the six unhybridized 2p atomic orbitals of benzene interact, forming six delocalized molecular orbitals, three of which are bonding and three of which are antibonding (Figure 5.8; cf. Chapter 1). The six electrons occupy the three bonding orbitals and the

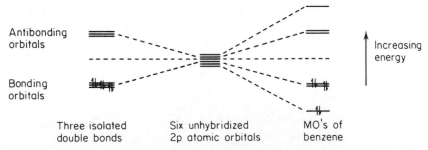

Figure 5.8 Relative energies of the atomic and molecular orbitals of benzene

molecule is more stable than it would be if it contained three isolated double bonds.

The stabilities of other cyclic conjugated molecules, relative to their acyclic counterparts, can also be predicted in a similar manner. Such considerations have led to a general rule known as Hückel's $(4n+2)$ rule which states that 'planar conjugated cyclic systems containing $(4n+2)$ π electrons possess enhanced stability' where $n = 1$, 2, 3, etc. These are the compounds which are nowadays described as aromatic, and benzene derivatives are not the only compounds which fulfil this criterion.

Question 5.1 Which of the following compounds are aromatic:

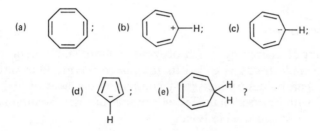

In pyridine (Figure 5.9) one of the CH groups of benzene has been replaced by an N atom. The nitrogen atom is sp^2 hybridized, and one of the sp^2 hybrid orbitals contains a lone pair of electrons. Otherwise the electronic structure is the same as in benzene and pyridine is therefore aromatic. In pyrrole the nitrogen atom is again sp^2 hybridized but all three sp^2 hybrid orbitals are involved in σ bonding. The lone pair of electrons occupies the p orbital and can interact with the four p electrons on the carbon atoms, giving a total of six π electrons. Pyrrole is therefore also aromatic.

One way of investigating whether or not a compound is aromatic makes use of nuclear magnetic resonance spectroscopy (Chapter 3). Thus, when the benzene molecule is placed in a magnetic field the π electrons circulate around the ring. This induced electric current gives rise to an induced magnetic field H$_i$ which reinforces the applied magnetic field H$_0$ in the region of the hydrogen atoms (Figure 5.10). Consequently the protons absorb at a slightly lower applied field (i.e. they are deshielded) and have a higher chemical shift than normal alkene protons. Thus the protons of an

Pyridine   Pyrrole

Figure 5.9 Two heterocyclic aromatic compounds

D

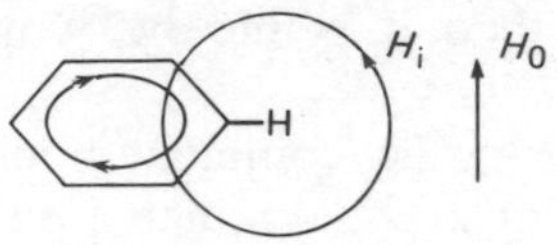

Figure 5.10 The use of n.m.r. to
investigate aromaticity

alkene appear around $\delta = 5.7$ p.p.m. whereas the protons of benzene absorb
at $\delta = 7.3$ p.p.m. Other aromatic compounds (e.g. pyridine and pyrrole) also
give n.m.r. spectra in which protons outside the aromatic ring are
deshielded.

5.3 Resonance energy

The amount of energy by which benzene is more stable than it would be
if it were a simple triene is called its *resonance energy*. One way in which
this quantity can be determined is by comparing the heat of hydrogenation
of benzene with an estimated value based on the assumption that it
contains three isolated double bonds:

Heat of hydrogenation of cyclohexene	$= -120$ kJ mol^{-1},
Estimated value for benzene	$= -3 \times 120$
	$= -360$ kJ mol^{-1},
Observed heat of hydrogenation of benzene	$= -209$ kJ mol^{-1},
Resonance energy of benzene	$= 360 - 209$
	$= 151$ kJ mol^{-1}.

The resonance energies of other aromatic compounds can be determined
in the same way (Table 5.1).

5.4 Electrophilic substitution

The electron cloud above and below the benzene ring shields it from
attack by nucleophiles and makes it susceptible to attack by electrophiles.
Furthermore the stability conferred by delocalization explains why the
characteristic reactions of aromatic compounds are substitution rather than
addition reactions, since substitution can take place with overall retention
of aromaticity, whereas addition cannot (Figure 5.11).

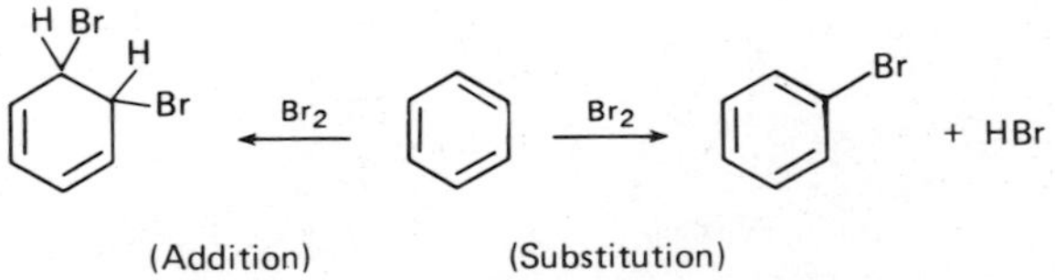

(Addition)   (Substitution)

Figure 5.11 Comparison of addition and substitution
reactions

Table 5.1 Resonance energies of aromatic compounds

Formula	Compound	Resonance energy (kJ mol^{-1})
	Benzene	151
	Naphthalene	255
	Anthracene	349
	Phenanthrene	382
	Pyridine	96
	Pyrrole	89
	Thiophene	120
	Furan	66

The process of electrophilic substitution involves attack by an electrophile and liberation of a proton. However this is not a simultaneous process but involves an intermediate carbonium ion (sometimes called a benzenonium ion or σ complex) which is itself a resonance hybrid of structures I, II and III (Figure 5.12). The structure of the intermediate is probably best represented by IV but it is often more convenient to show just one canonical form. The intermediate is relatively stable, for a carbonium ion, because the charge is spread over five carbon atoms. They have in some cases been detected spectroscopically and their salts have on occasions been isolated. Loss of a proton from these intermediates is favoured because stable aromatic products are formed.

(a) Nitration

Treatment of benzene with a mixture of concentrated nitric and sulphuric acids brings about nitration giving nitrobenzene as the product. The

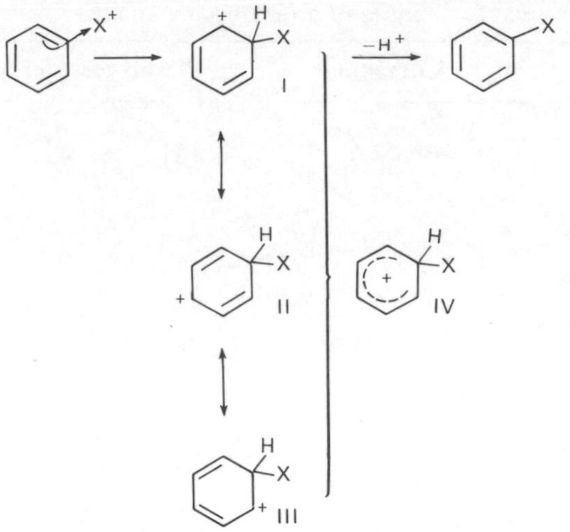

Figure 5.12 Substitution of benzene, showing the resonance structures of the benzenonium ion intermediate

species which actually brings about nitration is the electrophilic nitronium ion (NO_2^+) which is generated *in situ* (Figure 5.13(a)). There is very good evidence for the presence of the NO_2^+ ion. Firstly, a solution of nitric acid in sulphuric acid exhibits a freezing point depression consistent with the presence of four ions in the solution. Secondly, salts of the NO_2^+ ion can be prepared (e.g. $NO_2^+BF_4^-$) and they are themselves powerful nitrating agents. The function of the sulphuric acid is to protonate the nitric acid to give H_3O^+ and NO_2^+ ions, and indeed other very strong acids such as perchloric acid can be used in place of sulphuric acid. However, nitration of benzene using nitric acid alone is very slow because nitric acid itself is not sufficiently strong to give an appreciable concentration of NO_2^+ ions. Other reagents which generate NO_2^+ ions (e.g. N_2O_5 or $CH_3CO_2NO_2$) are also effective nitrating agents.

Nitration is an important reaction because nitro compounds can be easily reduced to give amino compounds which are themselves important synthetic intermediates (see Chapter 7).

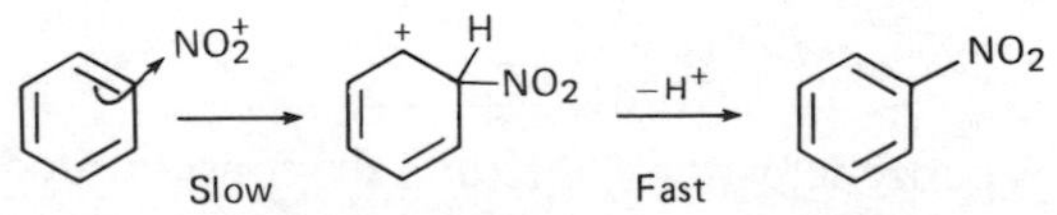

Figure 5.13 The nitration of benzene

Question 5.2 The fact that the second step of the nitration reaction is faster than the first (Figure 5.13(b)) was established by comparing the rates of nitration of C_6H_6 and C_6D_6 (where D represents deuterium). Given that the energy required to break a C—D bond is greater than that required to break a C—H bond, would you expect the nitration of C_6D_6 to proceed faster or slower than the nitration of C_6H_6?

(b) Halogenation

Halogenation of aromatic compounds is brought about by treatment with a mixture of the halogen and a Lewis acid. The electrophile involved is assumed to be a polarized complex of the halogen molecule and the Lewis acid (Figure 5.14), this being a stronger electrophile than the free Br_2 molecule.

In biological systems halogenation reactions are brought about by enzymes. For example, a chlorine atom is incorporated into an aromatic ring during the fungal biosynthesis of the antibiotic chlorotetracycline (Figure 5.15).

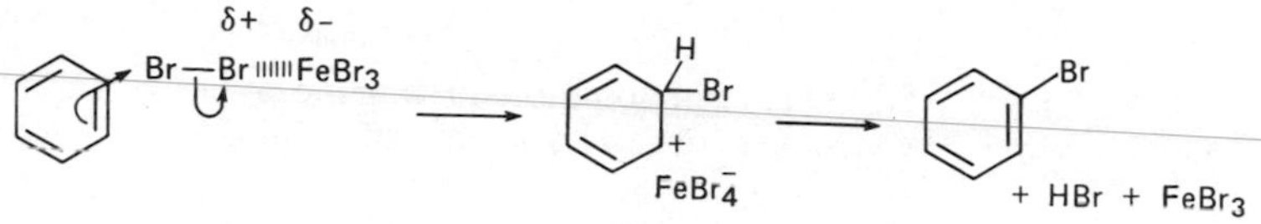

Figure 5.14 The bromination of benzene

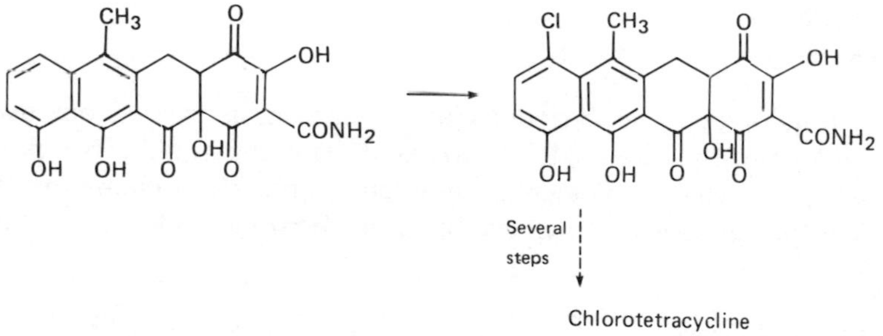

Figure 5.15 The chlorination of an aromatic ring during the biosynthesis of chlorotetracycline

(c) Sulphonation

Sulphonation occurs when aromatic compounds are heated with fuming sulphuric acid. The electrophile involved is probably sulphur trioxide or the HSO_3^+ ion (Figure 5.16).

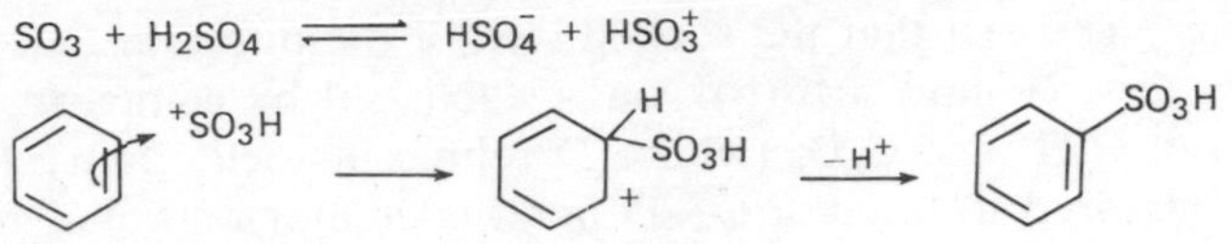

$$SO_3 + H_2SO_4 \rightleftharpoons HSO_4^- + HSO_3^+$$

Figure 5.16 The sulphonation of benzene

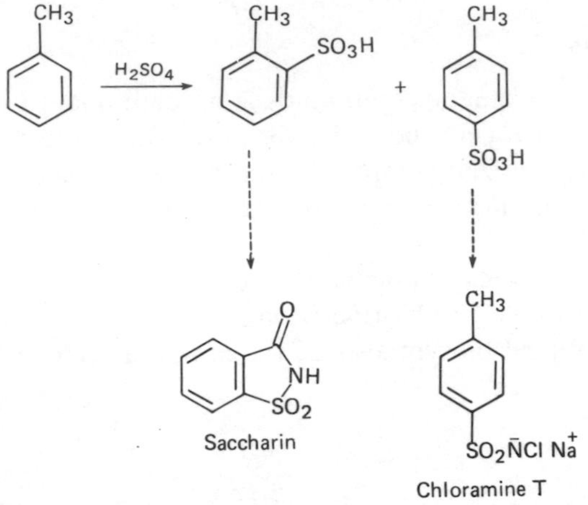

Figure 5.17 The sulphonation of toluene as a
step in the syntheses of saccharin and chloramine T

Sulphonation of toluene (methylbenzene, Figure 5.17) is important because it gives a mixture of two sulphonic acids which are precursors for the synthesis of the artificial sweetener, saccharin, and the antiseptic, chloramine T.

(d) Alkylation

Alkylation of benzene (Friedel Crafts reaction) is carried out using an alkyl halide or an alkene and a Lewis acid. The electrophile is a polarized alkyl halide or alkene molecule, or even a free carbonium ion (Figure 5.18). The production of ethylbenzene from benzene and ethene is a most

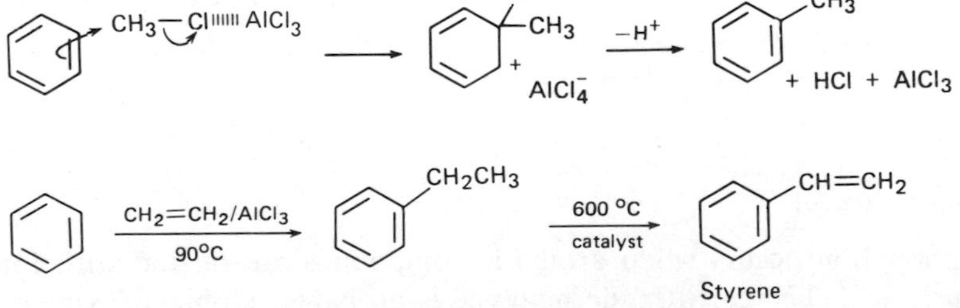

Figure 5.18 Alkylation of benzene

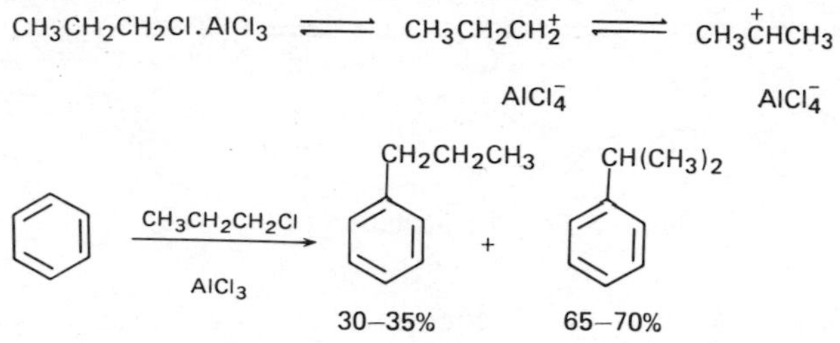
CH3CH2CH2Cl.AlCl3
CH3CH2CH2+
CH3CHCH3
AlCl4-
AlCl4-
CH2CH2CH3
CH(CH3)2
CH3CH2CH2Cl
AlCl3
+
30–35%
65–70%

Figure 5.19 Alkylation of benzene with partial re-arrangement

important industrial process because ethylbenzene is readily converted into styrene.

From the point of view of laboratory synthesis this reaction has a number of disadvantages. Firstly, the alkylbenzene produced is more reactive than benzene itself (see Section 5.5) so that polyalkylation occurs. Secondly, isomeric products are often formed since any carbonium ion formed may rearrange to give a more stable secondary or tertiary carbonium ion (Figure 5.19).

Question 5.3 Explain why the Friedel Crafts reaction of 1-chloro-2-methylpropane with benzene yields only *t*-butylbenzene, $PhC(CH_3)_3$.

The electrophiles responsible for alkylation of aromatic compounds in living systems are generated from such compounds as the amino acid methionine (for transfer of methyl groups) and isopentenyl pyrophosphate (for transfer of C_5 units), as in the biosynthesis of the fungal product auroglaucin (Figure 5.20, see Chapter 15).

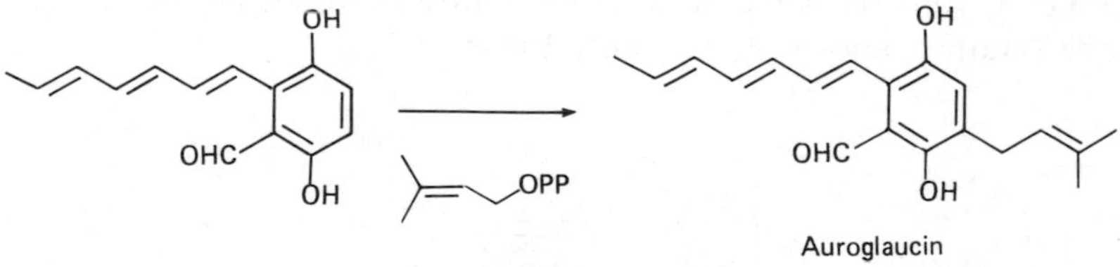
OH
OHC
OH
OPP
OH
OHC
OH
Auroglaucin

Figure 5.20 Alkylation by *iso*pentenyl pyrophosphate in the biosynthesis of auroglaucin

(e) Acylation

The acylation of benzene (also called the Friedel Crafts reaction) requires an acyl halide or anhydride and a Lewis acid. Thus the electrophile is the acylium ion $R—\overset{+}{C}{=}O$ or a complex of the acyl halide or anhydride and $AlCl_3$ (Figure 5.21).

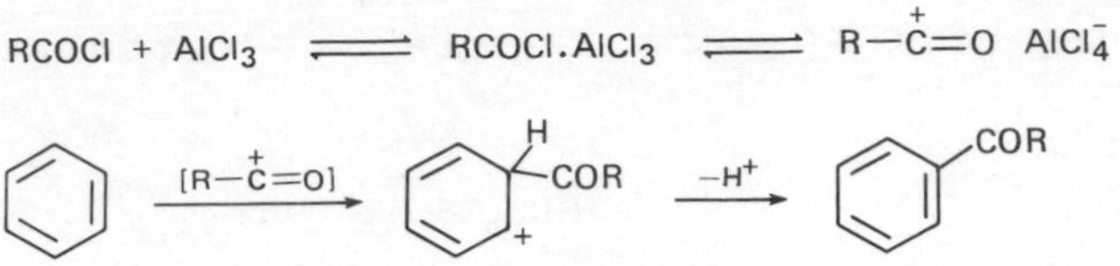

$$RCOCl + AlCl_3 \rightleftharpoons RCOCl.AlCl_3 \rightleftharpoons R\overset{+}{-}C=O \ AlCl_4^-$$

The acylation of benzene figure:

Figure 5.21 The acylation of benzene

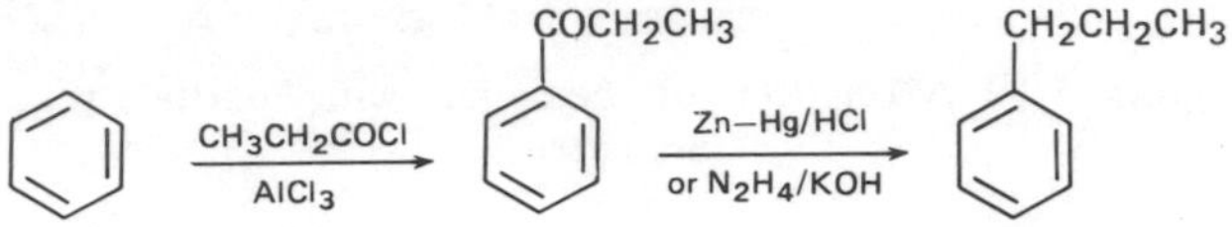

Figure 5.22 Acylation of benzene followed by either Clemmenson or Wolff–Kishner reduction

Unlike polyalkylation, polyacylation does not present a problem since the first formed acylbenzene is less reactive than benzene itself (see Section 5.5). Furthermore there is no possibility of isomerization. Consequently the Friedel Crafts acylation reaction is more useful for laboratory synthesis than alkylation, even when alkylated derivatives are required, since a carbonyl group (C=O) can easily be converted into a CH_2 group by either the Clemmensen or the Wolff–Kishner reaction (Figure 5.22, cf. Chapter 8).

Question 5.4 How could the above procedure be used to prepare a sample of (2-methylpropyl)benzene, $PhCH_2CH(CH_3)_2$ (cf. Question 5.3)?

It is also possible to introduce a formyl group (HCO) into an aromatic ring by reactions which closely resemble the Friedel Crafts reaction. The electrophile can be generated from carbon monoxide and hydrogen chloride (Gattermann–Koch reaction, Figure 5.23).

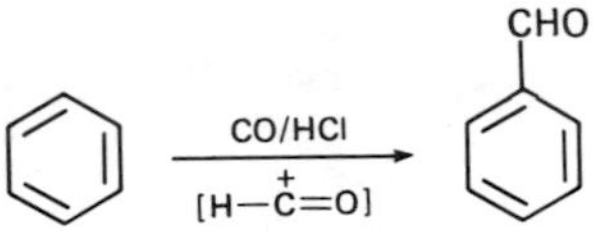

Figure 5.23 Formylation of benzene

5.5 Substituent effects in electrophilic substitution

(a) Influence on reactivity

Substituent groups exert a profound influence on the reactivity of the benzene nucleus. Broadly speaking any substituent which increases the

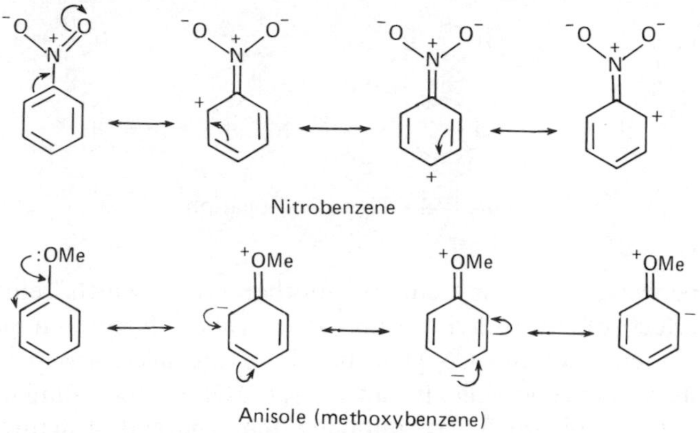

Nitrobenzene

Anisole (methoxybenzene)

Figure 5.24 Electron-withdrawing and electron-donating effects

electron density of the benzene ring increases the rate of electrophilic substitution, whereas any substituent which reduces the electron density decreases the rate of substitution.

There are two mechanisms by which a substituent can exert influence, viz. the inductive and mesomeric effects (see Chapter 1). Most groups (e.g. NO_2, OMe, OH, CN) exert an electron-withdrawing inductive effect, the main exception being alkyl groups which exert an electron-donating effect. When the substituent also exerts a mesomeric effect this may reinforce or be opposed to the inductive effect. For example the nitro group exerts an electron-withdrawing mesomeric effect whereas the methoxyl group exerts an electron-donating mesomeric effect (Figure 5.24).

Question 5.5 Draw diagrams to illustrate how the mesomeric effects of OH and CN groups influence the electron density of a benzene ring.

Where both inductive and mesomeric effects reinforce one another the overall effect is easy to predict. Thus the nitro group clearly exerts an electron-withdrawing effect. When the two effects oppose one another, in general the mesomeric effect predominates so that, for example, OMe and OH groups exert an overall electron-donating effect. The overall effects of the various substituents are shown in Table 5.2.

Table 5.2   Electronic effects of substituent groups

Electron-donating groups	Electron-withdrawing groups
OH, OMe	NO_2
NH_2, NHR, NR_2	$\overset{+}{N}R_3$
NHCOR	CN
Alkyl	CHO, COR
	CO_2H, SO_3H
	Halogen

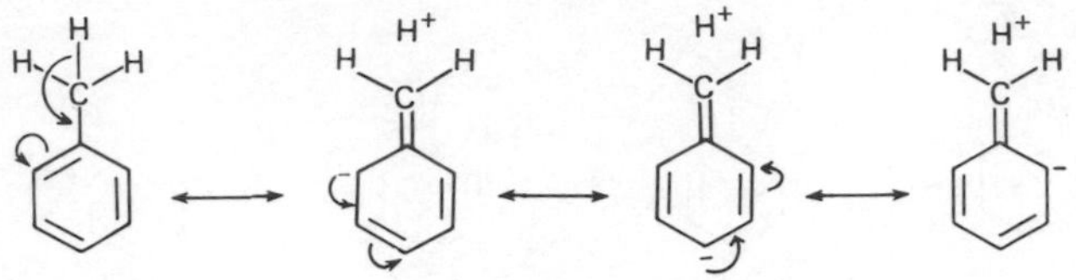

Figure 5.25 Hyperconjugation

In alkylbenzenes there is actually another effect which reinforces the inductive effect of the alkyl group. It is called hyperconjugation and involves interaction of the C—H σ bonds of the alkyl group with the π electrons of the benzene ring. It can be represented in a similar way to a mesomeric effect (Figure 5.25), implying that charged structures make a small contribution to the hybrid.

(b) Directing effects

The nature of the substituent also determines the position at which the benzene nucleus is attacked. Broadly speaking electron-donating groups promote attack at the 2 (*ortho*) and 4 (*para*) positions, whereas electron-withdrawing groups promote attack at the 3 (*meta*) position (Table 5.3). The main exceptions to this generalization are the halogenobenzenes which undergo *ortho* and *para* substitution, although the rates of their reactions are generally lower than those of benzene. This can be understood if their inductive and mesomeric effects are of comparable magnitude. Overall there is a slight reduction of electron density on the ring due to the electron-withdrawing inductive effect but the reduction of electron density is least at the *ortho* and *para* positions due to the electron-donating mesomeric effect.

Since the first step in electrophilic substitution is normally the rate-determining step, the rates of formation of the various possible

Table 5.3 Isomer distribution for nitration

	% ortho	% meta	% para
PhOMe	44	—	56
PhNHCOMe	20	—	80
PhMe	59	4	37
PhCl	30	1	69
PhBr	37	1	62
PhCO$_2$H	22	76	2
PhNO$_2$	6	94	—
PhNMe$_3^+$	—	89	11

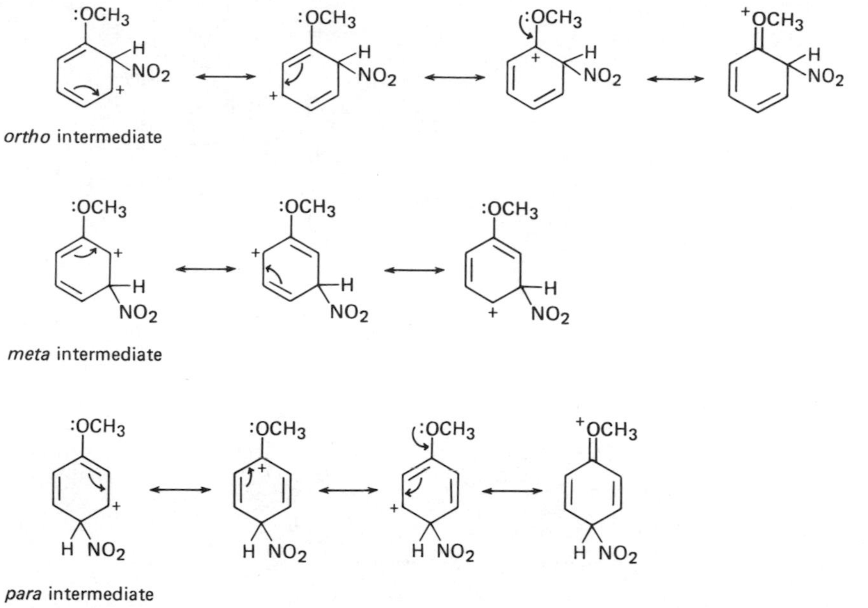

:OCH₃ ... :OCH₃ ... :OCH₃ ... ⁺OCH₃

ortho intermediate

:OCH₃ ... :OCH₃ ... :OCH₃

meta intermediate

:OCH₃ ... :OCH₃ ... :OCH₃ ... ⁺OCH₃

para intermediate

Figure 5.26 Relative stabilization of *ortho, meta,* and *para* intermediates by an electron-donating substituent

intermediate carbonium ions determine the relative proportions of the possible products. Thus an understanding of directing effects can be obtained by comparing the stabilities of these ions. Electron-donating groups (e.g. OMe) stabilize the intermediates involved in *ortho* and *para* substitution more than those involved in *meta* substitution. The lone pair of electrons on the methoxyl group assists in delocalizing the positive charge on the *ortho* and *para* intermediates, as represented by the extra canonical form in the appropriate resonance hybrids (Figure 5.26).

Where the substituent is electron-withdrawing (e.g. NO_2) all three possible intermediates are destabilized (Figure 5.27). However the *meta* intermediate is destabilized to a lesser extent than the *ortho* and *para* intermediates and so *meta* substitution predominates. The intermediates involved in *ortho* and *para* substitution are destabilized because one canonical form has positive charges on adjacent atoms. This form (underlined in Figure 5.27) has a high energy content and thus makes only a small contribution to the hybrid.

The size of a substituent may also influence the outcome of an electrophilic substitution reaction since a bulky substituent will tend to hinder *ortho* attack. Thus, whereas toluene (methylbenzene), for example, usually gives substantial quantities of both *ortho* and *para* substitution products, *t*-butylbenzene gives almost entirely *para* substitution products.

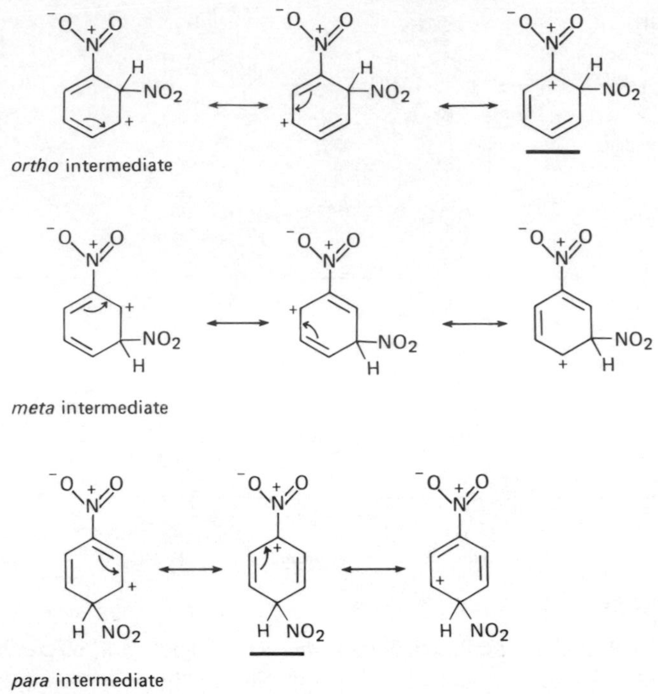

94

ortho intermediate

meta intermediate

para intermediate

Figure 5.27 Relative stabilization of *ortho, meta,* and *para* intermediates by an electron-withdrawing substituent

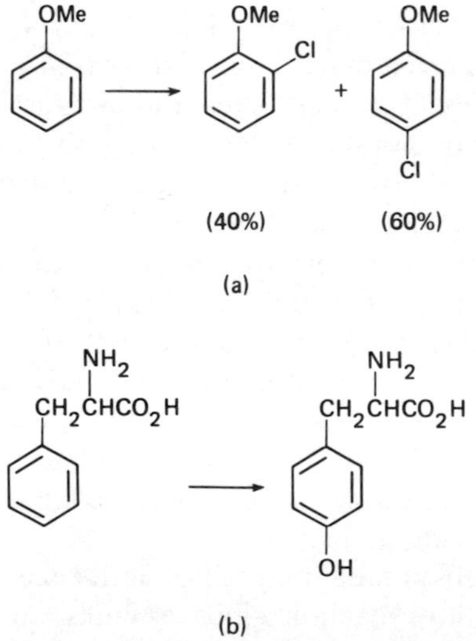

(40%) (60%)

(a)

(b)

Figure 5.28 (a) Chlorination of methoxybenzene and (b) hydroxylation of phenylalanine

Electrophilic substitution reactions can also be brought about by enzymes. It is interesting to note that chlorination of anisole (methoxybenzene) with the enzyme chlorinase gives the same ratio of products as the standard laboratory procedure (Figure 5.28(a)). Another example of enzymic electrophilic aromatic substitution is provided by the hydroxylation of phenylalanine to tyrosine; (Figure 5.28(b)); the intermediate in this reaction is believed to be an epoxide rather than the usual type of carbonium ion intermediate. Some babies are born with a defective gene for making the enzyme needed to convert phenylalanine to tyrosine with the result that a lot of phenylalanine is converted into phenylpyruvic acid which appears in the urine. This condition is called phenylketonuria and results in mental retardation.

Question 5.6 Predict the monosubstitution products formed when the following compounds are nitrated:

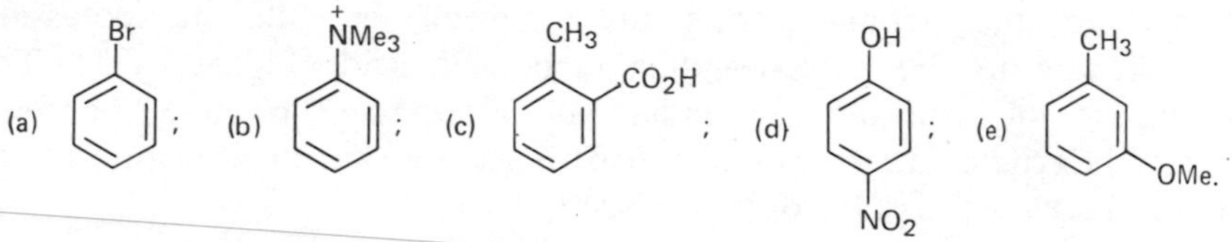

(a) ; (b) ; (c) ; (d) ; (e)

5.6 Polynuclear aromatic hydrocarbons

Naphthalene and azulene (Figure 5.29) both contain ten π electrons (i.e. $4n + 2$, where $n = 2$) and exhibit the properties of aromatic compounds. Similarly anthracene and phenanthrene contain fourteen π electrons ($n = 3$) and are also aromatic.

However such polynuclear aromatic hydrocarbons tend to undergo addition reactions as well as substitution reactions, as can be understood by comparing the resonance energies of the various compounds (Table 5.1). For example, the resonance energy of anthracene is 349 kJ mol^{-1} whereas that of the bromine addition product (Figure 5.30) is approximately the same as that of two benzene rings, i.e. 302 kJ mol^{-1}. Thus the loss of stability when anthracene undergoes addition is only about 47 kJ mol^{-1} whereas it would be approximately 151 kJ mol^{-1} in the case of benzene

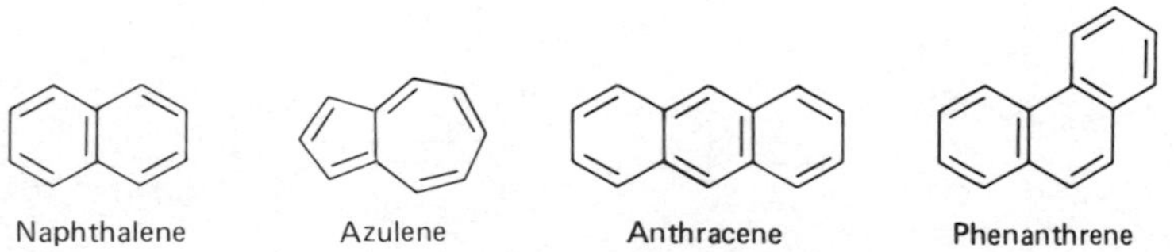

Naphthalene   Azulene   Anthracene   Phenanthrene

Figure 5.29 Some polynuclear aromatic hydrocarbons

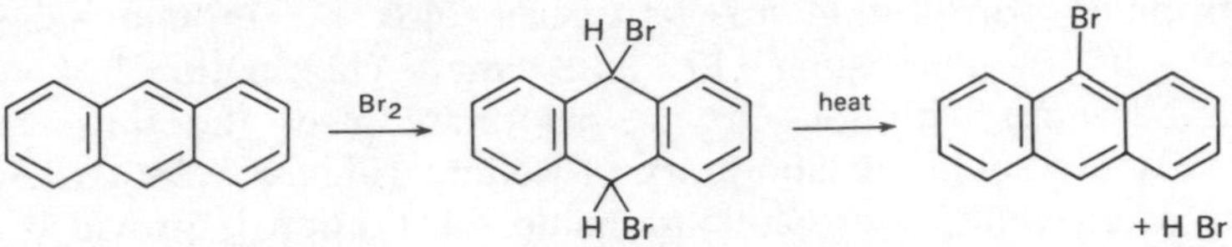

96

Figure 5.30 The bromination of anthracene

(see Section 5.4). Nevertheless, although the addition product can be isolated, the more stable substitution product is formed on heating.

5.7 Side chain reactions of alkylbenzenes

In addition to electrophilic substitution reactions of the aromatic ring, alkylbenzenes also undergo two important reactions involving the side chain. Thus, alkylbenzenes in which a primary or secondary alkyl group is attached to the aromatic ring are oxidized by alkaline potassium permanganate to the corresponding carboxylic acid (Figure 5.31). This reaction is of considerable industrial importance because benzene-1,4,-dicarboxylic acid (terephthalic acid) is used in the manufacture of the polymer Terylene® or Dacron® (see Section 9.7).

Similar oxidation occurs extensively in biological systems. Methyl groups can be oxidized through —CHO to —CO$_2$H with the help of enzymes. For example, 3-hydroxybenzene-1,2-dicarboxylic acid is biosynthesized from 6-methyl-2-hydroxybenzoic acid by fungi of the *Pencillium* species (Figure

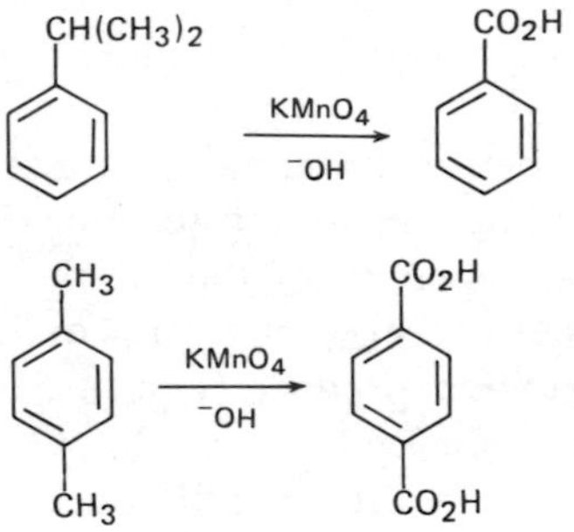

Figure 5.31 The oxidation of alkylbenzenes

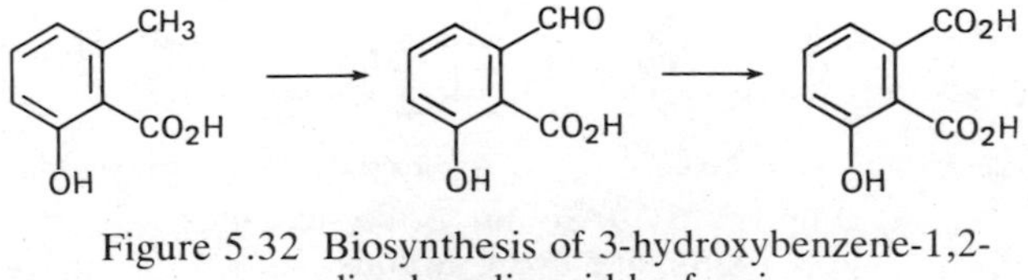

Figure 5.32 Biosynthesis of 3-hydroxybenzene-1,2-dicarboxylic acid by fungi

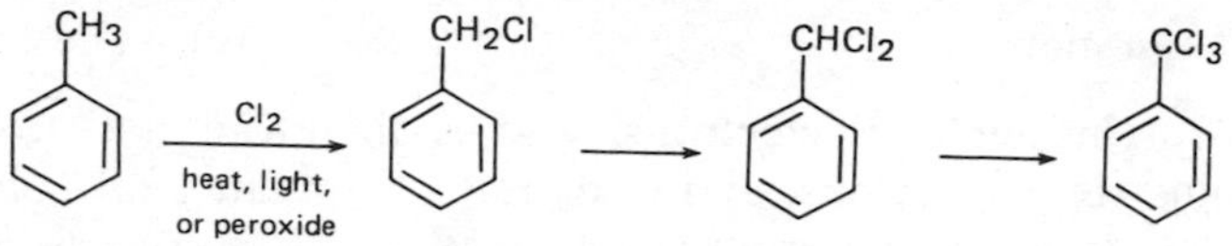

Figure 5.33 The free radical chlorination of toluene

5.32). Furthermore, biological oxidation of toluene to benzoic acid (not possible for benzene) provides a mechanism for the removal of toluene from the human body. Hence toluene is much less toxic than benzene.

Alkylbenzenes also undergo free radical chlorination of the side chain (Figure 5.33). The mechanism of this reaction is a chain mechanism analogous to the free radical chlorination of methane (Chapter 4). In order to carry out Lewis acid-catalysed ring substitution by halogens of alkylbenzenes, it is therefore necessary to exclude light.

5.8 Preparation of aromatic hydrocarbons

Many aromatic hydrocarbons are obtained directly from coal tar or indirectly from petroleum. Coal tar contains benzene, naphthalene, toluene, xylene, etc., which can be isolated by distillation, and indeed coal tar used to be the primary source of aromatic hydrocarbons. However during the Second World War a process (hydroformation) was developed for obtaining aromatic hydrocarbons from petroleum, and this is now the major source of supply. Petroleum itself consists mainly of aliphatic hydrocarbons such as heptane and octane which are converted into aromatic compounds such as toluene and xylene when passed over a metal oxide catalyst at high temperatures. In the laboratory alkylbenzenes can be prepared by Friedel Crafts alkylation, or by acylation followed by reduction (see Section 5.4).

In nature the cyclization of polyketide chains can lead to the formation of stable aromatic rings. For example, 6-methyl-2-hydroxybenzoic acid is produced from an eight-carbon polyketide (Figure 5.34; see Chapter 15). A second biological route to aromatic compounds is the shikimic acid pathway (Chapter 15).

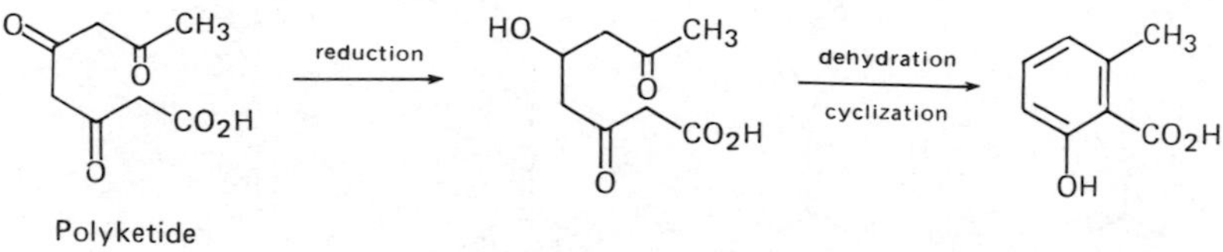

Figure 5.34 An example of polyketide cyclization

Answers to questions

5.1 (a) contains eight π electrons, and is therefore not aromatic; (b) contains six π electrons, and is therefore aromatic; (c) contains eight π electrons and is therefore not aromatic; (d) contains six π electrons and is therefore aromatic; (e) does not contain a fully conjugated cyclic system of overlapping p orbitals and is therefore not aromatic, although it contains six π electrons.

5.2 Nitration of C_6D_6 and C_6H_6 in fact proceed at the same rate since cleavage of the C—D bond is not involved in the rate-determining (slow) step of the reaction. If cleavage of the C—H or C—D bond were involved in the rate-determining step then we would expect C_6D_6 to react more slowly than C_6H_6.

5.3 The initially formed primary carbonium ion undergoes rearrangement to a more stable tertiary carbonium ion, and this then reacts with the benzene molecule:

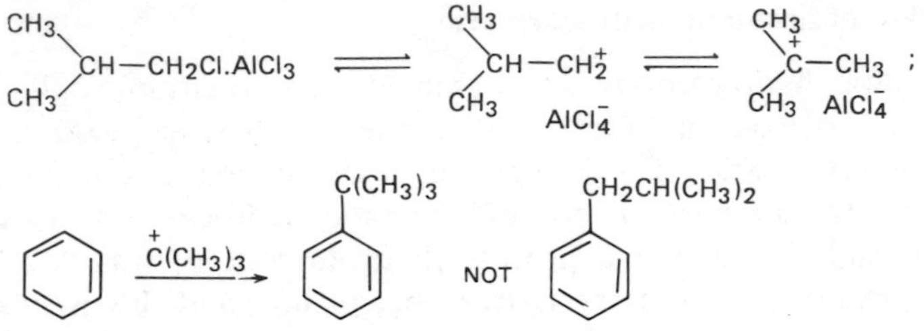

5.4

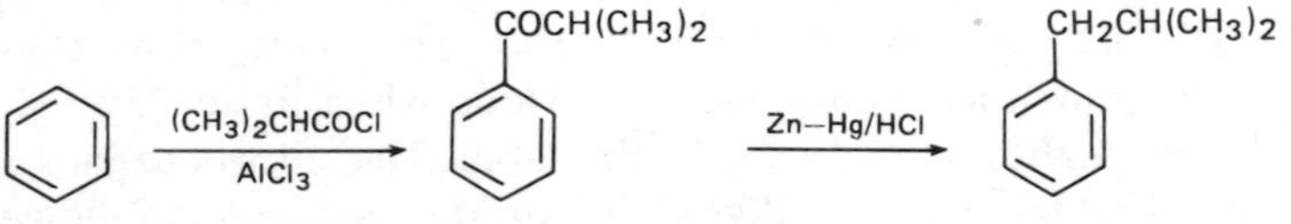

In this case, the electrophile, $(CH_3)_2CHCO^+$, does not undergo rearrangement and hence the expected product is obtained. This can then be converted into the required alkylbenzene by using the Clemmenson reaction. (Note that this compound cannot be prepared by direct alkylation, see Question 5.3.)

5.5

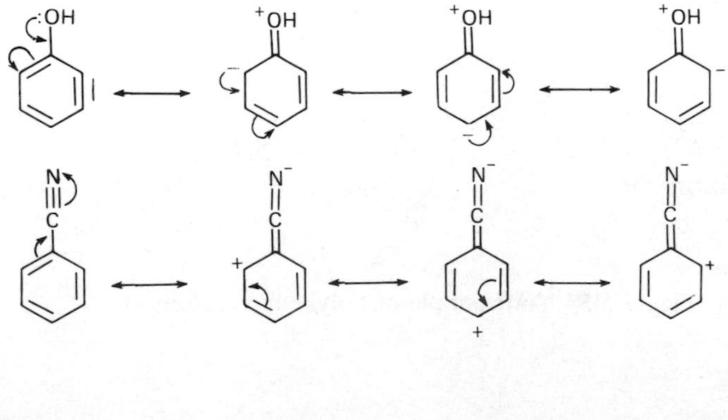

5.6(a)

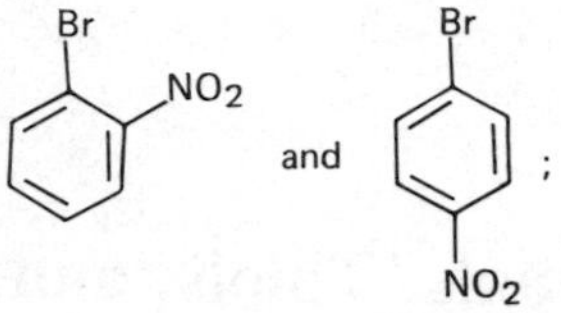
Br
NO2
and
Br
;
NO2

(b)

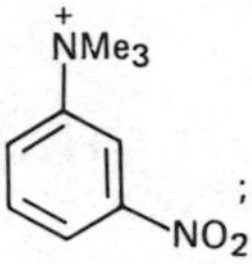
+
NMe3
;
NO2

(c)

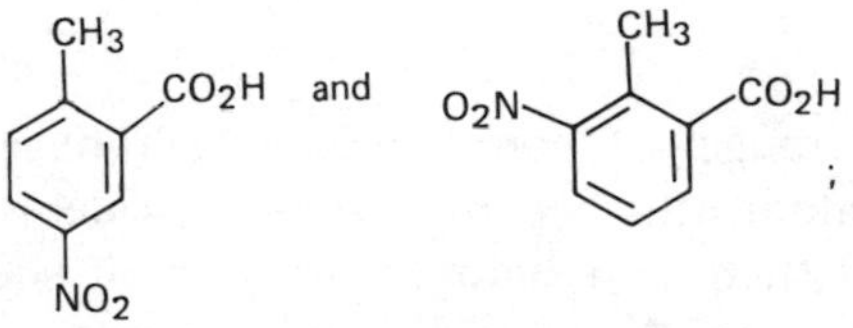
CH3
CO2H
and
CH3
O2N
CO2H
;
NO2

(d)

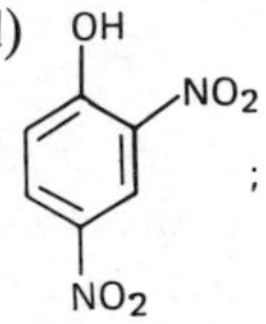
OH
NO2
;
NO2

(e)

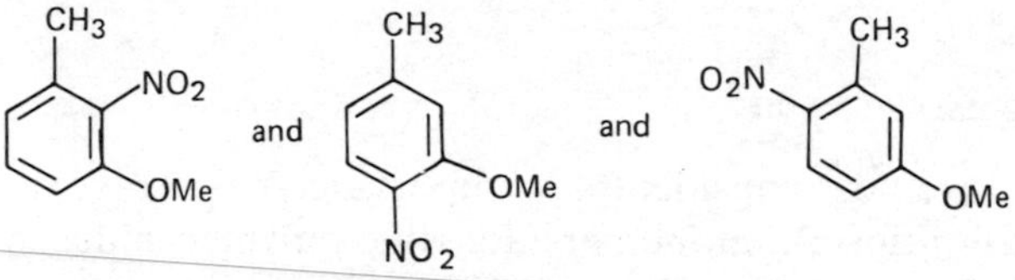
CH3
NO2
OMe
and
CH3
OMe
NO2
and
CH3
O2N
OMe

CHAPTER 6

Organic Halides, Alcohols, Phenols, Thiols, and Ethers

This chapter covers a diverse range of compounds which have one common characteristic: a carbon atom attached by a single covalent bond to a more electronegative element such as a halogen, oxygen, or sulphur. The common classes of compounds in this category are summarized in Table 6.1.

6.1 Occurrence and uses

Organic halides are of considerable importance as solvents (e.g. dichloromethane, chloroform), monomers for the polymer industry (e.g. vinyl chloride, tetrafluoroethene, Section 4.9), aerosol propellants, and refrigerants (e.g. dichlorodifluoromethane), and as pesticides (e.g. DDT) (Figure 6.1).

Examples of important alcohols (Figure 6.2) include methanol, ethanol, ethylene glycol, and glycerol. Ethanol is present in alcoholic drinks while ethanol and methanol are both extensively used as industrial solvents and raw materials. Ethylene glycol is used as antifreeze and for making the polyester, Terylene® (Section 9.7), while glycerol, prepared by hydrolysis of animal fats, is used in pharmaceutical preparations. Carbohydrates (Chapter 11) also possess many hydroxyl groups.

Table 6.1

General formula	Compound type
R^1-Hal	Alkyl or aryl halide
R^1-OH	Alcohol (R^1 = alkyl) or phenol (R^1 = aryl)
R^1-SH	Thiol
R^1-O-R^2	Ether
R^1-S-R^2	Thioether
$\begin{array}{c} R^1 \quad O \quad R^3 \\ \diagdown\diagup \\ C-C \\ \diagup\diagdown \\ R^2 \qquad R^4 \end{array}$	Epoxide (oxirane)

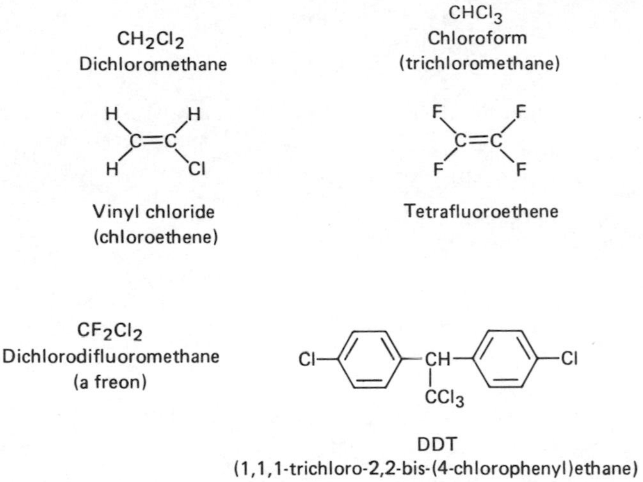

$$CH_2Cl_2$$
Dichloromethane

$$CHCl_3$$
Chloroform
(trichloromethane)

Vinyl chloride
(chloroethene)

Tetrafluoroethene

$$CF_2Cl_2$$
Dichlorodifluoromethane
(a freon)

DDT
(1,1,1-trichloro-2,2-bis-(4-chlorophenyl)ethane)

Figure 6.1 Some important organic halides

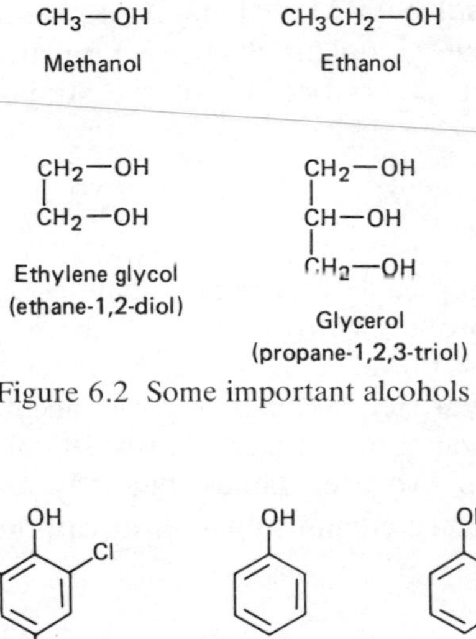

$$CH_3—OH$$
Methanol

$$CH_3CH_2—OH$$
Ethanol

Ethylene glycol
(ethane-1,2-diol)

Glycerol
(propane-1,2,3-triol)

Figure 6.2 Some important alcohols

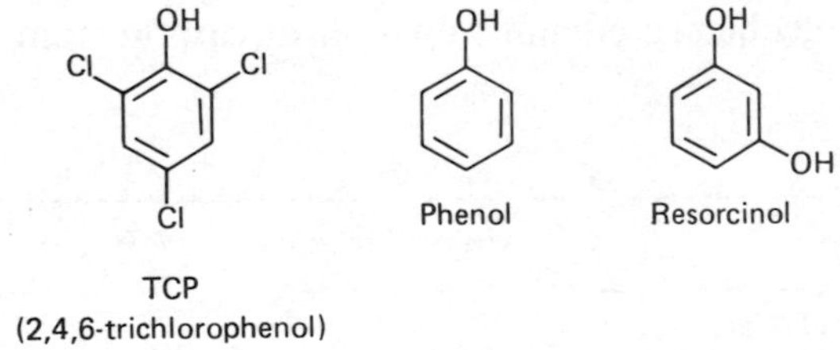

TCP
(2,4,6-trichlorophenol)

Phenol

Resorcinol

Figure 6.3 Three important phenols

Phenols are substances of varying degrees of toxicity and find application as antiseptics (e.g. TCP, Figure 6.3). Their reactivity enables phenol and resorcinol to polymerize with formaldehyde (methanal), forming bakelite and thermosetting adhesives respectively.

The thiol group (SH) is present in many biologically important com-

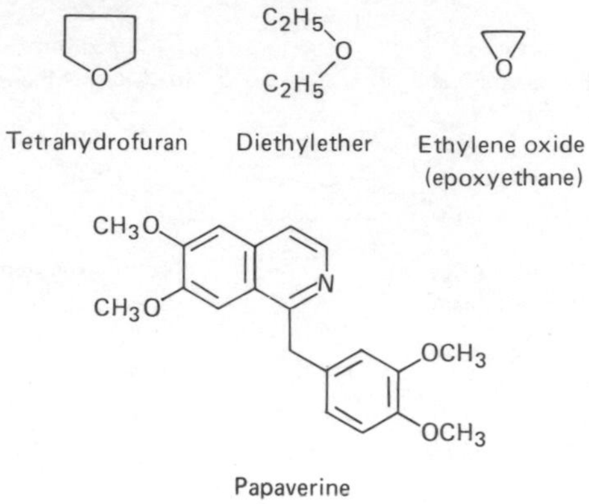

Tetrahydrofuran   Diethylether   Ethylene oxide
(epoxyethane)

Papaverine

Figure 6.4 Some important ethers

pounds such as the amino acid cysteine (Chapter 12) and coenzyme A (Chapter 14).

Ethers (Figure 6.4) such as diethyl ether and tetrahydrofuran are used in organic chemistry as solvents. Methyl ethers of phenols occur extensively in nature (e.g. papaverine, an opium alkaloid). Epoxides (e.g. ethylene oxide) are especially reactive ethers because of the strain present in the three-membered ring.

6.2 Physical properties

As can be seen from Table 6.2, organic halides and ethers have boiling points similar to those of hydrocarbons of comparable molecular mass. Alcohols and phenols, however, have much higher boiling points because of the formation of hydrogen bonds between molecules (Section 1.11).

Simple organic halides are not significantly soluble in water because of their inability to form hydrogen bonds, but they readily mix with hydrocarbons. Lower alcohols, containing up to three carbon atoms, are completely

Table 6.2

Compound	Molecular mass	Boiling point (°C)
$CH_3CH_2CH_2CH_3$	58	−0.5
$CH_3CH_2CH_2CH_2CH_3$	72	36.3
CH_3CH_2Cl	64.5	12.5
$CH_3CH_2CH_2OH$	60	97.2
$CH_3CH_2OCH_3$	60	7
$C_6H_5.C_2H_5$	106	136
C_6H_5Cl	112.5	132
$p\text{-}CH_3C_6H_4OH$	108	202
$C_6H_5.OCH_3$	108	154

miscible with water because of their ability to form hydrogen bonds with water molecules. As the size of the organic group in an alcohol increases solubility in water decreases whereas that in hydrocarbons increases. For example pentan-1-ol is only slightly soluble in water but appreciably soluble in hydrocarbons. Although unable to form hydrogen bonds with themselves, ethers can form hydrogen bonds with water molecules, and the lower members are therefore appreciably soluble in water. Indeed, a number of ethers (e.g. tetrahydrofuran and 1,2-dimethoxyethane) are completely miscible with water, while diethyl ether is soluble to the extent of 8 g per 100 cm^3 at 16 °C.

Alcohols and phenols show characteristic O—H absorption bands in their infra-red spectra, while chloroalkanes exhibit characteristic C—Cl bands (Chapter 3).

6.3 Nucleophilic substitution

A nucleophilic substitution reaction involves attack of an electron-rich species (nucleophile) on an electron-deficient carbon atom and loss of an atom or group from the carbon atom (Figure 6.5). Stable particles such as iodide ion are easily displaced, i.e. are good *leaving groups*.

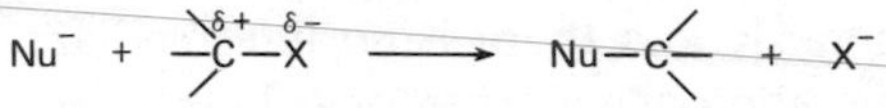

Figure 6.5 The principle of nucleophilic substitution

Nucleophilic substitution reactions can proceed by one of two mechanisms known as S$_N$1 (substitution, nucleophilic, unimolecular) and S$_N$2 (substitution, nucleophilic, bimolecular).

In S$_N$1 reactions the rate-determining step involves only one species, which dissociates forming a carbonium ion and a leaving group (Figure 6.6). The rate equation for such a reaction is of the type,

Rate=k [substrate],

i.e. it is independent of the concentration of the nucleophile (Section 1.9). Since carbonium ions are planar, attack by the nucleophile can take place from either side, so that if the original molecule is chiral the product should be a racemic mixture of the two possible enantiomers. In fact, in

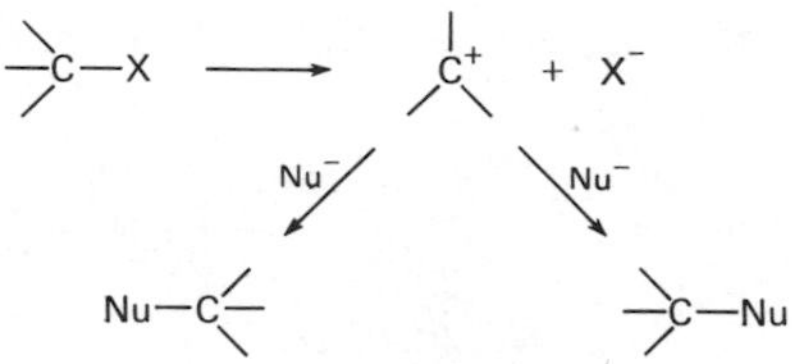

Figure 6.6 The S$_N$1 mechanism

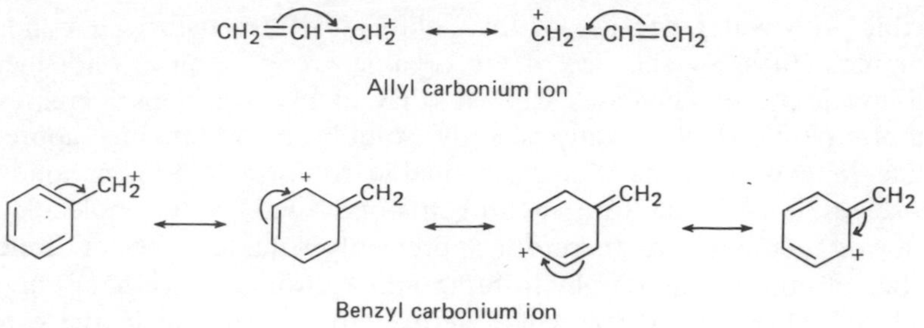

$$CH_2{=}\overset{\frown}{CH}{-}CH_2^+ \quad \longleftrightarrow \quad \overset{+}{CH_2}{-}\overset{\frown}{CH}{=}CH_2$$

Allyl carbonium ion

Benzyl carbonium ion

Figure 6.7 Stabilization of carbonium ions by delocalization

many cases only partial racemization is found because attack of the nucleophile from the side of the departing group (X^-) is somewhat restricted.

The tendency of a compound RX to undergo S_N1 reactions depends upon the nature of both R and X. For any given leaving group X, the order of reactivity in S_N1 reactions corresponds to the order of stability of the carbonium ions formed, i.e. allyl, benzyl > tertiary > secondary > primary > methyl (Section 4.5). Allyl and benzyl carbonium ions are particularly stable due to delocalization of the positive charge by interaction with the neighbouring π system (Figure 6.7).

In S_N2 reactions attack by the nucleophile and loss of the leaving group take place simultaneously and there is no intermediate carbonium ion. The incoming nucleophile approaches the molecule from the opposite side to the leaving group, and in the transition state the C—X bond is broken as the new bond is formed (Figure 6.8). This has the effect of 'turning the molecule inside out' and the reaction is therefore said to proceed with *inversion of configuration*. If the starting material is optically active the product is also optically active.

The rate equation for an S_N2 reaction is of the type,

Rate=k [substrate] [nucleophile],

i.e. the rate depends upon the concentrations of both the substrate and the nucleophile (Section 1.9). The tendency of a substrate RX to undergo S_N2 reactions depends upon the nature of both X and the organic group, and for a given leaving group X, reactivity decreases in the order R = methyl > primary > secondary > tertiary alkyl. This is because large substituents around the carbon atom hinder the approach of an incoming nucleophile.

Note that the order of reactivity for the S_N2 mechanism is the reverse of

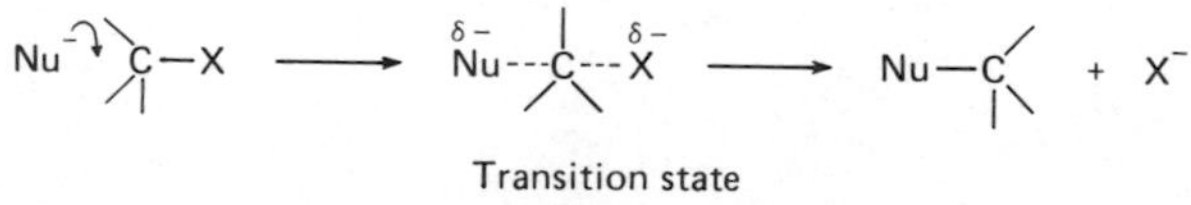

Transition state

Figure 6.8 The S_N2 mechanism

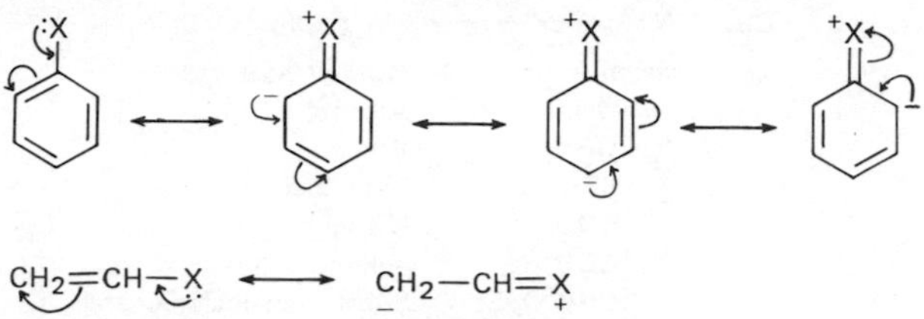

Figure 6.9 The resonance forms contributing to the structure of aryl and vinyl halides

that for S_N1. The net result is that primary alkyl compounds usually react by the S_N2 mechanism and tertiary compounds by S_N1. For secondary compounds the situation is less clear and they may react by either mechanism depending upon the nucleophile, the leaving group, and the solvent. A high concentration of a powerful nucleophile favours the S_N2 mechanism, whereas a polar solvent tends to favour S_N1 because solvation stabilizes the intermediate carbonium ions.

Aryl and vinyl compounds are relatively unreactive towards nucleophilic substitution and can only be made to undergo substitution reactions under very vigorous conditions. The reason for this lies in the contribution to the overall structure of resonance forms which involve double bonds between carbon and the substituent group (Figure 6.9). This increases the strength of the C—X bond, making it much more difficult to break.

An exception to this generalization occurs when strongly electron-withdrawing groups (e.g. NO_2) are attached to the 2, 4, or 6 positions of an aryl compound. This enables the formation of a stabilized intermediate which can then lose the leaving group. An example of such a reaction is shown in Figure 6.10. A similar reaction is used for determination of N-terminal amino acid residues in proteins (Section 12.7).

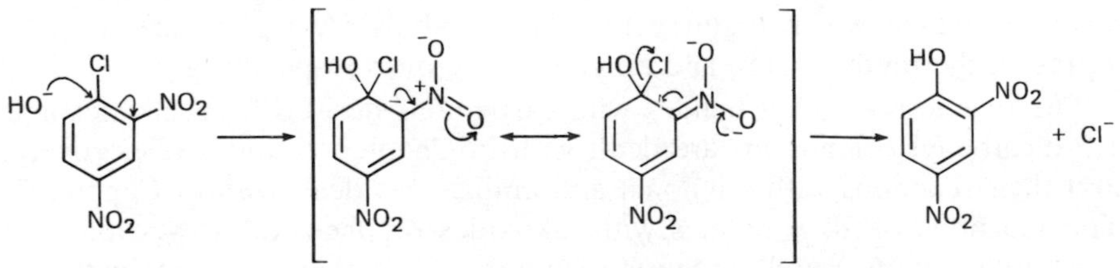

Figure 6.10 An example of a substitution reaction of an aryl compound containing two strongly electron-withdrawing groups

6.4 Examples of nucleophilic substitution reactions

Alkyl and benzyl halides undergo a large number of nucleophilic substitution reactions which are of considerable synthetic importance. Alcohols and ethers are less susceptible to nucleophilic substitution due to the

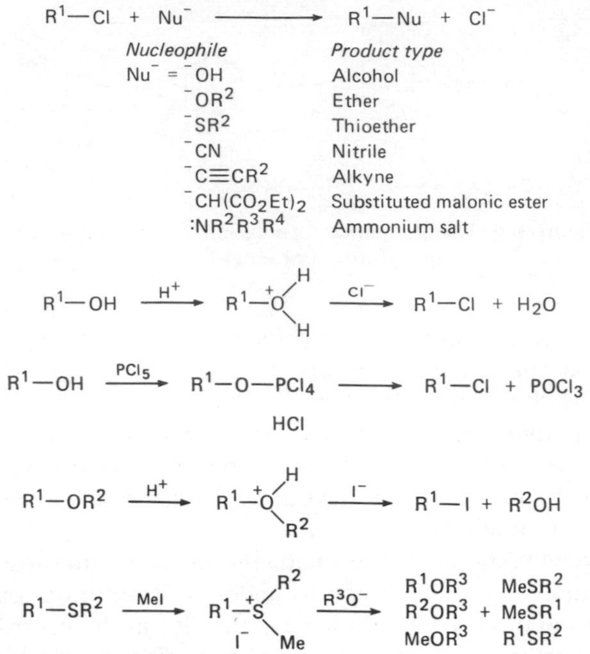

$$R^1-Cl \ + \ Nu^- \longrightarrow R^1-Nu \ + \ Cl^-$$

Nucleophile	Product type
$Nu^- = \ ^-OH$	Alcohol
$^-OR^2$	Ether
$^-SR^2$	Thioether
^-CN	Nitrile
$^-C\equiv CR^2$	Alkyne
$^-CH(CO_2Et)_2$	Substituted malonic ester
$:NR^2R^3R^4$	Ammonium salt

$$R^1-OH \xrightarrow{H^+} R^1-\overset{+}{\underset{H}{O}}{\diagup}^H \xrightarrow{Cl^-} R^1-Cl \ + \ H_2O$$

$$R^1-OH \xrightarrow{PCl_5} R^1-O-PCl_4 \longrightarrow R^1-Cl \ + \ POCl_3$$
$$HCl$$

$$R^1-OR^2 \xrightarrow{H^+} R^1-\overset{+}{\underset{R^2}{O}}{\diagup}^H \xrightarrow{I^-} R^1-I \ + \ R^2OH$$

$$R^1-SR^2 \xrightarrow{MeI} R^1-\overset{R^2}{\underset{I^- \ \diagdown Me}{S}} \xrightarrow{R^3O^-} \begin{array}{cc} R^1OR^3 & MeSR^2 \\ R^2OR^3 \ + & MeSR^1 \\ MeOR^3 & R^1SR^2 \end{array}$$

Figure 6.11 Some examples of nucleophilic substitution reactions

poorer leaving groups (HO^- and RO^- are less stable than Cl^-). These compounds do however undergo nucleophilic substitution reactions if the OH or OR group is first converted into a better leaving group, for example by protonation or conversion into a sulphonate or phosphate ester. Thiols and thioethers are converted into good leaving groups by alkylation or complexation to a heavy metal. Some examples of nucleophilic substitution reactions are shown in Figure 6.11. (The symbols Me and Et are used to represent the methyl, CH_3, and ethyl, C_2H_5, groups respectively.)

The reactions of alkyl halides with carbanions derived from alkynes and 1,3-dicarbonyl compounds are dealt with in Chapters 4 and 9 respectively, and their reactions with ammonia and amines are dealt with in Chapter 7. The reactions of alkyl halides with alkoxides or phenoxides (Section 6.7) constitute a very useful synthesis of ethers, while their reactions with hydroxide, RS^- and cyanide anions afford useful syntheses of alcohols, thioethers, and nitriles respectively.

Question 6.1 Starting from an organic halide devise syntheses for the following compounds:

$$OH$$
$$|$$
(a) $C_6H_5OCH_2CH_3$; (b) $CH_3CHCH_2CH_3$; (c) $C_6H_5C\equiv CCH_3$.

$$ROH \xrightarrow[\text{SOCl}_2]{\text{PCl}_5 \text{ or}} RCl$$

$$ROH \xrightarrow{\text{PBr}_3} RBr$$

Figure 6.12 Some methods for the synthesis of alkyl halides from alcohols

The acid catalysed nucleophilic substitution of hydroxide by halide is important for conversion of alcohols into alkyl halides. The reaction is usually carried out by treating the alcohol with dry hydrogen halide. An alternative method of converting alcohols to halides involves reaction with the corresponding phosphorus halides (e.g. PCl_5, PBr_3) or thionyl chloride (sulphur dichloride oxide, $SOCl_2$) as illustrated in Figure 6.12. The latter reagent has the advantage that the by-products (SO_2, HCl) are gaseous and therefore easily removed.

Reaction of ethers with concentrated hydrobromic or hydriodic acid results in cleavage. The mechanism involves initial protonation of the ether oxygen to give a relatively good leaving group, an alcohol. The alcohol formed reacts further to give a second alkyl halide molecule (Figure 6.13(a)). Note that when an aryl alkyl ether is cleaved the products are an alkyl halide and a phenol (Figure 6.13(b)). Phenols do not react further because even when protonated they are not susceptible to nucleophilic substitution (Section 6.3).

Nucleophilic substitution reactions of thioethers are not of particular importance in the laboratory because a complex mixture of products may

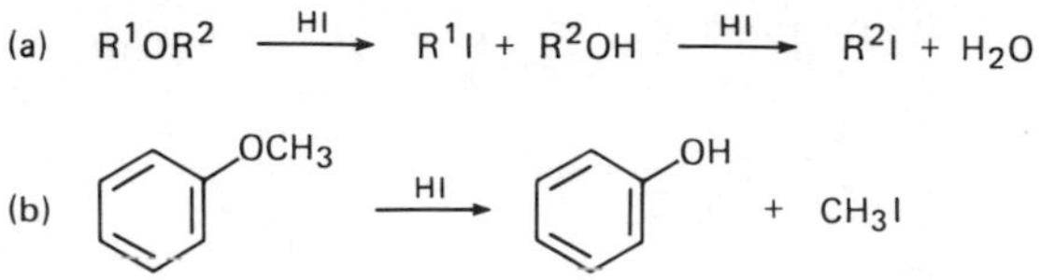

(a) $\quad R^1OR^2 \xrightarrow{\text{HI}} R^1I + R^2OH \xrightarrow{\text{HI}} R^2I + H_2O$

(b)

Figure 6.13 The cleavage of ethers using hydriodic acid

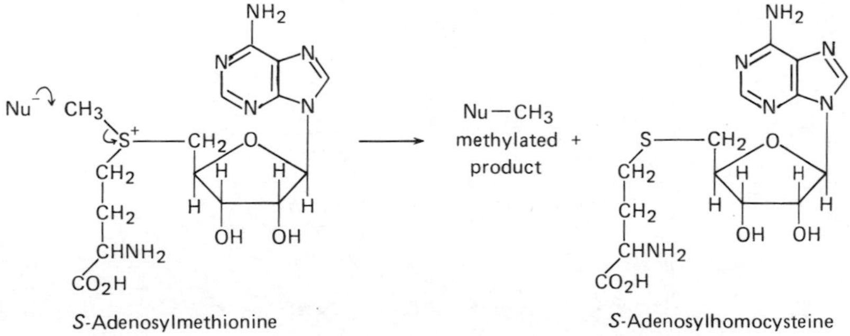

Figure 6.14 S-Adenylmethionine as a methyl group donor

be obtained. However enzyme catalysed reactions of this type are much more specific and such reactions are therefore of much greater significance in nature. For example, S-adenosylmethionine (Figure 6.14) is responsible for the biological methylation reactions involved in the biosynthesis of the antibiotic, tetracycline, and the vitamin, B_{12}.

6.5 Elimination reactions

Alkyl halides and alcohols undergo elimination reactions (Figure 6.15) and these reactions often compete with substitution. The elimination of hydrogen halide from an alkyl halide is brought about by treatment with a base (e.g. HO⁻, RO⁻), whereas elimination of water from an alcohol occurs under strongly acidic conditions (e.g. concentrated sulphuric or phosphoric acid).

As with substitution there are two possible mechanisms, one unimolecular (E1), the other bimolecular (E2). In the E1 mechanism the rate-determining step is loss of the leaving group forming a carbonium ion, from which H⁺ is subsequently lost to give the alkene (Figure 6.16). As in the S_N1 reaction the order of reactivity corresponds to the order of stability of the carbonium ions produced, i.e. tertiary > secondary > primary > methyl.

The E2 reaction involves just one step (Figure 6.17), the abstraction of a proton by a base (B⁻) occurring simultaneously with the loss of the leaving group.

In general the greater the number of alkyl groups surrounding a C=C bond the greater the stability of the alkene. For this reason, when there are alternative elimination pathways from alkyl halides or alcohols, the more substituted alkene predominates (e.g. Figure 6.18).

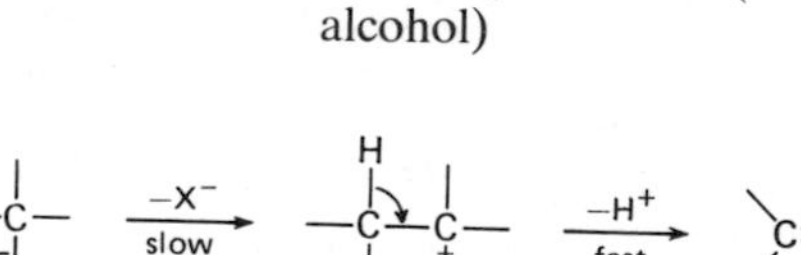

X = halogen or OH

Figure 6.15 Elimination of hydrogen halide (or the elements of water) from an alkyl halide (or alcohol)

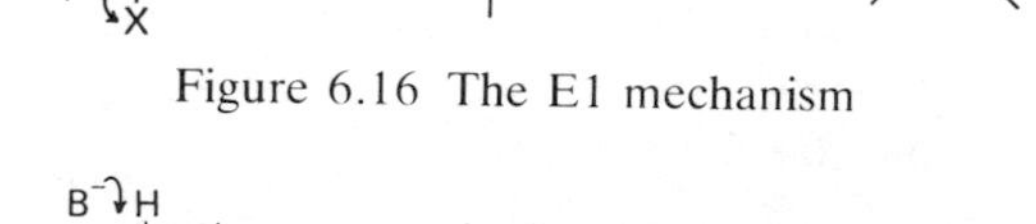

Figure 6.16 The E1 mechanism

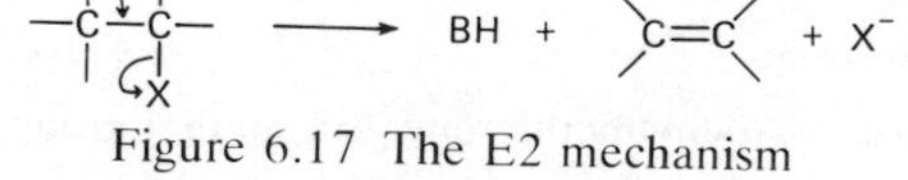

Figure 6.17 The E2 mechanism

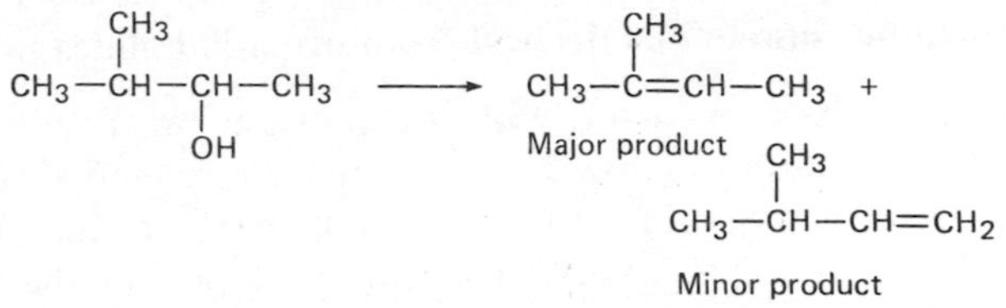

Figure 6.18 The formation of two products in an elimination reaction

Usually for tertiary alkyl halides and alcohols, elimination occurs readily and competes with substitution, while for primary compounds substitution tends to predominate over elimination. For a given substrate, elimination tends to be favoured relative to substitution by conditions which favour a bimolecular mechanism rather than a unimolecular one. With alkyl halides, for example, a solvent of relatively low polarity and a high concentration of a strong base favour elimination. Conversely the best conditions for promoting substitutions are those which favour a unimolecular mechanism, i.e. a highly polar solvent to promote ionization and a weakly basic nucleophile which will not remove H^+ from the carbonium ion.

The conditions favouring the various competing reactions of alkyl halides are summarized in Table 6.3.

Table 6.3 Conditions favouring S_N1, S_N2, E1, and E2 reactions

S_N1	S_N2	E1	E2
Tertiary halide	Primary halide	Tertiary halide	Primary halide
Poor nucleophile	Good nucleophile	Weak base	Strong base
Polar solvent	Non-polar solvent	Polar solvent	Non-polar solvent

Question 6.2 Predict the major alkene formed in each case when the following alkyl halides are heated with KOH in ethanol:

(a) $CH_3 - CH_2 - CH_2 - CH_2 - Br$;

(b) $CH_3 - CH_2 - \underset{\underset{Br}{|}}{\overset{\overset{CH_3}{|}}{C}} - CH_3$;

(c) $CH_3 - CH_2 - \underset{\underset{Br}{|}}{CH} - CH_3$.

What is the order of reactivity of these three alkyl halides?

6.6 Organometallic compounds derived from organic halides

Alkyl and aryl halides react with magnesium in dry ether to give organometallic compounds known as a Grignard reagents (Figure 6.19). Since the electronegativity of carbon is much greater than that of magnesium the Mg—C bond is polarized in such a way that the carbon atom bears a partial negative charge.

The carbon atom attached to the magnesium in a Grignard reagent is nucleophilic and attacks substrates possessing an electron deficient centre. For example Grignard reagents are sensitive to moisture and react with other weakly acidic compounds such as alcohols and alk-1-ynes (Figure 6.20). The latter reaction can be used to prepare alkynyl Grignard reagents (Chapter 4).

Some of the most important reactions of Grignard reagents are those with carbonyl compounds (Chapter 8). Thus Grignard reagents react with formaldehyde (methanal), other aldehydes, and ketones to give primary, secondary, and tertiary alcohols respectively (Figure 6.21). Esters react with two moles of a Grignard reagent to give tertiary alcohols whereas

$$RX \xrightarrow{\text{Mg}} \overset{\delta-}{R}{-}\overset{\delta+}{MgX}$$

$$(X = Cl, Br, I)$$

Figure 6.19 The formation of Grignard reagents

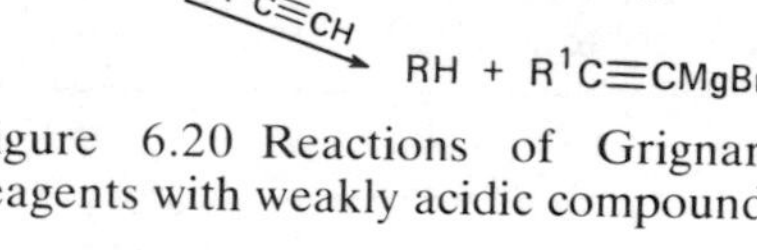

$$\overset{\delta-}{R}{-}\overset{\delta+}{MgBr} \quad \begin{array}{c} \xrightarrow{H_2O} \quad RH + HOMgBr \\ \xrightarrow{R^1OH} \quad RH + R^1OMgBr \\ \xrightarrow{R^1C\equiv CH} \quad RH + R^1C\equiv CMgBr \end{array}$$

Figure 6.20 Reactions of Grignard reagents with weakly acidic compounds

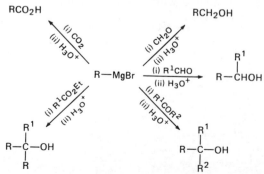

Figure 6.21 Reactions of Grignard reagents with carbonyl compounds

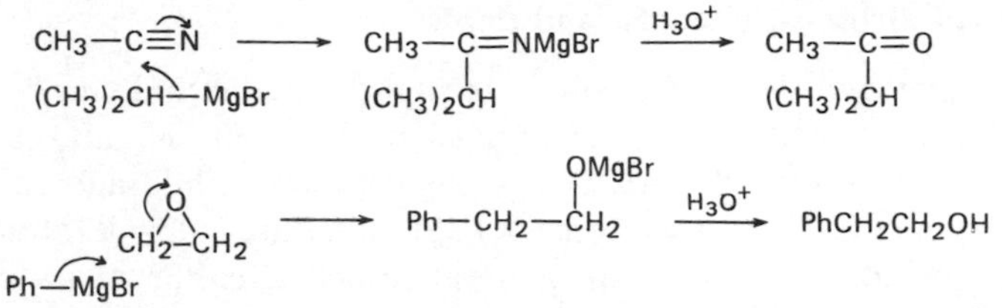

CH$_3$—C≡N $\longrightarrow$ CH$_3$—C=NMgBr $\xrightarrow{H_3O^+}$ CH$_3$—C=O

(CH$_3$)$_2$CH—MgBr   (CH$_3$)$_2$CH   (CH$_3$)$_2$CH

Ph—MgBr + CH$_2$—CH$_2$ (O) $\longrightarrow$ Ph—CH$_2$—CH$_2$ (OMgBr) $\xrightarrow{H_3O^+}$ PhCH$_2$CH$_2$OH

Figure 6.22 Reactions of Grignard reagents with nitriles and epoxides

carbon dioxide reacts with Grignard reagents to give carboxylic acids. In all of these cases it is necessary to decompose the magnesium salts formed initially by treating with dilute mineral acid.

Two other useful reactions of Grignard reagents are those with nitriles (Chapter 7) and epoxides (Section 6.11). These reactions are useful for the synthesis of ketones and primary alcohols respectively (e.g. Figure 6.22).

Question 6.3 Suggest syntheses for each of the following compounds using Grignard reagents:

(a) (CH$_3$)$_3$CCH$_2$CH$_2$OH from (CH$_3$)$_3$CCl;
(b) 4-CH$_3$OC$_6$H$_4$CO$_2$H from 4-CH$_3$OC$_6$H$_4$Br;
(c) (CH$_3$CH$_2$CH$_2$)$_2$COH from CH$_3$CH$_2$CH$_2$Cl.
$\qquad\qquad$ |
$\qquad\qquad$ CH$_3$

Grignard reagents can be used to prepare other organometallic compounds by reacting them with inorganic halides, e.g. BCl$_3$, CdCl$_2$. Organoboranes are more conveniently prepared however by hydroboration of alkenes (Chapter 4). Organocadmium reagents, R$_2$Cd, are much less reactive than Grignard reagents and react with acid chlorides to give ketones which do not react further (e.g. Figure 6.23).

Organometallic derivatives of other electropositive metals such as Li and Zn can be prepared in the same way as Grignard reagents from an alkyl halide and the metal. The most important organozinc reagent is the Reformatsky reagent, BrZnCH$_2$CO$_2$Et, which reacts with aldehydes and ketones to give β-hydroxyesters (Chapter 8). Organolithium reagents behave very much like Grignard reagents but are somewhat more reactive.

Organometallic reagents are widely used in the laboratory synthesis of naturally occurring and physiologically important compounds. Coenzyme B$_{12}$, which is related to vitamin B$_{12}$ (Chapter 13), is an important naturally occurring organocobalt compound.

(PhCH$_2$)$_2$Cd + CH$_3$COCl $\longrightarrow$ PhCH$_2$COCH$_3$

Figure 6.23 Reaction of an organocadmium compound with an acid chloride

6.7 Acidity of alcohols, phenols, and thiols

The hydroxyl hydrogen of alcohols is slightly acidic and can be replaced by direct reaction with reactive metals such as Na, K, and Li (Figure 6.24(a)). The resulting alkoxide ion is a good nucleophile and can be used for preparation of ethers (Section 6.4). Phenols are considerably stronger acids than alcohols, though in general much weaker than carboxylic (Chapter 8) and mineral acids. They neutralize strong alkalis such as sodium hydroxide solution forming salts (Figure 6.24(b)). Thiols resemble alcohols and phenols in that they possess an acidic hydrogen atom.

The nature of substituents on the aromatic ring of a phenol influences its strength as an acid. Electron-withdrawing groups such as —NO_2 delocalize the charge on the phenoxide ion and so increase its stability and the strength of the corresponding acid. Indeed picric acid (2,4,6-trinitrophenol, Figure 6.25) is such a strong acid that it is almost completely dissociated in dilute aqueous solution. On the other hand electron-donating groups such as —OR tend to destabilize the phenoxide ion and therefore lower the acid strength.

(a) $ROH + Na \longrightarrow RO^- Na^+ + \frac{1}{2} H_2$

(b) $ArOH + NaOH \longrightarrow ArO^- Na^+ + H_2O$

Figure 6.24 The acidic behaviour of alcohols and phenols

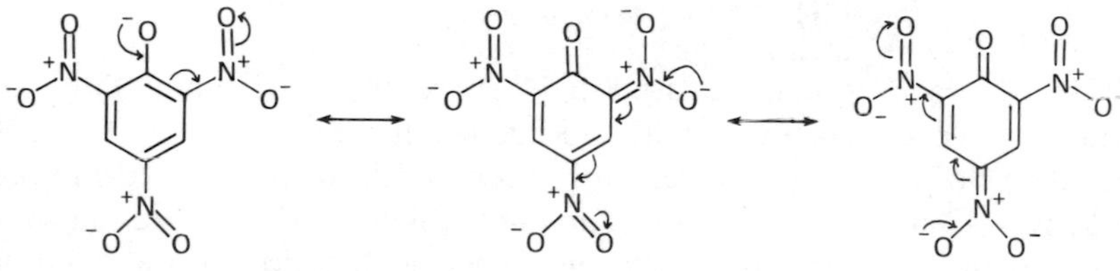

Figure 6.25 Stabilization of the picrate ion by delocalization

6.8 Esterification of alcohols and phenols

Alcohols and phenols can be converted into esters using a variety of reagents (Figure 6.26).

Direct esterification of alcohols and phenols by carboxylic acids is carried out by refluxing with a mineral acid catalyst. The reaction is

$R^1OH + R^2CO_2H \rightleftharpoons R^2CO_2R^1 + H_2O$

$R^1OH + (R^2CO)_2O \longrightarrow R^2CO_2R^1 + R^2CO_2H$

$R^1OH + R^2COCl \longrightarrow R^2CO_2R^1 + HCl$

$R^1OH + R^2SO_2Cl \longrightarrow R^2SO_2R^1 + HCl$

Figure 6.26 Esterification reactions

reversible and the yields therefore depend upon the equilibrium position. Yields can be increased by removing the water as it is formed.

An alternative method of preparing esters from alcohols and phenols is to react them with an acid anhydride or acyl chloride in the presence of an acidic or basic catalyst (Chapter 8). Sulphonate esters are prepared in a similar way from sulphonyl chlorides.

6.9 Oxidation of alcohols, phenols, and thiols

Primary and secondary alcohols are oxidized by a variety of reagents (e.g. chromic acid, $H_2Cr_2O_7$) to give aldehydes and ketones (Figure 6.27). Aldehydes readily undergo further oxidation to carboxylic acids (Chapter 8) so that if an aldehyde is the desired product of the oxidation of a primary alcohol the apparatus must be arranged so that it is removed as it is formed. Tertiary alcohols are not so readily oxidized.

Phenols, particularly those containing two or more hydroxyl groups,

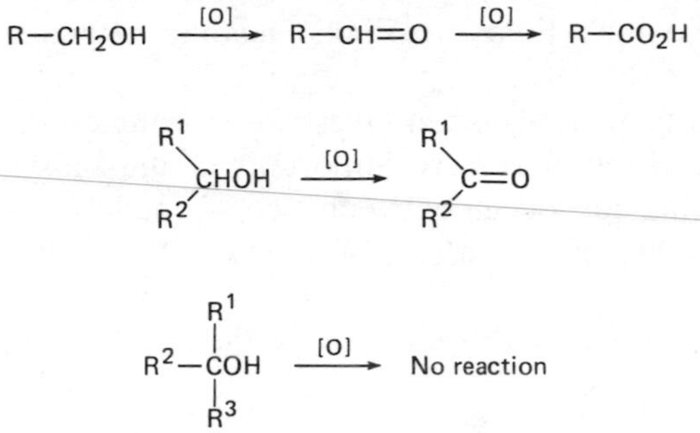

$$R-CH_2OH \xrightarrow{[O]} R-CH=O \xrightarrow{[O]} R-CO_2H$$

Figure 6.27 The oxidation of alcohols

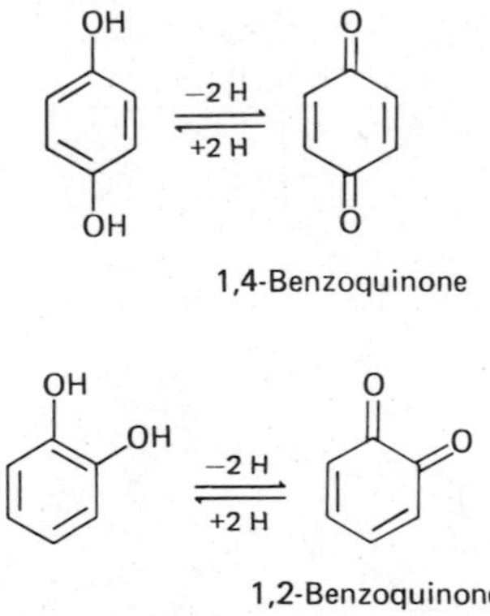

1,4-Benzoquinone

1,2-Benzoquinone

Figure 6.28 The oxidation of 1,2- and 1,4-dihydroxybenzenes to quinones

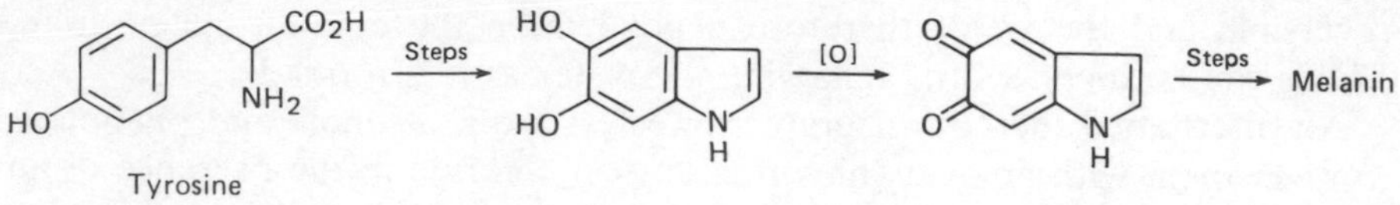

Figure 6.29 The formation of melanin via a quinone

are susceptible to oxidation by a wide variety of reagents, including atmospheric oxygen. Indeed, it is difficult to store phenol in a pure form in the laboratory and it is usually coloured due to the presence of oxidation products. The 1,2- and 1,4-dihydroxybenzenes (Figure 6.28) are particularly readily oxidized to cyclic $\alpha\beta$-unsaturated diketones known as quinones (Chapter 9). Interconversion of the oxidized and reduced species is achieved by such mild oxidizing agents as Fe^{3+} and reducing agents such as SO_2

Naturally occurring quinones with slightly different redox potentials take part in oxidation–reduction reactions in the respiratory cycle (Section 13.3). Formation of the skin pigment melanin also involves oxidation to give a quinone which then crosslinks, leading ultimately to the brown polymer (Figure 6.29).

The oxidation of phenols often gives rise to coupled or dimeric products and indeed this reaction is very important in the biosynthesis of natural products. Coupling can occur between two phenolic units within the same molecule, as for example in the biosynthesis of morphine from reticuline (Figure 6.30).

Lignin, which along with cellulose is the main structural material of

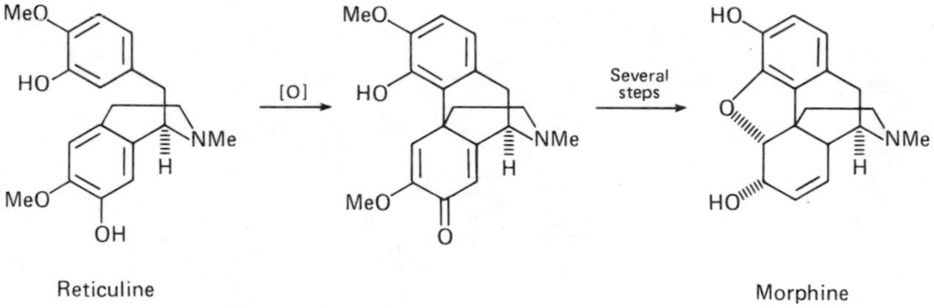

Figure 6.30 The biosynthesis of morphine

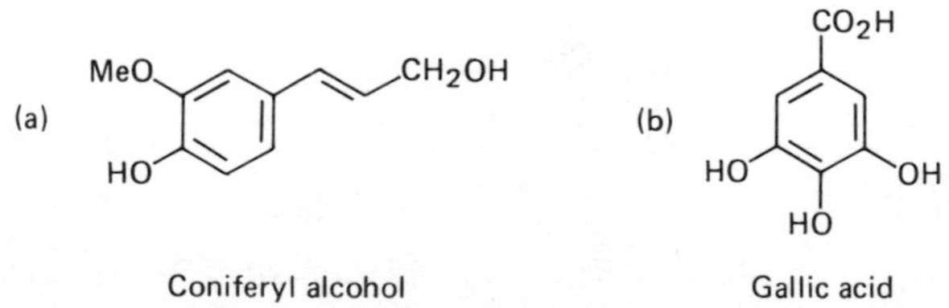

Figure 6.31 The structures of coniferyl alcohol and gallic acid

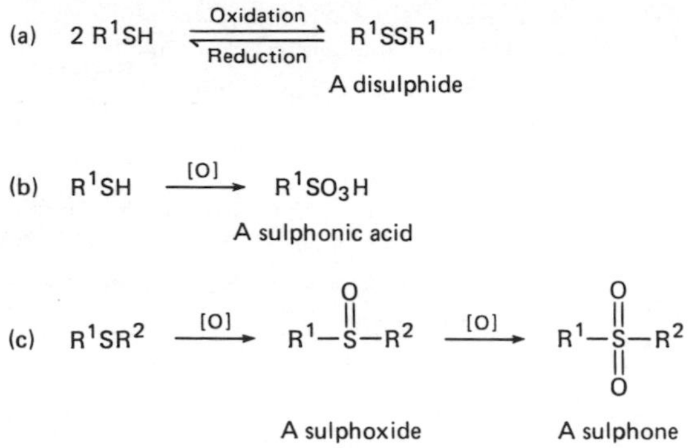

Figure 6.32 Oxidation of thiols and thioethers

wood, arises by oxidative coupling of the phenol, coniferyl alcohol (Figure 6.31(a)). Tannins, which occur in tea leaves, for example, and are used for converting animal hides into leather, are biosynthesized via phenolic coupling of gallic acid (Figure 6.31(b)).

Thiols are readily oxidized to disulphides, and disulphides are easily reduced back to thiols (Figure 6.32(a)). In nature disulphide bridges are important in maintaining the three-dimensional structure of proteins. (Section 12.8). With stronger oxidizing agents thiols can be oxidized (Figure 6.32(b)) to sulphonic acids (Chapter 8), while thiothers can be oxidized to sulphoxides or sulphones (Figure 6.32(c)).

6.10 Electrophilic substitution reactions of phenols

The electron-donating OH group of phenol activates the ring to attack by electrophiles (Chapter 5), and, in general, reactions of phenol proceed under considerably milder conditions than those required for the corresponding reactions of benzene. For example, nitration occurs readily at room temperature using dilute nitric acid (Figure 6.33(a)) and bromination

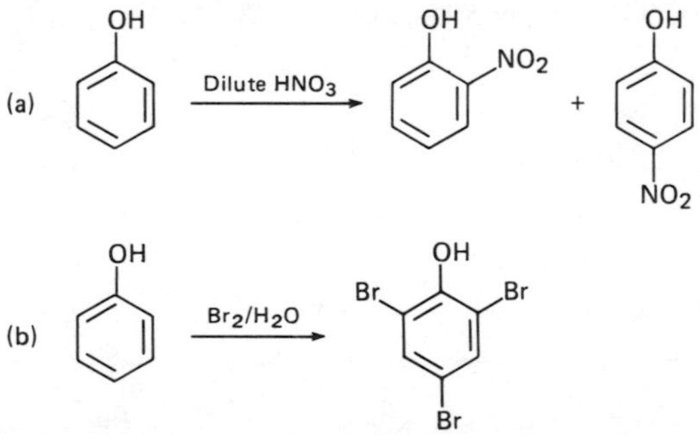

Figure 6.33 Examples of electrophilic substitution reactions of phenol

E

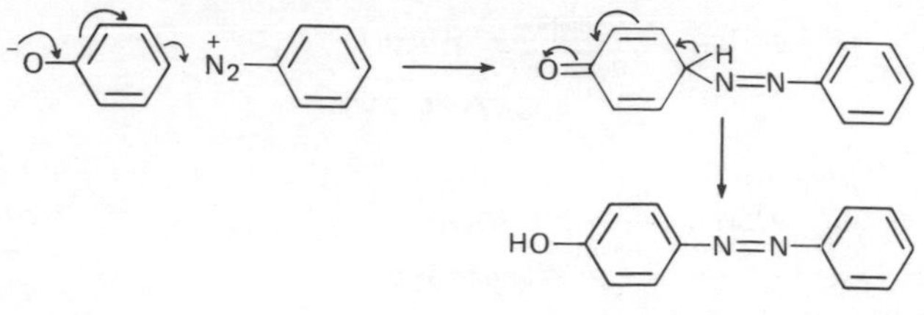

Hydroxyazobenzene

Figure 6.34 The mechanism of coupling reactions between phenoxide ions and diazonium salts

proceeds rapidly at room temperature even with bromine water, giving 2,4,6-tribromophenol (Figure 6.33(b)).

Coupling of phenoxide ions with diazonium salts gives rise to coloured compounds, some of which are important as dyestuffs (Chapter 7). For example, a solution of sodium phenoxide reacts with benzenediazonium chloride to yield hydroxyazobenzene (Figure 6.34), which is yellow.

Question 6.4 Draw curly arrows to indicate why the electrophilic attack of the NO_2^+ ion (Section 5.4(a)) on phenol gives only 2- and 4-nitrophenols.

6.11 Reactions of epoxides

Because of the strain present in the three-membered ring epoxides are much more reactive than other ethers. Attack by nucleophiles opens the ring, giving substituted alcohols. The reaction can occur under acidic or basic conditions (Figure 6.35).

Two important reactions of epoxides are their hydrolysis (Chapter 4) and their reaction with organometallic reagents (Section 6.6).

Introduction of hydroxyl groups into many natural products is thought to proceed via epoxidation of carbon–carbon double bonds by reaction with oxygen catalysed by a suitable oxygenase. For example, the biosynthesis of gliotoxin may proceed via epoxidation of a cyclic dipeptide (Figure 6.36).

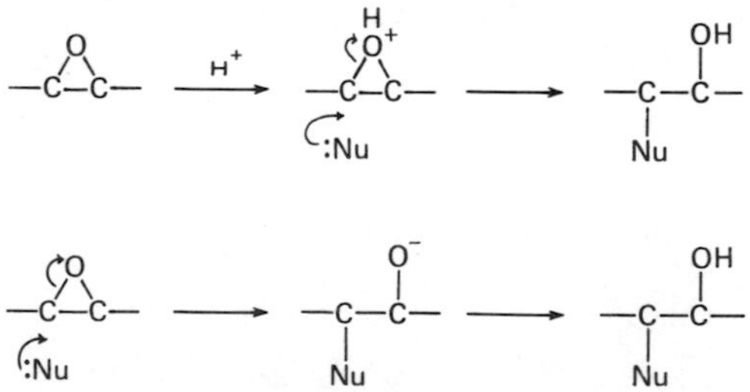

Figure 6.35 Nucleophilic cleavage of the epoxide ring

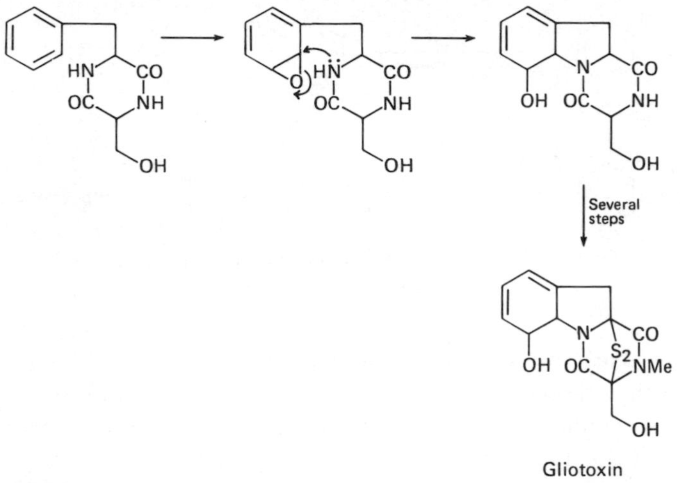

Gliotoxin

Figure 6.36 A possible mechanism for the biosynthesis of gliotoxin

6.12 Preparations

(a) Organic halides

Industrially, the most widely used halides are the chlorides, which are generally prepared by direct chlorination of hydrocarbons (Chapter 4). The most important laboratory methods for the preparation of alkyl halides are the addition of hydrogen halides or halogens to alkenes and alkynes (Chapter 4), and the reactions of alcohols with hydrogen halides, phosphorus halides, and thionyl chloride (Section 6.4). Aryl halides can be prepared by direct halogenation of arenes or by reactions of diazonium salts with copper(I) halides or potassium iodide (Chapter 5).

(b) Alcohols and phenols

Industrially, most of the simpler alcohols are prepared by hydration of alkenes. The alkene and steam are passed over a suitable catalyst such as phosphoric acid absorbed on an inert support. Ethanol in alcoholic drinks is produced by fermentation of sugars derived from fruit or grain. For example, during *beer* production partially sprouted *barley* (malt) is added to water where the enzyme diastase hydrolyses starch to the disaccharide maltose. Yeast is then added and the maltose is hydrolysed to glucose (Section 11.4). The glucose is converted to pyruvic acid (Chapter 15) which undergoes decarboxylation followed by reduction to give ethanol.

In the laboratory a wide range of methods are available for the production of alcohols. The most important are those involving hydration, oxidation, or hydroboration followed by oxidation of alkenes (Chapter 4),

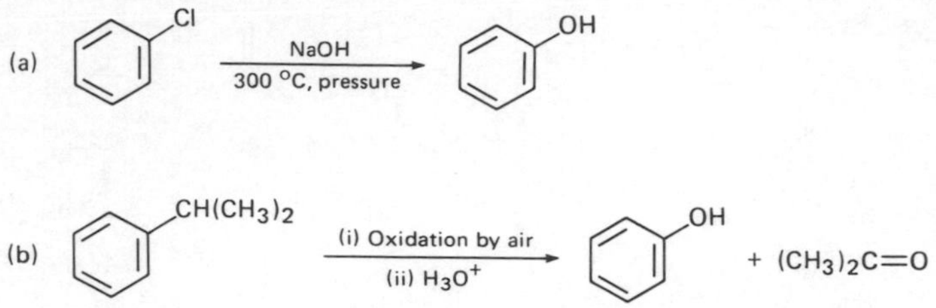

(a)

(b)

Figure 6.37 Reactions for the manufacture of phenol

reduction of carbonyl compounds (Chapter 8), and reactions of Grignard reagents with carbonyl compounds and epoxides (Section 6.6).

Phenol is manufactured on a large scale by hydrolysis of chlorobenzene (Figure 6.37(a)) or oxidation of cumene (Figure 6.37(b)).

(c) Ethers

Industrially, symmetrical ethers are produced by partial dehydration of alcohols using concentrated sulphuric acid. In the laboratory, preparation is by treatment of alkyl halides with sodium alkoxide or phenoxide (Section 6.4).

(d) Epoxides

Ethylene oxide is produced on a large scale by direct oxidation of ethylene using a silver catalyst at 250 °C.

Other synthetic methods include the reaction of an alkene with a peracid (Chapter 4), and the treatment of a halohydrin (Chapter 4) with a base (Figure 6.38).

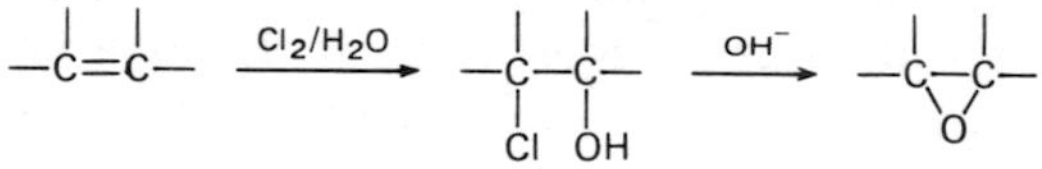

Figure 6.38 The synthesis of epoxides from alkenes via the halohydrin

(e) Thiols and thioethers

Thiols and thioethers are usually made by reacting alkyl halides with HS^- or RS^- (Section 6.4).

Answers to Questions

6.1 (a) $C_6H_5O^- + CH_3CH_2Br$; (b) $CH_3\overset{\overset{\displaystyle Cl}{|}}{C}HCH_2CH_3 + HO^-$;

    (c) $C_6H_5C \equiv C^- + MeI$

6.2 (a) $CH_3CH_2CH = CH_2$; (b) $CH_3CH = \overset{\overset{\displaystyle CH_3}{|}}{C}CH_3$;(c) $CH_3CH = CHCH_3$.
    Order of reactivity is (b) > (c) > (a)

6.3 (a) $(CH_3)_3CCl \xrightarrow{Mg} (CH_3)_3CMgCl \xrightarrow[\text{(ii) }H_3O^+]{\text{(i) }\triangle} (CH_3)_3CCH_2CH_2OH$;

    (b) $4 - CH_3OC_6H_4Br \xrightarrow{Mg} 4 - CH_3OC_6H_4MgBr \xrightarrow[\text{(ii) }H_3O^+]{\text{(i) }CO_2}$

                              $4 - CH_3OC_6H_4CO_2H$;

    (c) $CH_3CH_2CH_2Cl \xrightarrow{Mg} CH_3CH_2CH_2MgCl \xrightarrow[\text{(ii) }H_3O^+]{\text{(i) }CH_3CO_2Et}$

                             $(CH_3CH_2CH_2)_2\overset{\overset{\displaystyle}{}}{C}OH.$
                                      $\overset{|}{CH_3}$

6.4

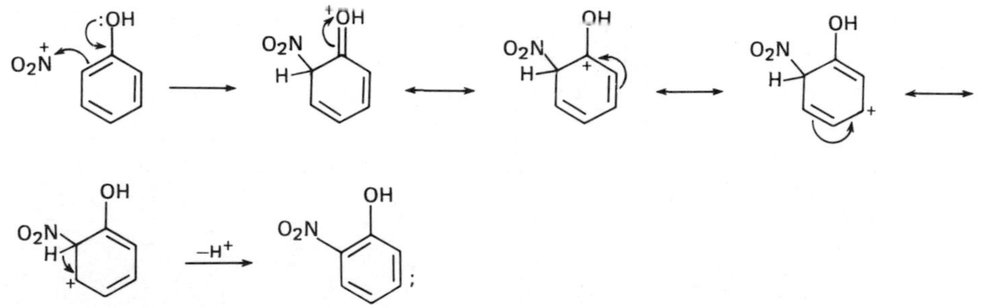

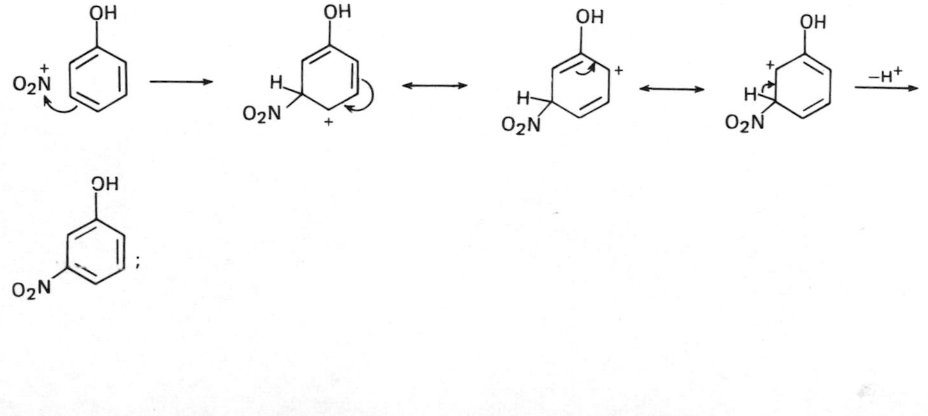

The intermediate involved in *meta* substitution has fewer canonical forms; in particular it lacks the important contribution from the form in which the positive charge resides on oxygen.

CHAPTER 7

Organic Nitrogen Compounds

The element nitrogen occurs in many organic compounds including many biologically important molecules such as amino acids, proteins, and nucleic acids (Chapter 12). Nitrogen compounds also play an important role in the manufacture of local anaesthetics, drugs, and many other medicinals (see below).

Some of the most important types of organic nitrogen compounds are shown in Figure 7.1(a). In addition there is a whole host of *heterocyclic* nitrogen compounds in which one or more hetero atoms (i.e. not carbon), in this case nitrogen, form part of a cyclic system. Some examples of such heterocyclic systems are shown in Figure 7.1(b).

Pyridine and purine rings are present in nicotinamide adenine dinucleotide (NAD⁺, Figure 7.2(a)), a coenzyme involved in oxidation and reduction reactions in plants and animals (Chapter 15). The pyrrole ring is

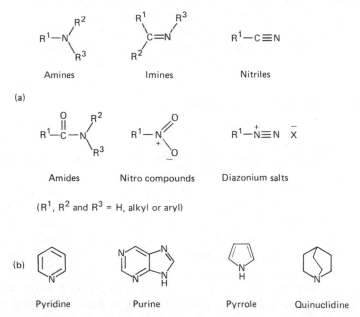

Figure 7.1 Some important types of organic nitrogen compounds and some specific examples of heterocyclic nitrogen compounds

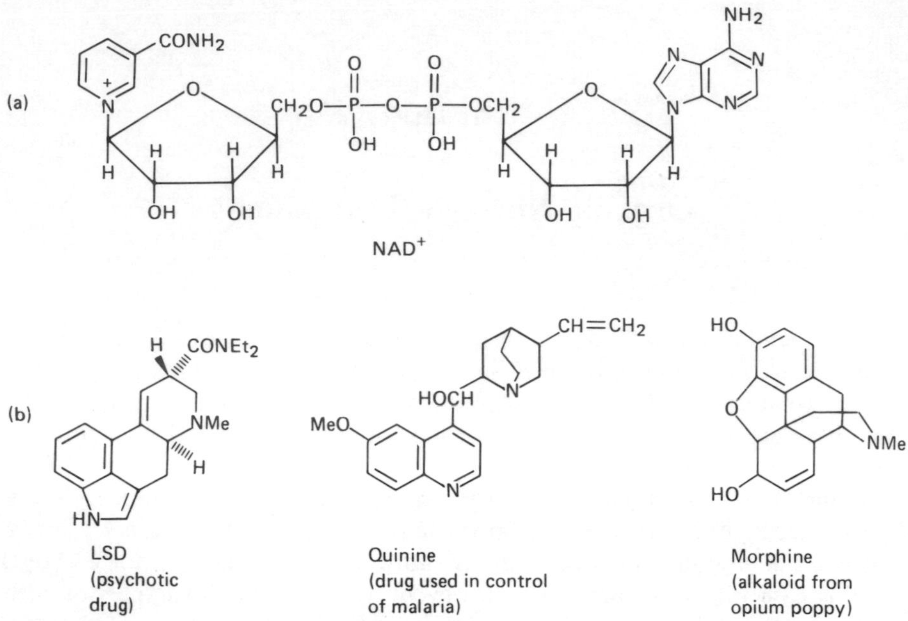

122

Figure 7.2 Some physiologically active heterocyclic nitrogen compounds

present in two important natural colouring materials, chlorophyll and haem (Chapter 13). Some plants produce physiologically active nitrogen heterocycles called alkaloids, many of which are drugs (e.g. Figure 7.2(b)).

Since a knowledge of the chemistry of amines is a prerequisite to understanding the chemistry of other nitrogen containing compounds, amines are dealt with first and in greater detail than the rest.

7.1 General characteristics of amines

Amines are organic derivatives of ammonia. They contain alkyl or aryl groups attached to nitrogen. Amines are classified as primary, secondary, or tertiary depending upon whether one, two, or three of the hydrogen atoms of ammonia have been replaced by organic groups (Figure 7.3).

Primary and secondary amines form intermolecular hydrogen bonds of the type N—H······N which are not as strong as O—H······O bonds because nitrogen is less electronegative than oxygen. Thus the boiling points of amines are lower than those of comparable alcohols (e.g. b.p. of CH_3NH_2 is -6.5 °C, b.p. of CH_3OH is 65 °C). Volatile amines have odours similar to ammonia but less pungent and more fishlike. Amines also form hydrogen bonds of the type O—H······N, for example with water, and many amines are therefore soluble in water. Hydrogen bonds involving nitrogen are very important in determining the structures of proteins and nucleic acids (Chapter 12).

H—N⟨H,H⟩	R—N⟨H,H⟩	R—N⟨R,H⟩	R—N⟨R,R⟩
Ammonia	Primary amine	Secondary amine	Tertiary amine

CH_3NH_2	$(CH_3)_2NH$	$(CH_3)_3N$
Methylamine	Dimethylamine	Trimethylamine
(methanamine)	(N-methylmethanamine)	(N,N-dimethylmethanamine)
$C_6H_5NH_2$	$C_6H_5NHCH_3$	$C_6H_5N(CH_3)_2$
($PhNH_2$)	(PhNHMe)	($PhNMe_2$)
Aniline	N-Methylaniline	N,N-Dimethylaniline
(benzenamine	(N-methylbenzenamine	(N,N-dimethylbenzenamine
or phenylamine)	or N-methylphenylamine)	or N,N-dimethylphenylamine)

Figure 7.3 The general structures of amines and some specific examples

Primary, secondary, and tertiary amines can be readily distinguished by their infrared spectra. Thus primary amines exhibit two N—H absorption bands between 3300 and 3500 cm^{-1}, secondary amines exhibit a single N—H band in this region, and tertiary amines exhibit no such band (Chapter 3).

Although only three groups are attached to nitrogen in amines, spectroscopic studies indicate that the bond angles (*ca.* 108°) are close to the tetrahedral angle expected for sp³ hybridization. However the energy barrier to inversion at nitrogen is low so that at ordinary temperatures most amines interconvert from one pyramidal form to the other (mirror image form) many times per second (Figure 7.4). As a result amines are not normally optically active (Chapter 2).

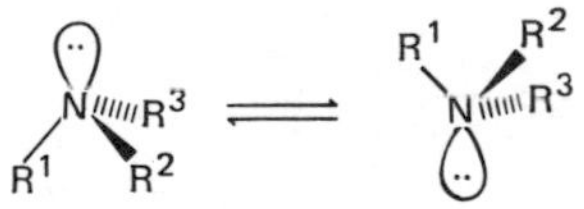

Figure 7.4 Inversion of the pyramidal form of amines

7.2 Basicity of amines

The unshared electron pair on the nitrogen atom of an amine can be shared with a proton to form a positively charged ammonium ion. Amines are therefore bases and dissolve in water to give alkaline solutions (Figure 7.5). The equilibrium constant for this process, called the *base dissociation constant* (K_b), gives a measure of the base strength of the amine. The larger K_b the stronger the base (Table 7.1). Base strength is sometimes expressed in terms of pK_b values, where $pK_b = -\log_{10}K_b$.

$$R_3N + H_2O \rightleftharpoons R_3\overset{+}{N}H \quad \overset{-}{O}H$$

$$K_b = \frac{[R_3\overset{+}{N}H][\overset{-}{O}H]}{[R_3N]}$$

Figure 7.5 The basicity of amines

Table 7.1 Basicities of amines

Amine	Formula	K_b	pK_b
Ammonia	NH_3	2×10^{-5}	4.70
Methylamine	CH_3NH_2	44×10^{-5}	3.36
Dimethylamine	$(CH_3)_2NH$	51×10^{-5}	3.29
Aniline	$PhNH_2$	4.2×10^{-10}	9.38
N-Methylaniline	$PhNHCH_3$	7.1×10^{-10}	9.15
N,N-Dimethylaniline	$PhN(CH_3)_2$	11×10^{-10}	8.96

Differences in the basicities of amines can usually be understood by considering the availability of the lone pairs of electrons on nitrogen and the relative stabilities of ammonium ions. Thus aliphatic amines are stronger bases than ammonia because alkyl groups are electron releasing and stabilize the positive charge on the ammonium ion. Aromatic amines on the other hand are weaker bases because the availability of the lone pair is reduced by resonance interaction with the aromatic π system (Figure 7.6).

Because they are bases, amines react with a wide variety of acids to form crystalline water soluble salts. Salt formation is routinely used to separate amines from neutral and acidic compounds (Section 3.3).

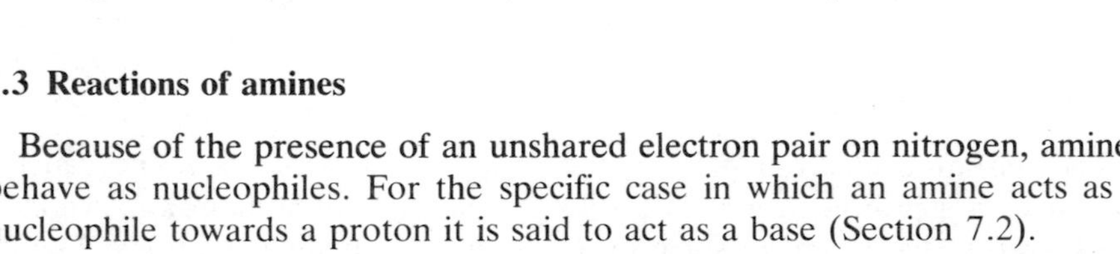

Figure 7.6   The resonance structures of aniline and its protonated form

7.3 Reactions of amines

Because of the presence of an unshared electron pair on nitrogen, amines behave as nucleophiles. For the specific case in which an amine acts as a nucleophile towards a proton it is said to act as a base (Section 7.2).

(a) Alkylation

The reactions of ammonia and amines with alkyl halides are examples of nucleophilic substitution reactions (Chapter 6). When the amine is tertiary

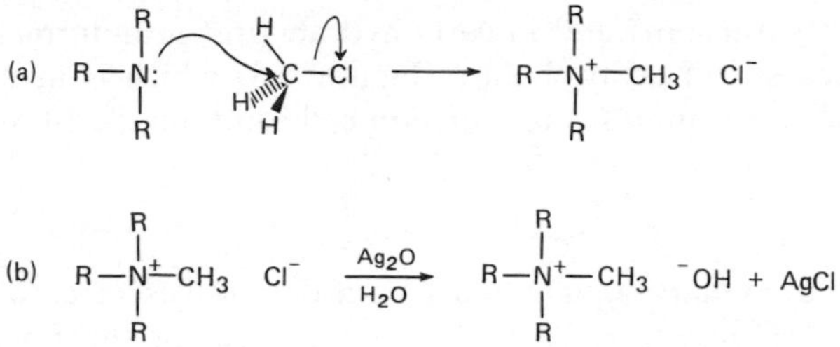

(a)

(b)

Figure 7.7 The synthesis of quaternary ammonium salts and
their conversion to quaternary ammonium hydroxides

the product is a *quaternary ammonium salt* (Figure 7.7(a)). Quaternary
ammonium halides can be converted into the corresponding quaternary
ammonium hydroxides by treatment with moist silver oxide (Figure
7.7(b)). These are strong bases similar to sodium and potassium hydroxides, and on heating undergo elimination to form an alkene. This reaction
is called the Hofmann elimination reaction. Because identification of the
alkene helps in locating the position of the nitrogen atom in the molecule,
the reaction has been extensively used in the structural determination of
complex natural products.

Question 7.1 $CH_3CH_2CH_2\overset{+}{N}(CH_3)_3\bar{O}H$ undergoes an elimination reaction
on heating. Predict the products and propose a mechanism for this reac-
tion.

Several quaternary ammonium salts have physiological activity. For
example, choline forms part of the lecithins (Figure 7.8) which make up
part of the brain and spinal cord tissue. Choline is essential to growth and

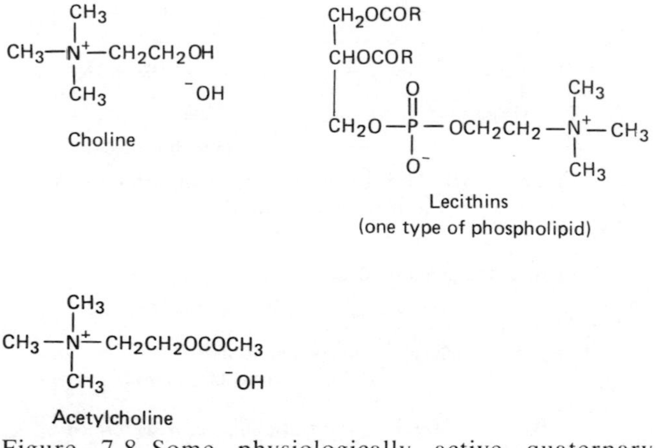

Choline

Lecithins
(one type of phospholipid)

Acetylcholine

Figure 7.8 Some physiologically active quaternary
ammonium compounds

is involved in fat transport and in carbohydrate and protein metabolism. It is also the precursor of acetylcholine (Figure 7.8) which is involved in the transmission of nerve impulses to ganglion cells and muscle fibres.

(b) Acylation

Primary and secondary amines react with derivatives of carboxylic acids to form *amides* (Figure 7.9). When the by-product of the reaction would be an acid this reacts with a further molecule of the amine to form a salt (e.g. Figure 7.9(a)).

Primary and secondary amines also form amides of other types of acids. For example, sulphonyl and phosphonyl chlorides react to form sulphonamides and phosphonamides respectively (Figure 7.10). Sulphonamides form the basis for an historically important class of drugs (Section 8.9).

The *Hinsberg test* which can be used to distinguish between primary, secondary, and tertiary amines is based upon their reactions with benzenesulphonyl chloride. Primary amines give sulphonamides which are sol-

(a) $CH_3COCl + 2\,PhNH_2 \longrightarrow PhNHCOCH_3 + Ph\overset{+}{N}H_3\overset{-}{Cl}$

(b) $CH_3CO_2CH_3 + CH_3CH_2NH_2 \longrightarrow CH_3CH_2NHCOCH_3 + CH_3OH$

Figure 7.9 The formation of amides

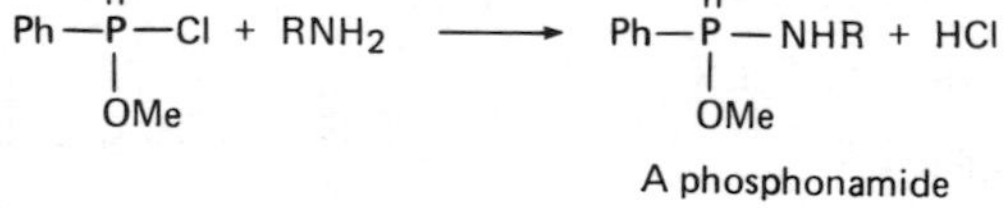

A sulphonamide

A phosphonamide

Figure 7.10 The formation of sulphonamides and phosphonamides

$RNH_2 + PhSO_2Cl \longrightarrow PhSO_2NHR$
(Soluble in alkali)

$R_2NH + PhSO_2Cl \longrightarrow PhSO_2NR_2$
(Insoluble in alkali)

$R_3N + PhSO_2Cl \longrightarrow$ No sulphonamide formed

Figure 7.11 The reactions of amines with benzenesulphonyl chloride

uble in alkali, whereas secondary amines give sulphonamides which are insoluble in alkali. Tertiary amines do not form amides or sulphonamides on reaction with acyl or sulphonyl chorides because they have no hydrogen atoms which can be replaced (Figure 7.11).

Question 7.2 Suggest an explanation for the fact that primary amines form sulphonamides which are soluble in alkali whereas those formed from secondary amines are insoluble in alkali.

(c) Reaction with nitrous acid

Another test for distinguishing between primary, secondary, and tertiary amines is based upon their reactions with nitrous acid. Nitrous acid is stable only in solution at low temperatures and is prepared by mixing solutions of sodium nitrite and a strong acid. In the presence of the latter nitrous acid generates nitrosonium ions ($\overset{+}{N}O$), and this is probably the reactive species (Figure 7.12(a)).

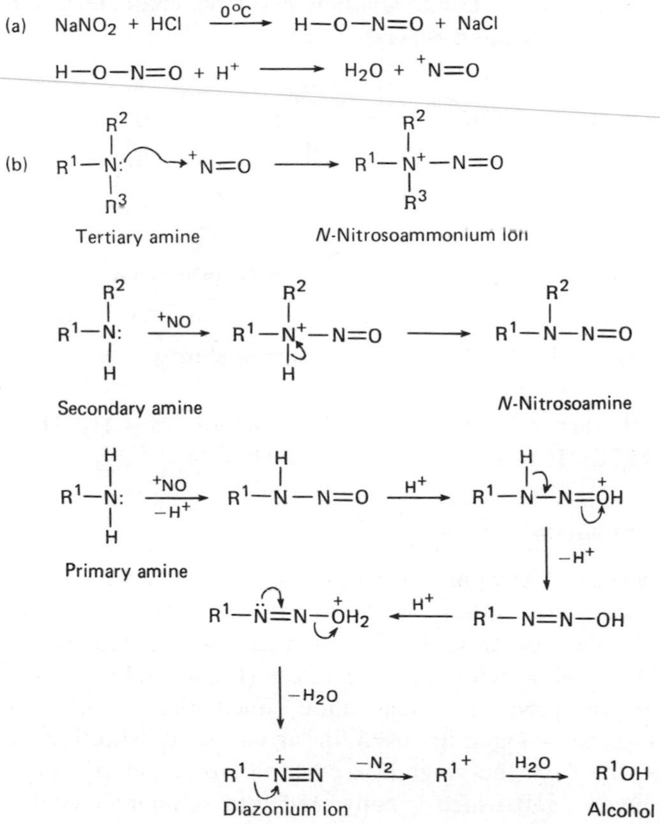

Figure 7.12 Reactions of amines with nitrous acid

The mode of reaction with amines is shown in Figure 7.12(b). Tertiary amines simply form an *N*-nitrosoammonium ion with no visible change. Secondary amines react further to form an *N*-nitrosoamine which separates from the aqueous solution as a yellow oil. Primary amines react further still by rearrangement and elimination of water to form a diazonium salt. Alkyl diazonium salts are unstable and decompose, forming the corresponding alcohol with liberation of nitrogen gas. Aromatic diazonium salts are moderately stable (<10 °C) and are usually detected by formation of coloured compounds on addition of an alkaline solution of 2-hydroxynaphthalene (*β*-naphthol). Aromatic diazonium salts are widely used for the synthesis of other aromatic compounds (see Section 7.8).

(d) Imine formation

Ammonia and its derivatives also act as nucleophiles towards aldehydes and ketones (Section 8.4(c)). The products contain a carbon-nitrogen double bond and are called imines or Shiff's bases (Section 7.4). Some such derivatives, e.g. oximes and hydrazones, are used to identify carbonyl compounds because they are crystalline solids which have characteristic melting points (Figure 7.13, see Section 8.4(c)).

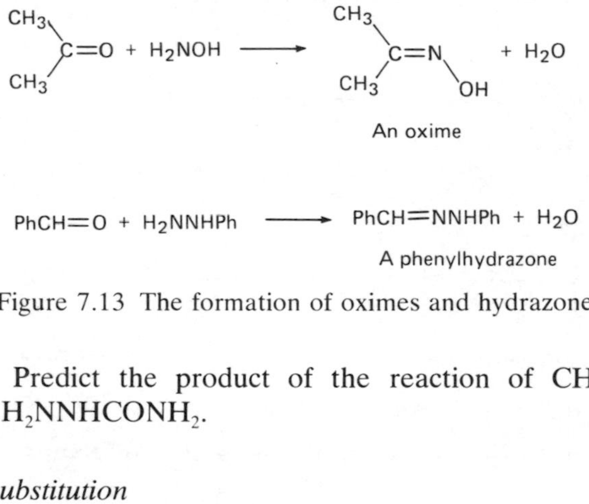

$$CH_3\text{-}C(CH_3)=O + H_2NOH \longrightarrow CH_3\text{-}C(CH_3)=N\text{-}OH + H_2O$$

An oxime

$$PhCH=O + H_2NNHPh \longrightarrow PhCH=NNHPh + H_2O$$

A phenylhydrazone

Figure 7.13 The formation of oximes and hydrazones

Question 7.3 Predict the product of the reaction of $CH_3CH=O$ with semicarbazide $H_2NNHCONH_2$.

(e) Aromatic substitution

Arylamines undergo electrophilic substitution on the aromatic ring. The electron-donating amino group is *ortho/para* directing and activating (Chapter 5). Because of this, 2,4,6-tribromoaniline is readily prepared simply by shaking aniline with bromine water (Figure 7.14(a)). However, since the nitrogen atom is itself quite nucleophilic, many of the electrophilic reagents commonly used in aromatic substitution react at nitrogen rather than on the ring. For example, reaction of aniline with sulphuric acid gives a salt which is converted into sulphanilic acid only on strong heating (Figure 7.14(b)).

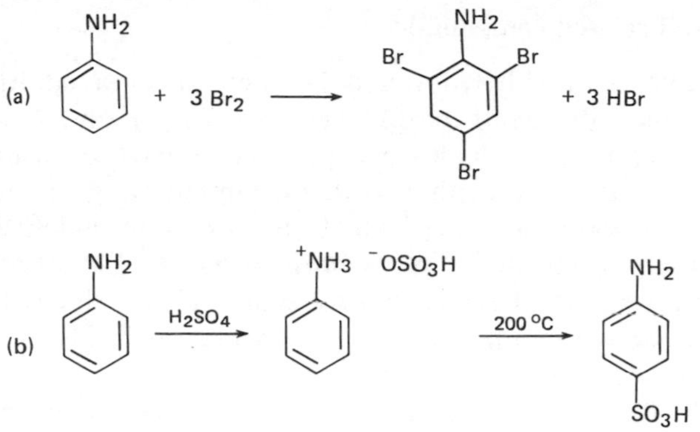

Figure 7.14 Two substitution reactions of aniline

For similar reasons the reaction of aniline with an acid chloride gives an amide rather than a ring substituted derivative (Section 7.3(b)). Indeed this reaction provides a useful means of moderating the reactivity of aniline towards electrophilic substitution. The acetamido group (CH_3CONH-) is less strongly electron-donating than the amino group because the carbonyl substituent reduces the availability of the lone pair on nitrogen. Thus bromination of acetanilide (N-phenylethanamide) gives a mixture of the 2- and 4-monobromo derivatives (Figure 7.15(a)). Furthermore the reaction of acetanilide with chlorosulphonic acid provides a useful method for introducing a sulphonyl group on to the aromatic ring, an important step in the synthesis of the drug sulphanilamide (Figure 7.15(b)).

The last step in the synthesis of sulphanilamide illustrates the regeneration of the amino group of hydrolysis of the amide. Indeed substitution of acetanilide followed by hydrolysis provides a general route to monosubstituted aniline derivatives.

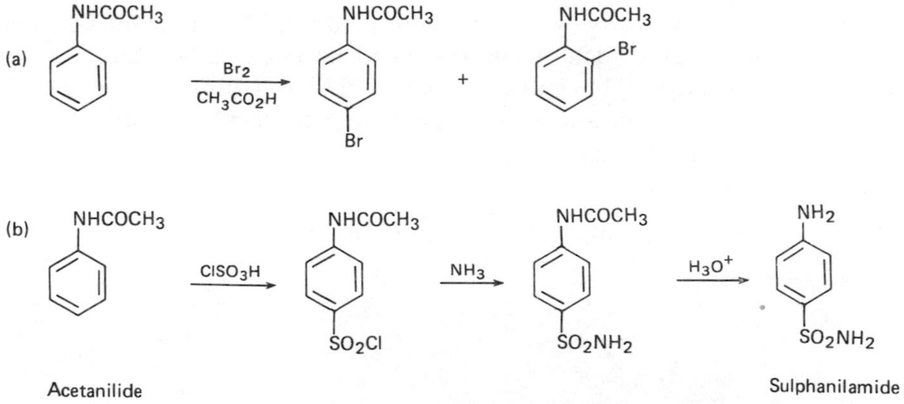

Figure 7.15 Two substitution reactions of acetanilide

130

7.4 Imines and related compounds

Amines react with aldehydes and ketones to form addition products known as imines or Schiff bases (Sections 7.3(d) and 8.4(c)). Imines formed from aliphatic aldehydes and primary amines are usually unstable and polymerize or react further with the amine to give more complex products. However imines derived from aromatic aldehydes such as benzaldehyde are often stable, as are most oximes and hydrazones (Section 7.3(d)). The formation of imines is reversible and the aldehyde or ketones may often be regenerated by hydrolysis with aqueous acid.

Question 7.4 Propose a mechanism for the hydrolysis of an imine (e.g. PhCH=NPh) by aqueous acid.

Imines are important in many biological reactions, particularly in the interconversion of amino acids and carbonyl compounds derived from protein and carbohydrate metabolic pathways (Section 8.4(c) and Chapter 15).

Imines and related compounds such as oximes can be reduced to give primary or secondary amines (Figure 7.16).

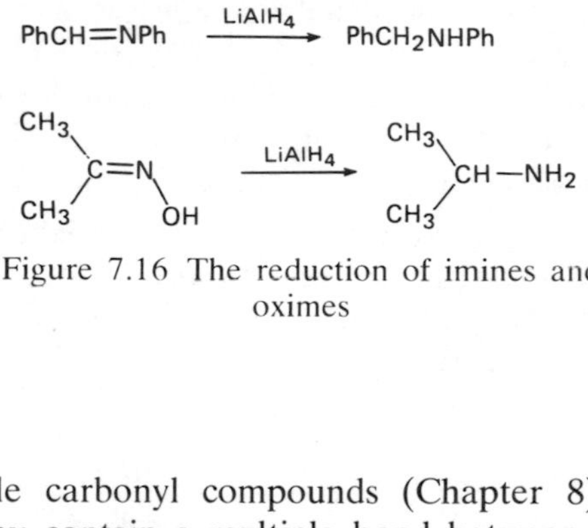

$$PhCH=NPh \xrightarrow{LiAlH_4} PhCH_2NHPh$$

Figure 7.16 The reduction of imines and oximes

7.5 Nitriles

Nitriles resemble carbonyl compounds (Chapter 8) in many of their reactions since they contain a multiple bond between carbon and a more electronegative atom. Thus nitriles contain acidic α-hydrogen atoms which can readily be removed to form carbanions which react with esters to give substituted ketones (Figure 7.17).

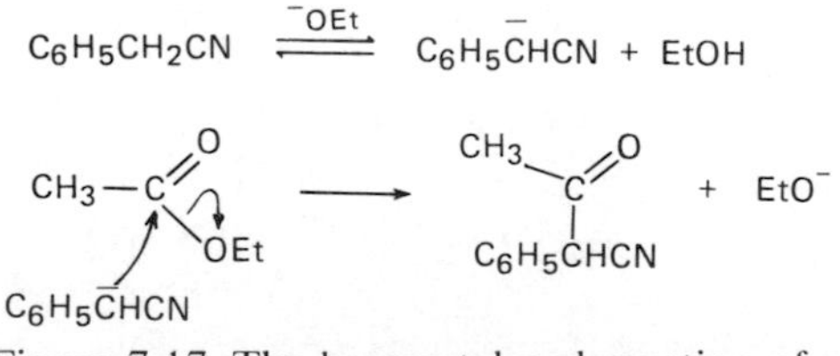

$$C_6H_5CH_2CN \underset{}{\overset{^-OEt}{\rightleftharpoons}} C_6H_5\overset{-}{C}HCN + EtOH$$

Figure 7.17 The base-catalysed reaction of a nitrile with an ester

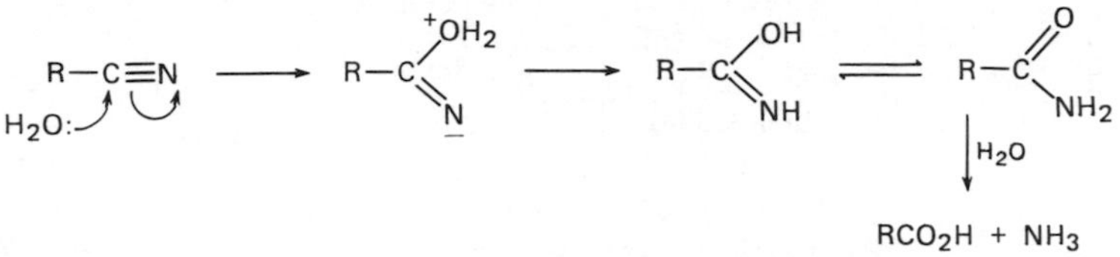

Figure 7.18 The hydrolysis of a nitrile

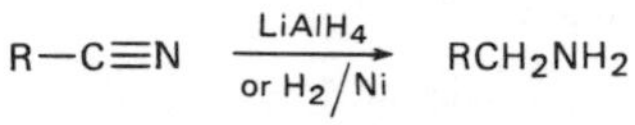

Figure 7.19 The reduction of a
nitrile

Hydrolysis of nitriles, either by acid or base, converts them into the corresponding carboxylic acids. The mechanism of this reaction involves nucleophilic attack on the carbon-nitrogen triple bond and is therefore analogous to the reactions of aldehydes and ketones with nucleophilic reagents. The nitrile is first converted into the corresponding amide, which can usually be isolated, but under more forcing conditions the amide is converted into the acid (Figure 7.18).

Nitriles can be reduced to primary amines by means of lithium aluminium hydride or by catalytic hydrogenation (Figure 7.19).

7.6 Amides

Simple amides are obtained by the reaction of ammonia or amines with acid derivatives such as esters, acyl chlorides, and anhydrides (Section 7.3(b)). They are also intermediates in the hydrolysis of nitriles (Section 7.5). Amides are the least susceptible of carboxylic acid derivatives to nucleophilic attack, because the electron density at the carbon atom is increased by interaction of the nitrogen lone pair with the carbonyl group (Figure 7.20).

Many of the properties of amides are dealt with in Chapter 8 along with those of other carbonyl compounds. Only those properties which specifically depend upon the presence of the nitrogen atom are included here. The C—N bonds of amides have partial double bond character due to interaction of the nitrogen lone pair with the carbonyl group (Figure 7.20). This tends to restrict rotation about the C—N bond and causes the amide unit to be planar, with important consequences for the structures of proteins.

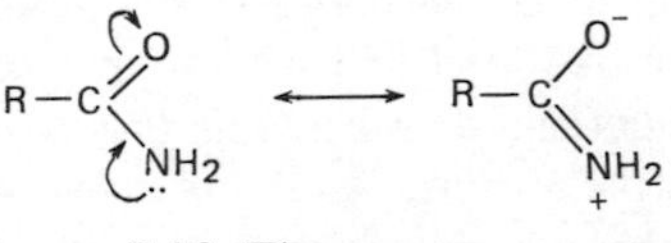

Figure 7.20 The resonance structures of an amide

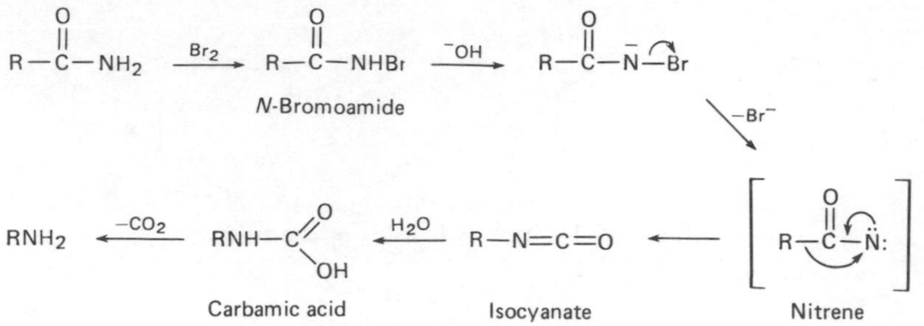

Figure 7.21 The Hofmann rearrangement reaction

One of the most interesting reactions of amides is the *Hofmann rearrangement* by which an amide is converted into an amine containing one less carbon atom (Figure 7.21). The mechanism of this reaction has been extensively studied, much of the evidence being derived from isolation of intermediates (e.g. the *N*-bromoamide) under appropriate conditions. If the reaction is carried out under anhydrous conditions the end product is an isocyanate presumed to be formed by rearrangement of a short-lived nitrene. Nitrenes are very reactive and extremely difficult to isolate. Hydrolysis of isocyanates is known to give carbamic acids which readily decarboxylate to give amines.

Urea is the diamide of carbonic acid (Figure 7.22). It is the normal end product of human metabolism of nitrogen containing compounds and is excreted to the extent of 30 g per day in the urine of the average adult. Urea is used as a fertilizer and also for the manufacture of urea–formaldehyde plastics (Section 9.4) and barbiturates. When urea is heated gently ammonia is evolved and biuret produced. An alkaline solution of biuret gives a violet pink coloration when copper sulphate is added due to the formation of a complex ion.

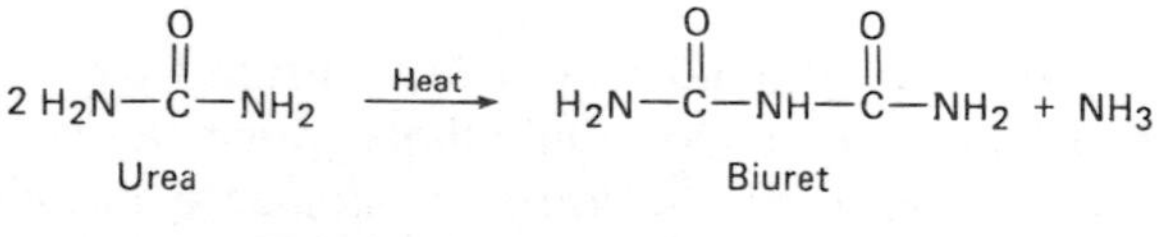

Figure 7.22 The formation of biuret from urea

7.7 Nitro compounds

In aliphatic nitro compounds, as in carbonyl compounds and nitriles, the α-hydrogen atoms are activated by the nitro group. The nitro group is in fact a more powerful electron-withdrawing group than the carbonyl group with the result that nitromethane is a stronger acid than acetone. The ionization of the α-hydrogen atoms occurs because the carbanion formed is stabilized by resonance (Figure 7.23(a)). The anions derived from

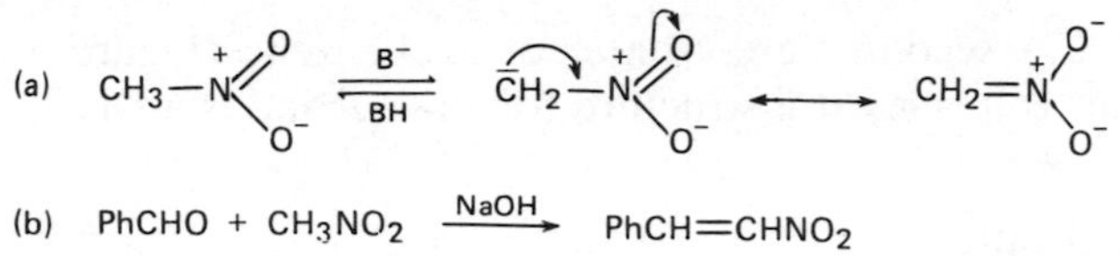

(a)

(b) $PhCHO + CH_3NO_2 \xrightarrow{NaOH} PhCH=CHNO_2$

Figure 7.23 (a) Resonance stabilization of the carbanion derived from nitromethane, and (b) the reaction of nitromethane with benzaldehyde

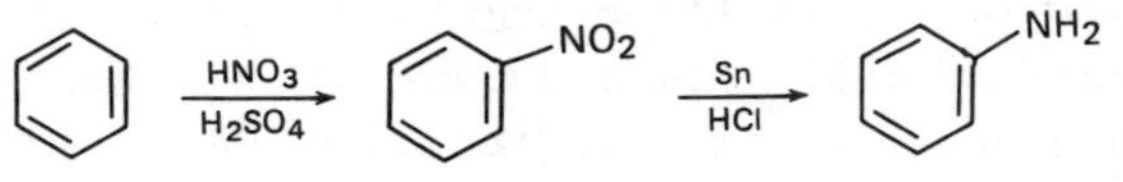

Figure 7.24 The formation and reduction of nitrobenzene

nitroalkanes add to aldehydes and ketones (Figure 7.23(b)) in a reaction analogous to the aldol condensation (Section 8.6(a)).

The importance of aromatic nitro compounds can be attributed to their ready formation by direct nitration of benzene derivatives and the ease with which they can be reduced to aromatic amines (e.g. Figure 7.24). The amino group of aromatic amines can in turn be transformed via diazonium salts into several other functional groups (Section 7.8).

The reduction of nitrobenzene under alkaline conditions leads to a number of different products depending upon the reaction conditions. The different products which can be obtained are shown in Figure 7.25.

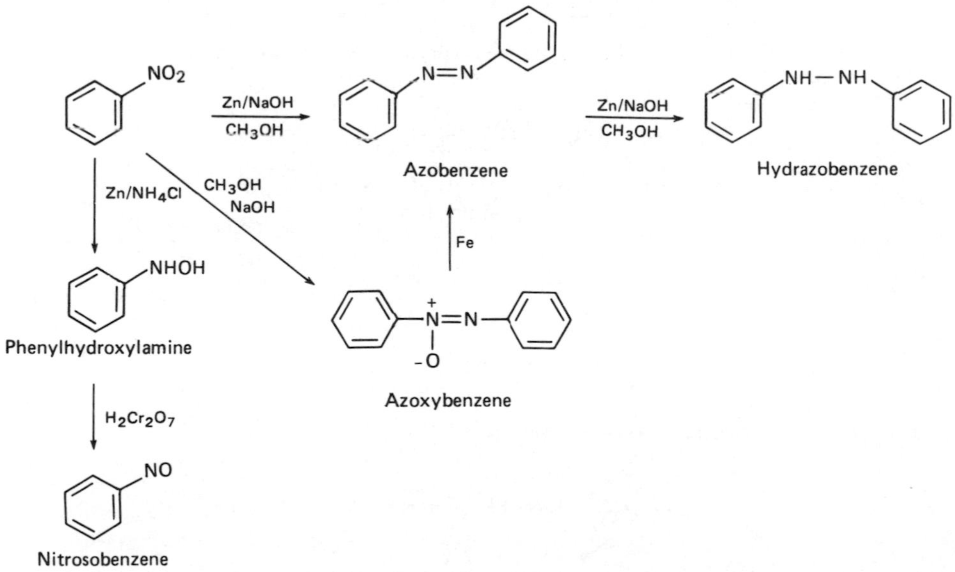

Figure 7.25 Alkaline reduction reactions of nitrobenzene

134

Question 7.5   A second stereoisomer of azobenzene (Figure 7.25) can be prepared. Can you suggest a structure for this second isomer?

7.8 Diazonium salts

Diazonium salts can be readily prepared by *diazotization* of primary aromatic amines (Section 7.3(c)) and are extremely useful and versatile in the synthesis of other aromatic compounds. For example, aniline reacts with nitrous acid and hydrochloric acid at 0 °C to yield a solution of benzenediazonium chloride. Unlike alkyl diazonium ions which usually decompose spontaneously even at very low temperatures, aromatic diazonium ions are moderately stable (< 10 °C) due to delocalization of the positive charge round the aromatic ring (Figure 7.26).

Nevertheless the great stability of the nitrogen molecule results in its ready displacement even from aryl diazonium salts. Indeed the solid salts explode when heated or subjected to mechanical shock. If an aqueous solution of a diazonium salt is heated, nitrogen is evolved and phenol is formed (Figure 7.27). The reaction is analogous to the reaction of primary aliphatic amines with nitrous acid and may proceed by an S_N1 mechanism.

Heating a solution of the diazonium salt with copper(I) chloride or

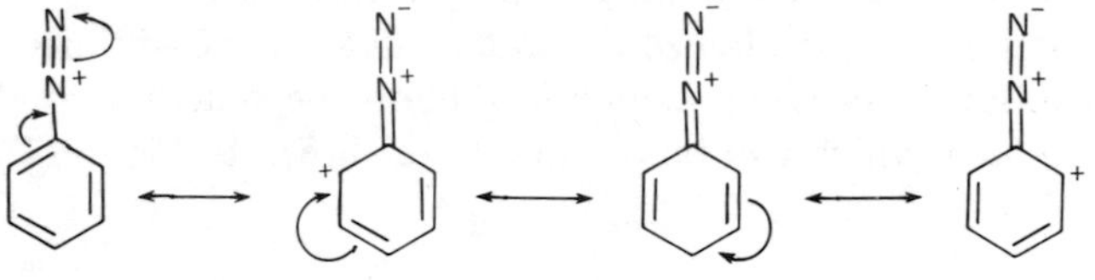

Figure 7.26 The resonance structures of a diazonium ion

$$Ph-\overset{+}{N}\equiv N \xrightarrow{H_2O} PhOH + N_2 + H^+$$

Figure 7.27 Hydrolysis of a diazonium ion

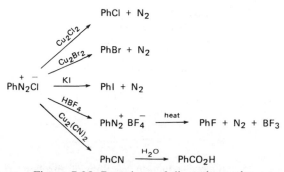

Figure 7.28 Reactions of diazonium salts

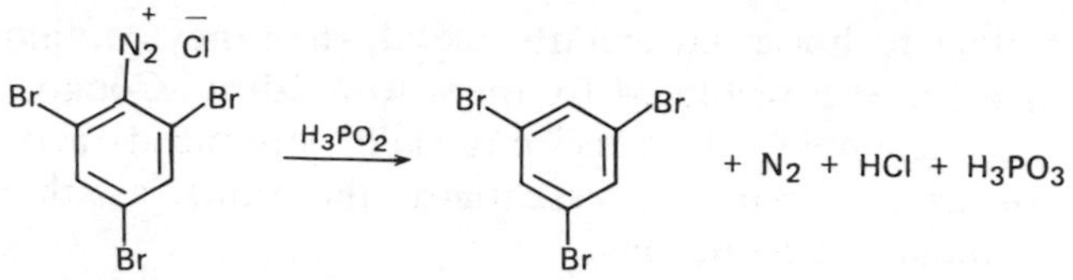

Figure 7.29 Reduction of 2,4,6-tribromobenzenediazonium chloride

bromide (Sandmeyer reaction), or with potassium iodide, produces the corresponding aryl halide. Aryl fluorides are prepared by heating the solid tetrafluoroborate salt (Figure 7.28). If cuprous cyanide is used in place of a halide in the Sandmeyer reaction the product is the aryl nitrile. Since nitriles are readily hydrolysed to carboxylic acids this reaction permits the conversion of an aromatic primary amine to an acid.

Certain reducing agents (e.g. H_3PO_2) enable the diazonium group to be replaced by a hydrogen atom. For example, diazotization of the readily available 2,4,6-tribromoaniline (Section 7.3(e)), followed by reduction, gives 1,3,5-tribromobenzene (Figure 7.29). In contrast other reducing agents (e.g. $SnCl_2$) convert aryldiazonium salts to aryl hydrazines (e.g. $PhNHNH_2$).

Question 7.6 Why can 1,3,5-tribromobenzene not be obtained by direct bromination of benzene?

The diazonium ion is a weak electrophile and can attack activated aromatic rings. On addition of phenols or aromatic amines to alkaline or neutral solutions of diazonium salts a coupling reaction occurs to give an azo compound (Figure 7.30(a)). Azo compounds are coloured and are

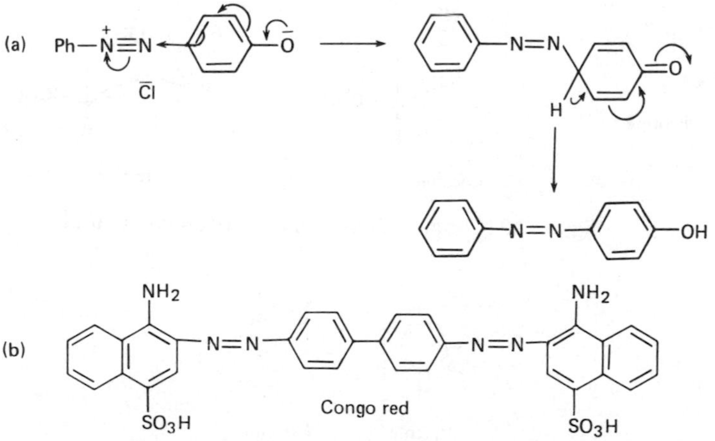

Figure 7.30 (a) The mechanism of the diazonium coupling reaction with phenol, and (b) the structure of Congo red

used as dyestuffs which are particularly useful when they contain functional groups which make it possible to fix them to a fabric. Congo red (Figure 7.30(b)) is an example of a commercially produced azo dye. The mechanism of azo coupling is essentially the same as that of other electrophilic aromatic substitutions.

Question 7.7 Predict the product that would be formed by reacting *N,N*-dimethylaniline with a solution of diazotized 4-aminobenzenesulphonic acid.

7.9 Heterocyclic Nitrogen Compounds

Heterocyclic nitrogen compounds contain at least one nitrogen atom incorporated in a ring. If the ring is saturated as in pyrrolidine or piperidine (Figure 7.31) the compounds are simply cyclic secondary amines. The pyrrolidine ring is present in two common amino acids, proline and hydroxyproline (Chapter 12), as well as in certain alkaloids such as nicotine, cocaine, and atropine.

Many other heterocyclic nitrogen compounds, e.g. pyrrole and pyridine (Figure 7.32), show properties characteristic of aromatic compounds (see Chapter 5). In pyrrole the lone pair on nitrogen together with the four π electrons from the two double bonds combine to form a planar aromatic

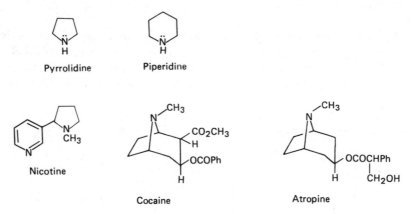

Figure 7.31 Some heterocyclic nitrogen compounds

Figure 7.32 Structures of pyrrole and pyridine

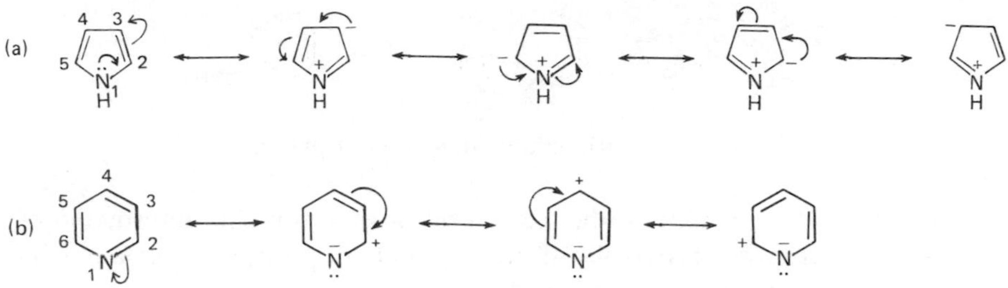

(a)

(b)

Figure 7.33 Resonance structures of pyrrole and pyridine

six π electron system similar to that in benzene. Since the lone pair on nitrogen is part of the aromatic system, it is not readily available to coordinate with a proton. Consequently pyrrole is an exceedingly weak base ($K_b = 2.5 \times 10^{-14}$) compared to pyrrolidine ($K_b = 2 \times 10^{-4}$).

Pyridine is the heterocyclic analogue of benzene with one CH group replaced by a nitrogen atom. The aromatic six π electron system is derived from the six p orbitals, one on each atom, so that unlike pyrrole, the lone pair on nitrogen is not part of the aromatic sextet and is available for accepting a proton. Pyridine ($K_b = 2.3 \times 10^{-9}$) is therefore more basic than pyrrole. It is however less basic than aliphatic amines because the electronegativity of an sp^2 hybridized nitrogen atom is greater than that of an sp^3 hybridized nitrogen atom.

Pyrrole and pyridine behave very differently from one another in electrophilic substitution reactions. Pyrrole is much more reactive than benzene whereas pyridine is much less reactive. In pyrrole the effect of the lone pair of electrons on the nitrogen atom is to increase the electron density on the carbon atoms (Figure 7.33(a)), thus increasing their reactivity towards electrophiles. In pyridine the electronegative nitrogen atom withdraws electrons from the carbon atoms of the ring (especially from the 2-, 4-, and 6- positions, Figure 7.33(b)), thus decreasing the reactivity towards electrophiles.

Athough all four carbon atoms of pyrrole are activated to electrophilic attack, in practice pyrrole itself is found to react preferentially at the 2-position (e.g. Figure 7.34(a)). As implied by the canonical forms for pyridine, it undergoes electrophilic substitution at the 3- position (e.g. Figure 7.34(b)).

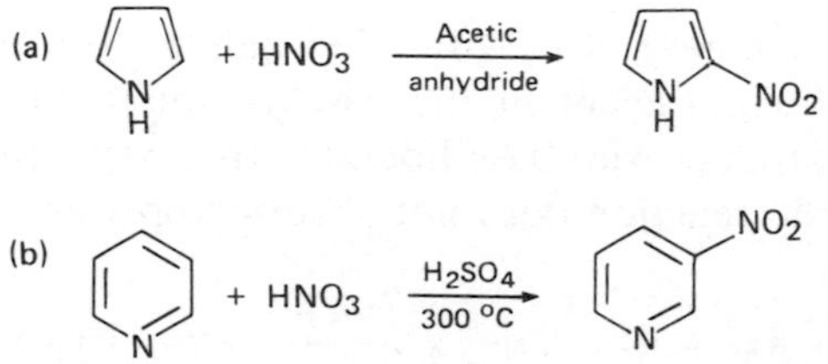

Figure 7.34 Nitration of pyrrole and pyridine

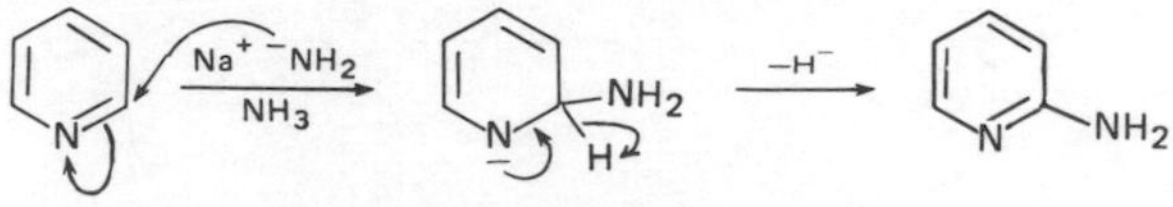

138

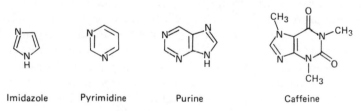

Figure 7.35 Nucleophilic amination of pyridine

Question 7.8 By examining the canonical forms for the intermediates involved in nitration of pyrrole at the 2- and 3- positions, explain why reaction at the 2- position is preferred.

Because of the reduced electron density at the 2-, 4-, and 6- positions of pyridine these positions are susceptible to attack by nucleophiles. For example, treatment of pyridine with sodium amide gives 2-aminopyridine (Figure 7.35).

Some other heterocyclic systems are shown in Figure 7.36. The imidazole ring occurs in the amino acid histidine (Chapter 12), the pyrimidine ring is present in vitamin B_1 (Chapter 14), and the purine ring forms part of certain alkaloids such as caffeine. Pyrimidine and purine derivatives are important constituents of nucleic acids (Chapter 12).

The detailed reactivity of these heterocycles is outside the scope of this book but, broadly speaking, since imidazole contains both pyrrole-like and pyridine-like nitrogen atoms it shows some of the characteristics of both systems. Pyrimidine on the other hand resembles only pyridine.

Imidazole   Pyrimidine   Purine   Caffeine

Figure 7.36 Structures of some heterocyclic nitrogen compounds
containing several nitrogen atoms

7.10 Preparation of organic nitrogen compounds

(a) Amines

Ammonia (and amines) react with alkyl halides to form alkylammonium salts (Section 7.3(a)). Except in the case of quaternary ammonium salts treatment of the product with base liberates the free amine (Figure 7.37).

Unfortunately the reaction does not always stop with the replacement of

$$NH_3 + RX \longrightarrow R\overset{+}{N}H_3 \overset{-}{X} \xrightarrow{NaOH} RNH_2 + H_2O + NaX$$

Figure 7.37 Reaction of alkyl halides with ammonia

$$NH_3 + R\overset{+}{N}H_3 \rightleftharpoons \overset{+}{N}H_4 + RNH_2$$

$$RNH_2 + RX \longrightarrow R_2\overset{+}{N}H_2 \ \overset{-}{X}$$

$$NH_3 + R_2\overset{+}{N}H_2 \rightleftharpoons \overset{+}{N}H_4 + R_2NH$$

$$R_2NH + RX \longrightarrow R_3\overset{+}{N}H \ \overset{-}{X}$$

$$NH_3 + R_3\overset{+}{N}H \rightleftharpoons \overset{+}{N}H_4 + R_3N$$

$$R_3N + RX \longrightarrow R_4N^+ \ X^-$$

Figure 7.38 Further alkylation of
amines

only one hydrogen atom. This is because the alkylammonium ion is in equilibrium with the corresponding amine and this amine can itself serve as a nucleophile, leading to further alkylation (Figure 7.38). A mixture of products is usually obtained, but by using an excess of ammonia or the starting amine, monoalkylation can be made to predominate.

An alternative and often more satisfactory route to amines is by reduction of other nitrogen containing compounds such as imines, nitriles, amides, and nitro compounds (Sections 7.4–7.7). The Hofmann rearrangement (Section 7.6) can also be used to prepare amines from amides.

(b) Imines

Imines, oximes, hydrazones, and similar compounds are prepared by reacting aldehydes or ketones with the appropriate amino compounds (Section 7.3d).

(c) Nitriles

Nitriles can be prepared (Figure 7.39) either by reacting alkyl halides with NaCN (Chapter 6), by dehydration of amides, or by reaction of $Cu_2(CN)_2$ with an aryl diazonium salt (Section 7.8).

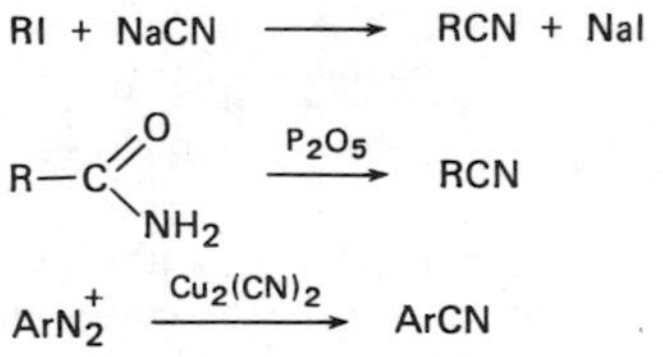

$$RI + NaCN \longrightarrow RCN + NaI$$

$$R-C\overset{\displaystyle O}{\underset{\displaystyle NH_2}{\big<}} \xrightarrow{P_2O_5} RCN$$

$$Ar\overset{+}{N}_2 \xrightarrow{Cu_2(CN)_2} ArCN$$

Figure 7.39 Methods of preparation
of nitriles

(d) Amides

Amides are obtained by reacting esters or other acid derivatives with ammonia or amines (Section 7.3(b)). They are also intermediates in the

$$RCO_2^-NH_4^+ \xrightarrow{\text{Heat}} RCONH_2 + H_2O$$

Figure 7.40 Preparation of amides

hydrolysis of nitriles (Section 7.5) and can be obtained by heating ammonium salts of carboxylic acids (Figure 7.40).

(e) Nitro compounds

Aromatic nitro compounds are prepared by nitration of the corresponding aromatic hydrocarbons (Chapter 5). Aliphatic nitro compounds are prepared commercially by nitration of aliphatic hydrocarbons (Figure 7.41).

$$CH_3CH_2CH_3 \xrightarrow{HNO_3} CH_3CH_2CH_2NO_2 + CH_3\underset{\underset{NO_2}{|}}{C}HCH_3 + CH_3CH_2NO_2 + CH_3NO_2$$

Figure 7.41 Commercial synthesis of aliphatic nitro compounds

(f) Diazonium salts

Aromatic diazonium salts are prepared by reacting aromatic amines with nitrous acid at low temperatures (Section 7.3(c)).

(g) Heterocyclic nitrogen compounds

Heterocyclic nitrogen compounds are usually prepared by reactions of amino compounds with bifunctional compounds (Chapter 9). especially those containing two carbonyl groups. Two examples are shown in Figure 7.42.

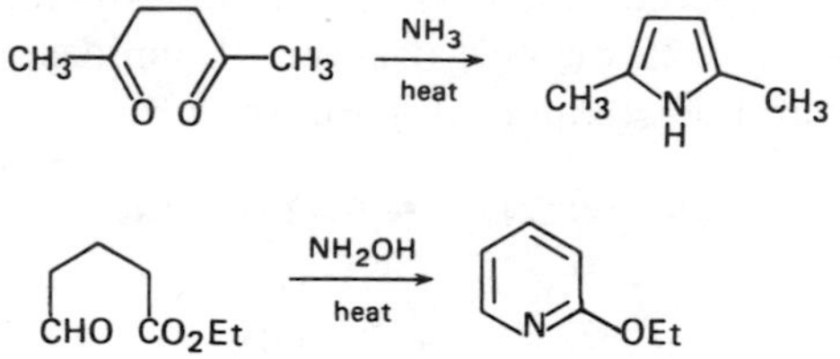

Figure 7.42 Two methods for the preparation of heterocyclic nitrogen compounds

Answers to questions

7.1

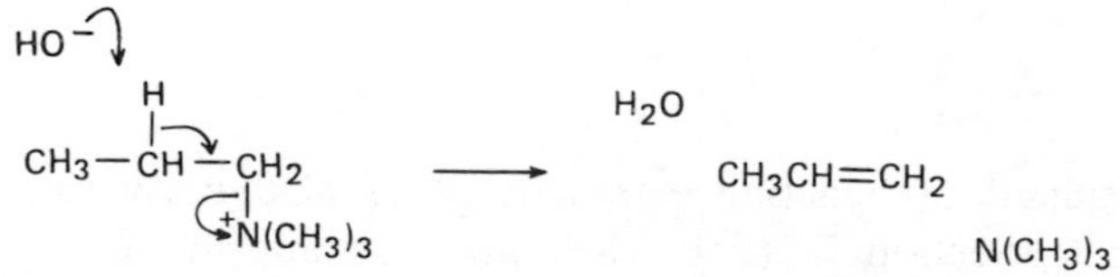

7.2 Since the sulphonyl group is powerfully electron-withdrawing it can stabilize a negative charge on an adjacent nitrogen atom. Sulphonamides containing an N—H group can therefore be deprotonated by base and are soluble in alkali.

$$PhSO_2NHR \xrightarrow{NaOH} PhSO_2\bar{N}R \ Na^+$$

7.3 The most electron-rich (nucleophilic) nitrogen atom of semicarbazide is the one furthest removed from the carbonyl group. The electron density on the other two nitrogen atoms is reduced by the C=O group.

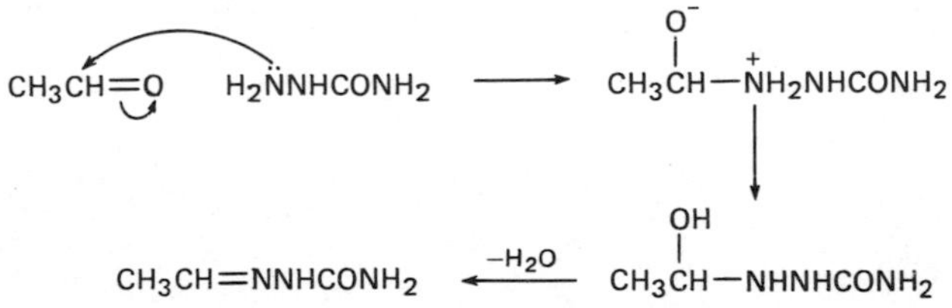

7.4

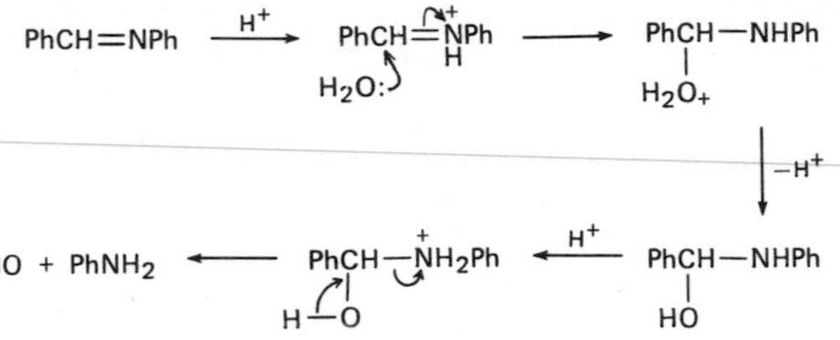

7.5

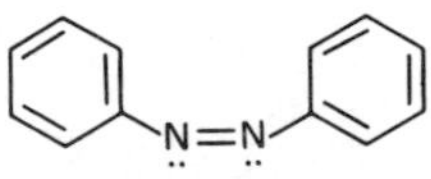

7.6 Bromination of benzene initially gives bromobenzene. Further bromination gives 1,2- and 1,4-dibromobenzenes because the bromo substituent is *ortho/para* directing. Production of 1,3,5-tribromobenzene would require *meta* substitution.

7.7

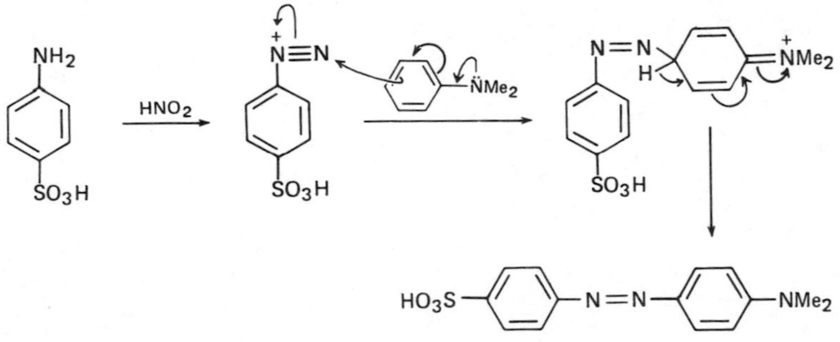

Methyl orange

142

7.8 Examination of the canonical forms of the intermediates involved in substitution at the 2- and 3- positions shows that the former is resonance stabilized to a greater extent than the latter (three canonical forms as opposed to two):

CHAPTER 8

Carbonyl Compounds

Many important types of natural products, including proteins, and many fats, flavouring substances, prostaglandins, and sex hormones, possess a carbon–oxygen double bond (C=O), known as the *carbonyl group*. A number of different classes of carbonyl compounds may be distinguished, depending upon the nature of the groups R^1 and R^2 attached to the carbonyl

carbon atom in R^1—$\overset{\overset{\displaystyle O}{\|}}{C}$—$R^2$ (Table 8.1). This chapter deals with the chemistry of carbonyl compounds and also includes a short section on the formally analogous sulphonyl compounds (containing the —SO_2— group).

There are many important carbonyl compounds, of which only a few well-known examples can be mentioned here. Thus, cinnamaldehyde is the compound largely responsible for the odour of cinnamon, muscone is a widely used perfume obtained from the musk deer, and methyl salicylate (oil of wintergreen) is a common constituent of household ointments (Figure 8.1). Animal and vegetable fats and oils (Chapter 10) consist largely of esters of glycerol, while the synthetic fibre Terylene® (or Dacron®) is a polymer containing many ester units, and is therefore described as a *polyester*. Similarly, nylon and proteins (Chapter 12) are *polyamides*. Acyl chlorides and acid anhydrides are not often encountered in everyday life because they are very reactive compounds, but they are widely used in chemical synthesis.

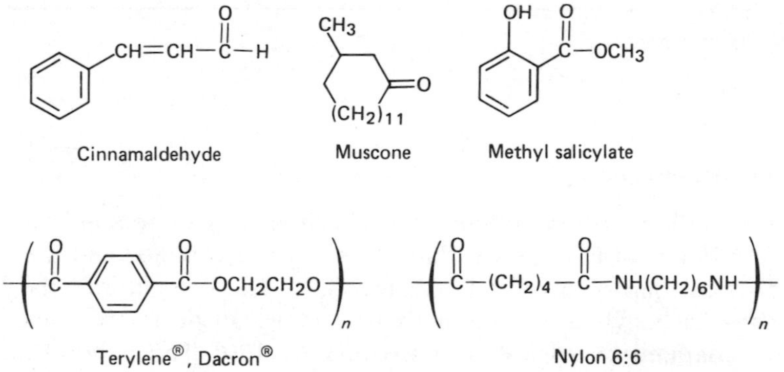

Figure 8.1 Some carbonyl compounds

Table 8.1 Classes of carbonyl compounds

Compound class	R^1	R^2	Example	Name
Aldehyde	H, R*	H	$CH_3CH_2CH_2-\overset{\overset{\displaystyle O}{\|}}{C}-H$	Butanal
Ketone	R	R	$CH_3-\overset{\overset{\displaystyle O}{\|}}{C}-CH_2CH_3$	Butanone
Carboxylic acid	H, R	OH	$CH_3-\overset{\overset{\displaystyle O}{\|}}{C}-OH$	Acetic (ethanoic) acid
Ester	H, R	OR	$CH_3-\overset{\overset{\displaystyle O}{\|}}{C}-OCH_2CH_3$	Ethyl acetate (ethyl ethanoate)
Thioester	H, R	SR	$CH_3-\overset{\overset{\displaystyle O}{\|}}{C}-SCH_2CH_3$	S-Ethyl ethanethioate (ethyl thioacetate)
Primary amide	H, R	NH_2	$C_6H_5-\overset{\overset{\displaystyle O}{\|}}{C}-NH_2$	Benzamide
Secondary amide	H, R	NHR	$CH_3CH_2-\overset{\overset{\displaystyle O}{\|}}{C}-NHCH_3$	N-Methylpropanamide
Tertiary amide	H, R	NR_2	$H-\overset{\overset{\displaystyle O}{\|}}{C}-N(CH_3)_2$	Dimethylformamide (N,N dimethyl-methanamide)
Acid anhydride	R	$O\overset{\overset{\displaystyle O}{\|}}{C}-R$	$CH_3-\overset{\overset{\displaystyle O}{\|}}{C}-O-\overset{\overset{\displaystyle O}{\|}}{C}-CH_3$	Acetic anhydride (ethanoic anhydride)
Acyl chloride	R	Cl	$C_6H_5-\overset{\overset{\displaystyle O}{\|}}{C}-Cl$	Benzoyl chloride

*R is an alkyl or aryl group.

8.1 The carbonyl group

As with carbon–carbon double bonds, carbon–oxygen double bonds consist of a σ bond, formed by overlap of sp^2 hybrid orbitals, and a π bond, formed by overlap of p orbitals (Figure 8.2). The σ bond and the bonds joining the carbonyl carbon atom to R^1 and R^2 all lie in the same plane and are separated by angles of approximately 120°. The π bond has two lobes, one above and one below this plane.

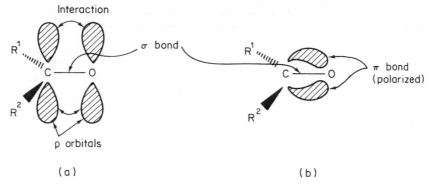

Figure 8.2 Bonding in the carbonyl group

Because oxygen is considerably more electronegative than carbon, the bonds joining the two atoms are highly polarized, especially the π bond, in which the electrons are less tightly held than in the σ bond. This means that the oxygen atom effectively carries a partial negative charge and the carbon a partial positive charge, which may be represented as in Figure 8.3.

Because of this polarization, the carbonyl carbon atom is susceptible to attack by nucleophiles. This tendency is often enhanced by the addition of acid, which protonates the oxygen atom and thereby increases the polarization. Furthermore, hydrogen atoms attached to the atoms adjoining the carbonyl group exhibit acidic properties because the anions formed by their abstraction as protons are stabilized by resonance (Figure 8.4). This property is most marked in the carboxylic acids (X=oxygen), but other carbonyl compounds also undergo important reactions which depend upon the acidity of *α-hydrogen atoms*.

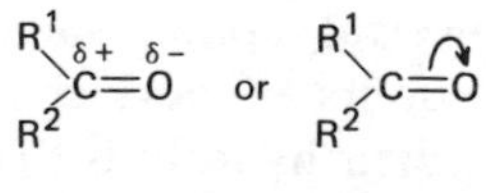

Figure 8.3 Polarization of the
carbonyl group

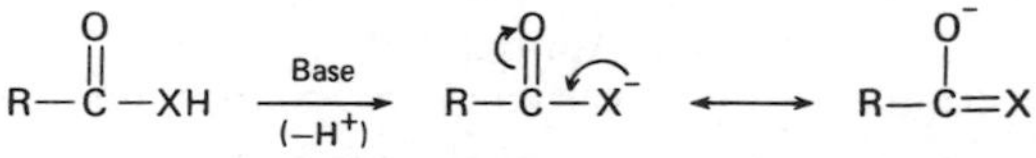

Figure 8.4 Ionization of carbonyl compounds

8.2 Physical properties of carbonyl compounds

The polarization of the carbonyl group leads to dipole–dipole attractive forces between molecules. These forces must be broken down before the

146

Table 8.2 Physical constants of some representative compounds

Compound	Formula	Molecular mass	Melting point (°C)	Boiling point (°C)
Pentane	$CH_3(CH_2)_3CH_3$	72	−130	36
Acetyl chloride (ethanoyl chloride)	CH_3COCl	78	−112	51
Methyl acetate (methyl ethanoate)	$CH_3CO_2CH_3$	74	−98	57
Butanal	$CH_3CH_2CH_2CHO$	72	−99	75
Butanone	$CH_3COCH_2CH_3$	72	−87	80
Butan-1-ol	$CH_3CH_2CH_2CH_2OH$	74	−90	118
Propanoic acid	$CH_3CH_2CO_2H$	74	−21	141
Benzoic acid	$PhCO_2H$	122	122	249
Propanamide	$CH_3CH_2CONH_2$	73	81	213

compounds can vaporize, so that carbonyl compounds have rather higher boiling points than hydrocarbons of comparable molecular mass (Table 8.2). Carbonyl compounds which in the liquid states are not strongly hydrogen bonded (i.e. acyl chlorides, esters, aldehydes, and ketones) boil at lower temperatures than alcohols of comparable molecular mass, but carboxylic acids, and especially primary amides, which form multiple hydrogen bonds, have very high boiling points. Indeed, whereas most low molecular mass aliphatic compounds are liquids, the primary amides are generally solids. In the aromatic series even the simple carboxylic acids are solids, as well as the primary amides.

Volatile carbonyl compounds generally have powerful odours. The odours of aliphatic aldehydes and carboxylic acids are pungent and are largely responsible for the unpleasantness associated with rancid butter and human perspiration. Perhaps not so paradoxically, some of these compounds also act as powerful aphrodisiacs to some animal species. Furthermore, the pungent odours of acyl chlorides and acid anhydrides may be at least partly due to their hydrolysis to carboxylic acids.

Table 8.3 Carbonyl group stretching frequencies

Compound class	Approximate position of carbonyl group absorption (cm^{-1})
Acid anhydride	1835, 1770 (two bands)
Acyl chloride	1800
Ester	1735
Aldehyde	1720
Acid	1720
Ketone	1705
Amide	1645

By contrast, many esters, ketones, and aromatic aldehydes are fragrant substances, responsible for the scents and flavours of many flowers and fruits, and extensively used in the perfume and artificial flavouring industries.

The carbonyl group strongly absorbs infrared radiation in the region $1850-1650$ cm^{-1} (Chapter 3). Furthermore, the actual carbonyl stretching frequency gives valuable information about the type of carbonyl compound present (Table 8.3). Of course, the frequency of the absorption also depends upon the structure of the rest of the molecule, so that the values recorded in the table are typical only of saturated aliphatic compounds.

8.3 Keto–enol tautomerism

The resonance stabilized anion (*enolate anion*) formed by abstraction of a proton from a carbon atom α- to a carbonyl group (Section 8.1), can be reprotonated either on the carbon atom, giving back the carbonyl compound, or on the oxygen atom, giving an isomeric compound known as an *enol* (Figure 8.5). Indeed, carbonyl compounds actually exist in equilibrium with their enol isomers under normal conditions. This type of isomerism, where the isomers are in equilibrium, is described as *tautomerism*, and the compounds are said to be *tautomers*.

Simple carbonyl compounds contain very little of the enol in the equilibrium mixture (e.g. 0.00025% enol in acetone). However, the enol form is stabilized if the carbon–carbon double bond is conjugated with a

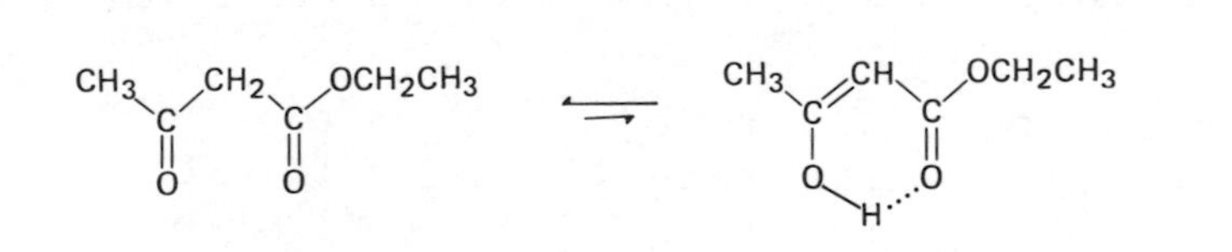

| Keto tautomer | Enolate anion | Enol tautomer |

Figure 8.5 Keto–enol tautomerism

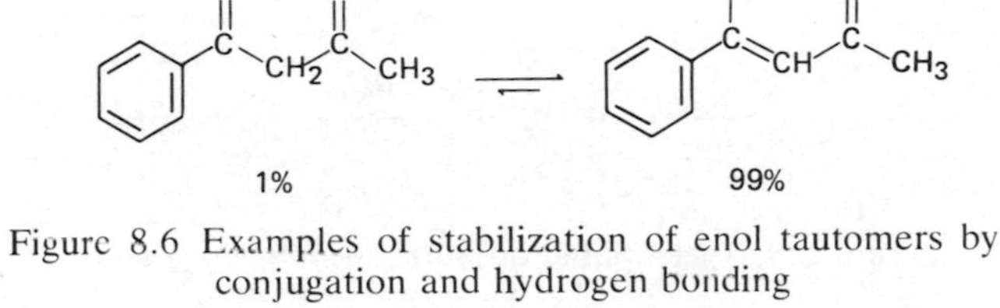

92.5%   7.5%

1%   99%

Figure 8.6 Examples of stabilization of enol tautomers by conjugation and hydrogen bonding

F

second π system or if the hydroxyl group is involved in intramolecular hydrogen bonding. In cases where such stabilization occurs the enol tautomer is much more abundant, and may even predominate (Figure 8.6).

As we shall see (Section 8.6), many reactions of carbonyl compounds, both in the laboratory and in living systems, involve enol forms or their simple derivatives.

8.4 Nucleophilic reactions of carbonyl compounds

By virtue of the polarization of the carbonyl group, carbonyl compounds are susceptible to attack by nucleophiles (Section 8.1). Addition of acid may catalyse the reaction by protonating the carbonyl oxygen atom and thereby enhancing the polarization. However, too much acid may prevent reaction by also protonating the nucleophile and rendering it unreactive.

Different types of carbonyl compounds undergo different reactions with nucleophiles, depending upon what happens after the initial nucleophilic attack. Thus, carboxylic acids and their derivatives usually undergo *nucleophilic substitution* reactions, whereas aldehydes and ketones usually undergo *nucleophilic addition* reactions.

The reason that carboxylic acids and their derivatives undergo nucleophilic substitution reactions is that the electronegative group attached to the carbonyl carbon atom in such compounds can act as a leaving group (Figure 8.7). The relative reactivities are in the same order as the stabilities of the displaced anions, i.e. acyl chloride > acid anhydride > ester $\simeq$ thioester > amide. Carboxylic acids themselves are anomalous because most nucleophiles cause deprotonation to form a salt (Section 8.5), which is then unreactive. However, under acidic conditions even carboxylic acids will undergo nucleophilic reactions.

Aldehydes and ketones undergo nucleophilic addition reactions (Figure 8.8) rather than substitution because they have no leaving group attached to the carbonyl carbon atom. In general, aldehydes are more reactive than

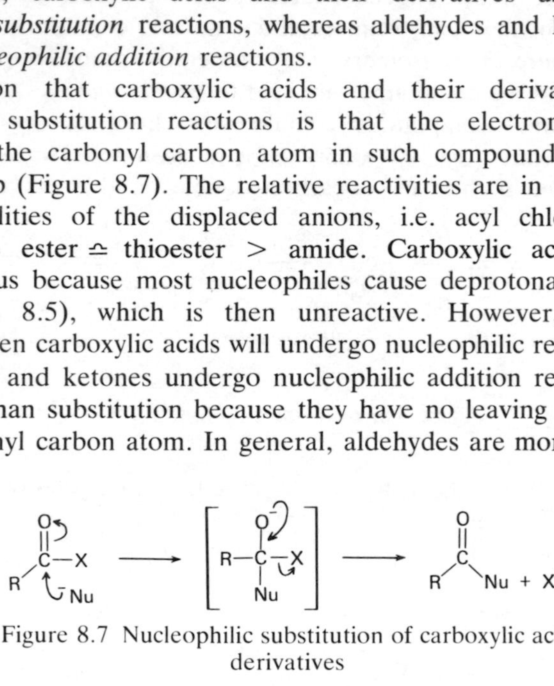

Figure 8.7 Nucleophilic substitution of carboxylic acid derivatives

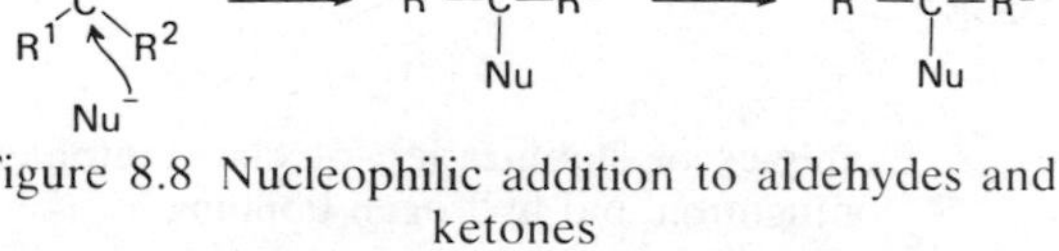

Figure 8.8 Nucleophilic addition to aldehydes and ketones

ketones because alkyl groups are both larger and more electron-donating than hydrogen. Thus, in ketones approach of the nucleophile is more hindered, and the partial positive charge on the carbonyl carbon atom is also reduced. If R^1, R^2, and the nucleophilic unit are different, a new asymmetric centre is generated by this reaction.

The reactivity of members of a particular class of carbonyl compounds depends upon the steric and electronic environment of the carbonyl group. In general, aromatic compounds are less reactive than their aliphatic counterparts because the partial positive charge on the carbonyl carbon atom can be delocalized around the benzene ring. For example, benzaldehyde is less reactive than acetaldehyde (ethanal). Electron-donating substituents (e.g. OH, NH_2) on the aromatic ring cause a further decrease in reactivity, whereas electron-withdrawing substituents (e.g. NO_2) cause an increase.

Question 8.1 Draw resonance structures to show why methyl 4-hydroxybenzoate is *less* susceptible, whereas methyl 4-nitrobenzoate is *more* susceptible to nucleophilic attack than methyl benzoate itself.

Some of the more important nucleophilic reactions of carbonyl compounds are summarized in Figure 8.9 (carboxylic acids and their derivatives) and 8.10 (aldehydes and ketones). They are discussed individually in the following sections.

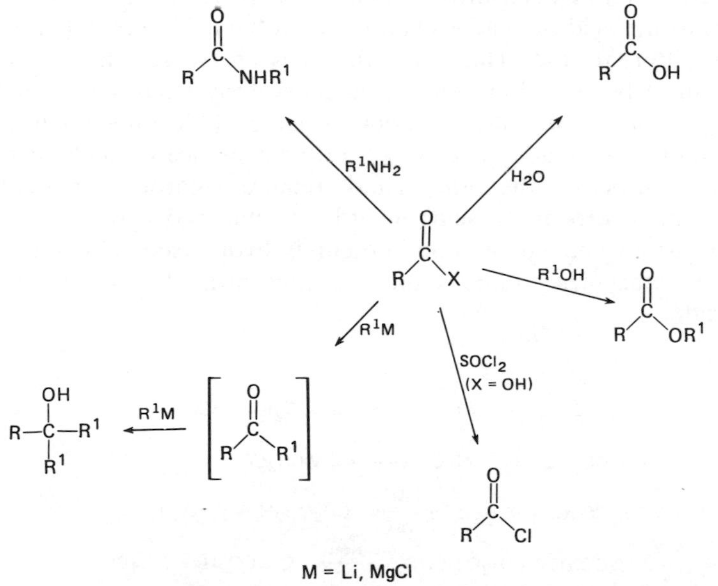

M = Li, MgCl

Figure 8.9 The important nucleophilic reactions of carboxylic acids and their derivatives

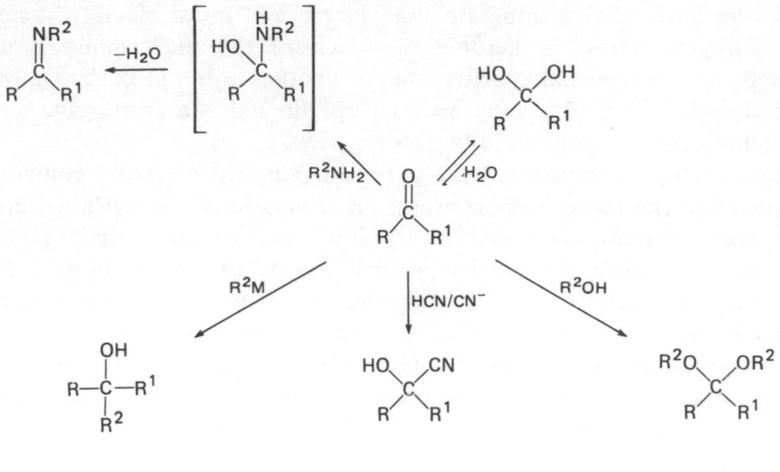

M = Li, MgBr, ZnBr

Figure 8.10 The important nucleophilic reactions of aldehydes and ketones

(a) Reactions with water

Hydrolyses of amides and esters are vitally important reactions in biological systems. For example, one of the most important reactions occurring during the process of digestion is the hydrolysis of food protein (a mixture of polyamides) into its constituent amino acids (Chapter 12).

All carboxylic acid derivatives can be hydrolysed (Figure 8.11), but with markedly different ease. Thus, acyl chlorides are so sensitive to moisture that they must be stored in sealed containers away from the water vapour in the atmosphere, while esters, though usually stable to water under normal conditions, are readily hydrolysed in the presence of acids or bases as catalysts. Amides, on the other hand, require vigorous conditions (e.g. refluxing with hydrochloric acid) in order to undergo hydrolysis. For this reason the stomach, where food protein is hydrolysed, albeit under the influence of enzymes, contains quite a concentrated solution of hydrochloric acid.

$$CH_3{-}\overset{O}{\overset{\|}{C}}{-}Cl + H_2O \longrightarrow CH_3CO_2H + HCl$$

$$CH_3CO_2COCH_3 + H_2O \longrightarrow 2\,CH_3CO_2H$$

$$CH_3CO_2CH_3 + H_2O \xrightarrow[\text{or } OH^-]{H_3O^+} CH_3CO_2H + CH_3OH$$

$$CH_3CONH_2 + H_2O + HCl \xrightarrow{\text{Heat}} CH_3CO_2H + NH_4Cl$$

Figure 8.11 Hydrolysis of carboxylic acid derivatives

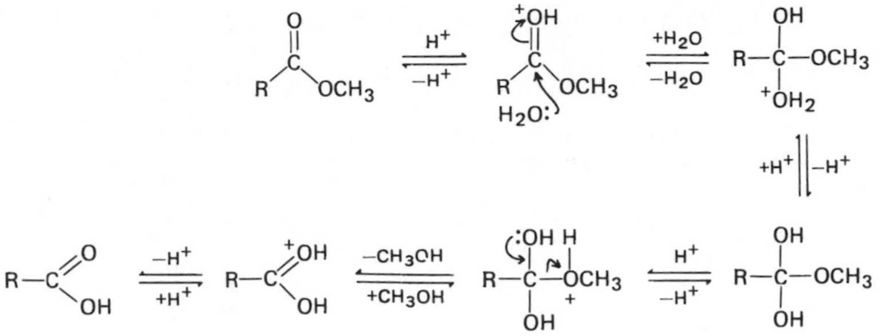

Figure 8.12 Hydrolysis of esters under acidic conditions

The hydrolysis of esters is one of the most widely studied of all organic reactions. A number of different mechanisms occur depending upon the nature of the ester and the reaction conditions, but the most common one for simple esters under acidic conditions is that depicted in Figure 8.12. This is an equilibrium reaction, but use of a large excess of water forces the equilibrium over in favour of the carboxylic acid.

Question 8.2 Propose a mechanism for hydrolysis of an ester under basic conditions.

Aqueous solutions of aldehydes contain the corresponding 1,1-diols in equilibrium (Figure 8.13). Formation of the 1,1-diol is particularly favourable in the case of formaldehyde (methanal, R=H), but much less favourable for higher aldehydes and for ketones.

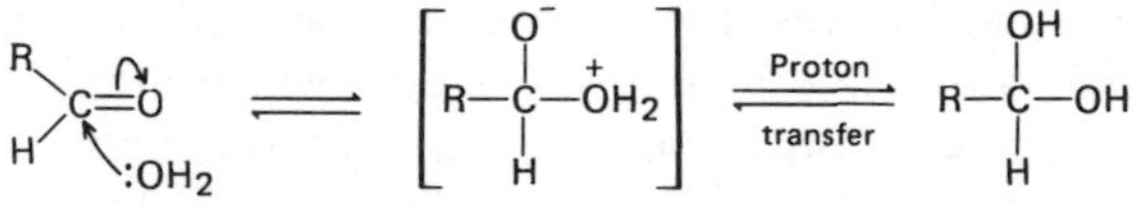

Figure 8.13 The equilibrium between aldehydes and 1,1-diols in aqueous conditions

(b) Reaction with alcohols

The reactions of carbonyl compounds with alcohols largely parallel the corresponding reactions with water. The reactions of alcohols with acyl chlorides and carboxylic acids are widely used as methods for synthesizing esters (Figure 8.14). As would be expected, acyl chlorides react very readily, whereas carboxylic acids require catalysis by acid and may require heating. Furthermore, the reactions of acyl chlorides with alcohols are irreversible, whereas those of carboxylic acids with alcohols are equilibrium reactions. Nevertheless, because in the latter reactions the ester is formed directly from the carboxylic acid, it is often convenient to adopt this method rather than to introduce the extra step of synthesizing the acyl chloride.

$$RCOCl + CH_3OH \longrightarrow RCO_2CH_3 + HCl$$

$$RCO_2H + CH_3OH \overset{H_3O^+}{\rightleftharpoons} RCO_2CH_3 + H_2O$$

Figure 8.14 Reactions of acids and acyl chlorides with alcohols

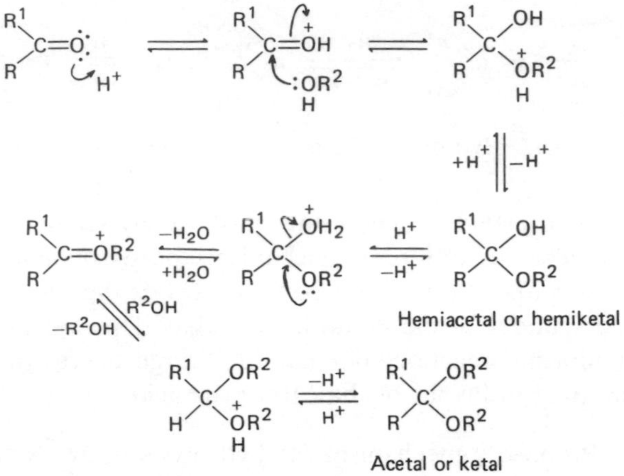

Figure 8.15 Formation of acetals and ketals

Question 8.3 Propose a mechanism for the acid catalysed esterification reaction between a carboxylic acid and an alcohol.

Aldehydes and ketones react with alcohols to give acetals and ketals, which are useful derivatives for protecting the carbonyl group because acetals and ketals are stable to many reagents which attack aldehydes and ketones. The reaction proceeds via an intermediate hemiacetal or hemiketal and under acidic or basic conditions an equilibrium is rapidly established (Figure 8.15).

For simple alcohols such as methanol and ethanol the equilibrium favours the free carbonyl compound, although acetals can still be prepared from such alcohols by removal of the water formed in the reaction by azeotropic distillation using toluene. However, the reactions of aldehydes and ketones with diols such as ethylene glycol (ethan-1,2-diol) give more favourable equilibrium concentrations of the cyclic acetals and ketals (Figure 8.16). Also, the cyclic derivatives are more rapidly formed because

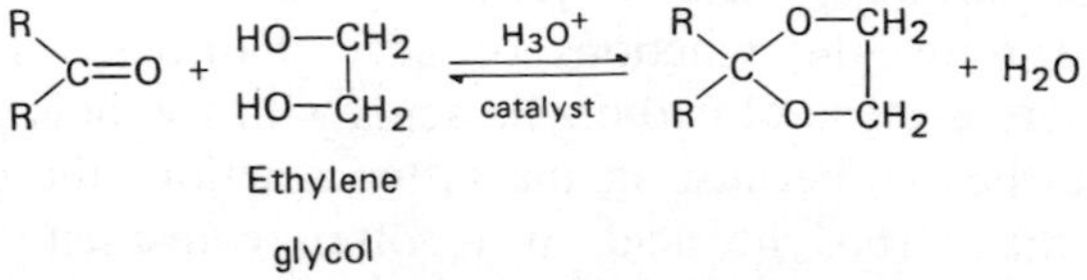

Figure 8.16 Reaction of aldehydes or ketones with ethylene glycol

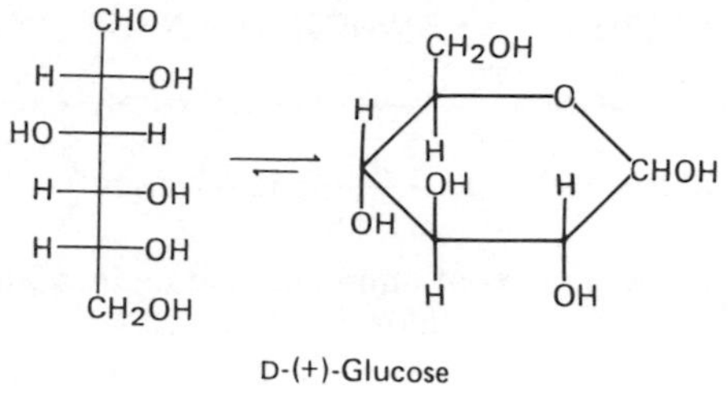

D-(+)-Glucose

Figure 8.17 The cyclization of D-(+)-glucose

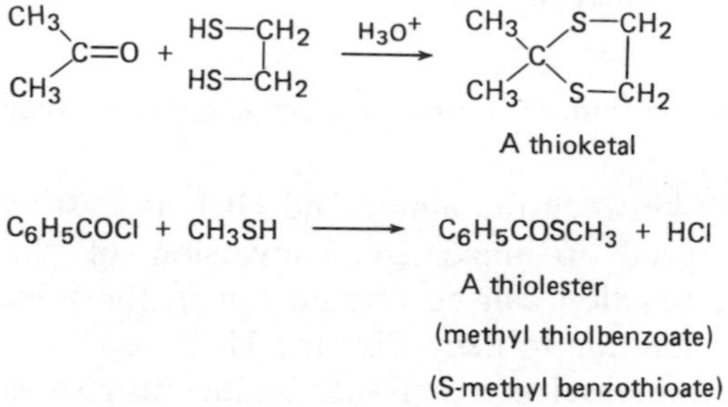

A thioketal

$$C_6H_5COCl + CH_3SH \longrightarrow C_6H_5COSCH_3 + HCl$$

A thiolester

(methyl thiolbenzoate)

(S-methyl benzothioate)

Figure 8.18 Reactions of thiols and dithiols with carbonyl compounds

the second step is *intramolecular* (i.e. takes place within the one molecule rather than between two molecules). There is, therefore, a higher probability of the two interacting groups coming together. Although acetals and ketals are stable compounds, they are readily hydrolysed to the parent carbonyl compounds on treatment with dilute solutions of acids and bases.

In general, hemiacetals and hemiketals cannot be isolated, as they readily eliminate a molecule of alcohol to give back the parent carbonyl compound. However, when the hydroxyl and carbonyl groups are in the same molecule, they can form a cyclic hemiacetal or hemiketal, and when there are five, six, or seven atoms in the ring, the equilibrium usually favours the cyclic derivative. Indeed, sugars, which are polyhydroxyaldehydes or ketones, usually exist as cyclic hemiacetals or hemiketals (Figure 8.17, see also Chapter 11).

Reactions of aldehydes and ketones with thiols give thioacetals and thioketals in a manner similar to the formation of acetals and ketals, while acyl chlorides react with thiols to give thiol esters (Figure 8.18).

(c) Reactions with ammonia and its derivatives

Ammonia and primary and secondary amines react with carboxylic acid derivatives to give primary, secondary, and tertiary amides respectively. As usual, acyl chlorides react most readily, and they are most widely used in laboratory preparations of simple amides (Figure 8.19). Due to the

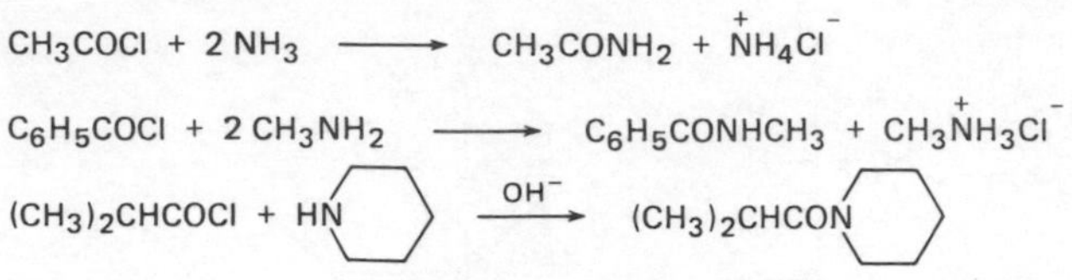

$$CH_3COCl + 2\,NH_3 \longrightarrow CH_3CONH_2 + \overset{+}{N}H_4Cl^-$$

$$C_6H_5COCl + 2\,CH_3NH_2 \longrightarrow C_6H_5CONHCH_3 + CH_3\overset{+}{N}H_3Cl^-$$

$$(CH_3)_2CHCOCl + HN\!\!\bigcirc \xrightarrow{OH^-} (CH_3)_2CHCON\!\!\bigcirc$$

Figure 8.19 Reactions of ammonia and amines with acyl chlorides

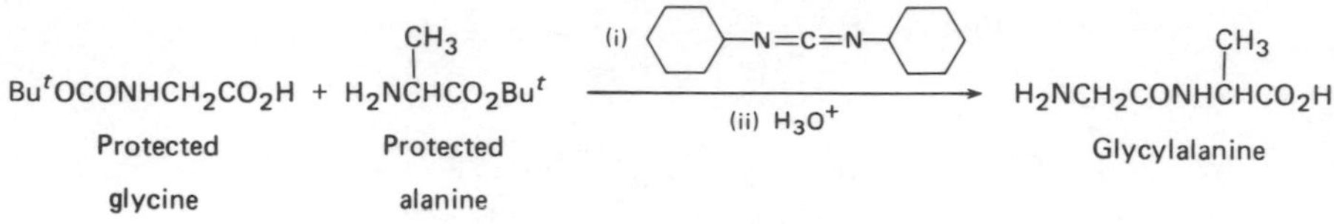

$$Bu^tOCONHCH_2CO_2H + H_2N\overset{\overset{\displaystyle CH_3}{|}}{CH}CO_2Bu^t \xrightarrow[\text{(ii) } H_3O^+]{\text{(i)}\;\bigcirc\!-N=C=N\!-\bigcirc} H_2NCH_2CONH\overset{\overset{\displaystyle CH_3}{|}}{CH}CO_2H$$

Protected   Protected   Glycylalanine

glycine   alanine

Figure 8.20 The laboratory synthesis of a peptide using protected amino acids

formation of a salt between the amine and HCl, at least two equivalents of amine must be used to maximize conversion of the acyl chloride. Alternatively, the reaction can be carried out in the presence of aqueous sodium hydroxide solution to neutralize the HCl.

Carboxylic acids themselves react with amines to give salts (Section 8.5) and these can be converted into amides only under vigorous conditions. However, under the influence of certain compounds (e.g. dicyclohexylcarbodiimide) which form activated derivatives of the carboxylic acids, direct formation of amides can be achieved. This technique is widely used for the laboratory synthesis of peptides and proteins from protected amino acids (Figure 8.20, see also Chapter 12.).

Question 8.4 The initial product of the reaction of a carboxylic acid with dicyclohexylcarbodiimide is shown below:

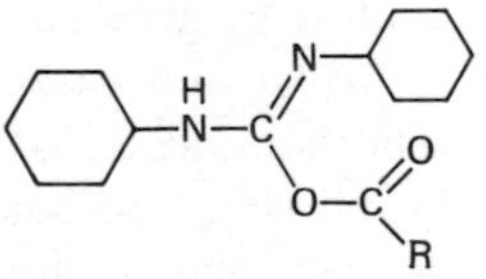

Draw curly arrows to show how this product is formed, and to rationalize its reaction with an amine to give an amide.

Primary amines add to aldehydes and ketones to give aminoalcohols as the initial products, although these products are not isolable. Water is then

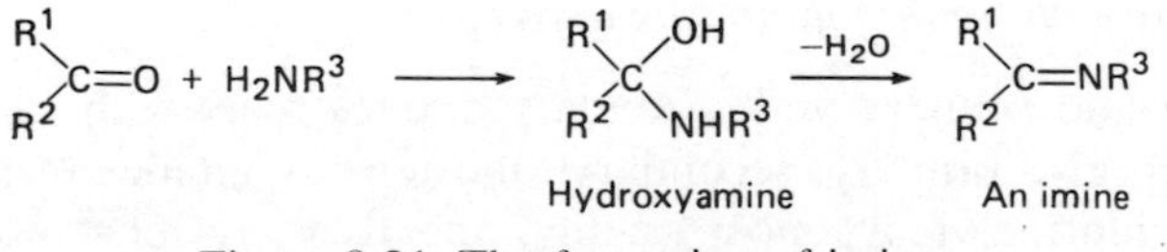

$$\overset{R^1}{\underset{R^2}{>}}C{=}O + H_2NR^3 \longrightarrow \overset{R^1}{\underset{R^2}{>}}\overset{\displaystyle OH}{\underset{\displaystyle NHR^3}{C}} \xrightarrow{-H_2O} \overset{R^1}{\underset{R^2}{>}}C{=}NR^3$$

Hydroxyamine   An imine

Figure 8.21 The formation of imines

Vitamin B6
(pyridoxine)

Pyridoxal
5-phosphate

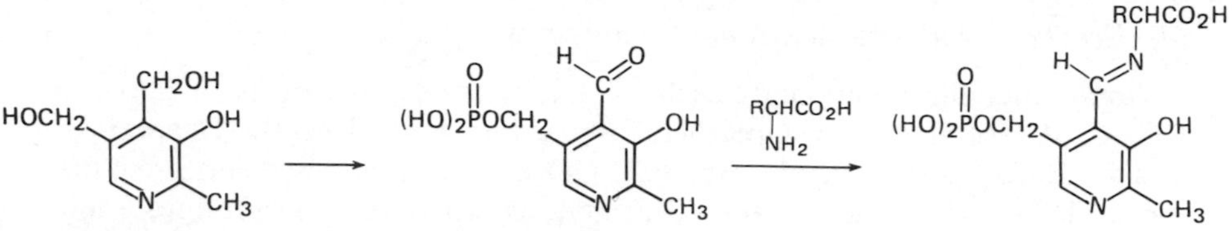

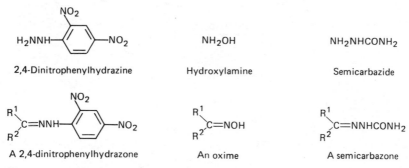

Figure 8.22 The involvement of pyridoxal-5-phosphate in biochemical transformations of amino acids

eliminated to give a compound with a carbon–nitrogen double bond, called an *imine* or *Schiff base* (Figure 8.21).

Imine formation is thought to be involved in a number of important biochemical transformations of amino acids, such as decarboxylation and oxidative deamination. A derivative of vitamin B_6, called pyridoxal-5-phosphate, interacts to give an imine which then undergoes further reactions (Figure 8.22).

In the laboratory, the formation of particular types of imines is used in the identification and characterization of aldehydes and ketones. Thus, formation of an orange or red precipitate on addition of 2,4-dinitrophenylhydrazine (Brady's reagent) to a compound indicates that it is an aldehyde or ketone. The precipitated compound is a 2,4-dinitrophenylhydrazone (Figure 8.23), and its melting point is characteristic of the carbonyl compound from which it was formed. Similarly, characteristic solid derivatives can be prepared with hydroxylamine (hydroxyamine) and semicarbazide, the products being oximes and semicarbazones respectively.

2,4-Dinitrophenylhydrazine　　Hydroxylamine　　Semicarbazide

A 2,4-dinitrophenylhydrazone　　An oxime　　A semicarbazone

Figure 8.23 The structures of some compounds used in the laboratory identification and characterization of aldehydes and ketones

Question 8.5 Two geometrical isomers of many oximes are known. They are called *syn* (or *cis*) and *anti* (or *trans*) isomers. Draw structures for the two isomers of the oxime of acetaldehyde.

(d) Reactions with organometallic compounds

Organometallic compounds such as alkali metal derivatives of alkynes, $RC\equiv C^- M^+$ (M=Na, Li; Chapter 4), Grignard reagents, RMgBr (Chapter 6), and the Reformatsky reagent, $BrZnCH_2CO_2Et$, are all highly polarized in the sense $R^{\delta-}$—$M^{\delta+}$, and attack carbonyl compounds to produce new carbon–carbon bonds. As such, these reactions are useful in the synthesis of complex molecules, such as may, for example, be required in relatively small quantities for pharmacological purposes.

Aldehydes and ketones undergo simple addition reactions with these organometallic compounds, yielding metal alkoxides, which are hydrolysed to the corresponding alcohols upon addition of dilute acid. Aldehydes give secondary alcohols, whereas ketones give tertiary alcohols (e.g. Figure 8.24).

An example of a Reformatsky reaction is involved in one approach to the synthesis of vitamin A (Figure 8.25).

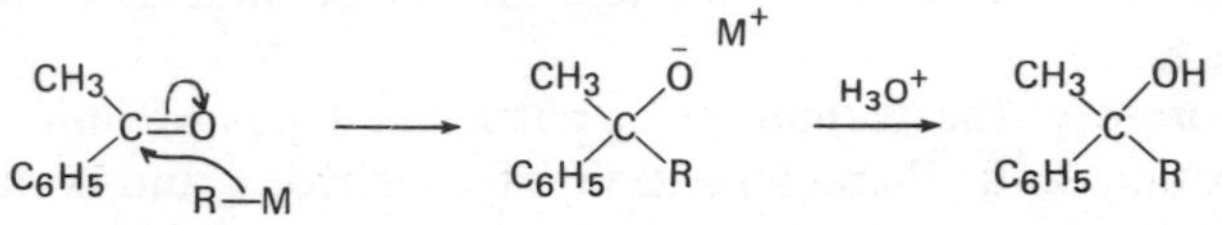

Grignard reagent:	RMgBr
Alkynyl metal:	$R^1C\equiv CLi$ or $R^1C\equiv CNa$ (R = $R^1C\equiv C$)
Reformatsky reagent:	$BrZnCH_2CO_2Et$ (R = CH_2CO_2Et)

Figure 8.24 Reaction of aldehydes and ketones with organometallic compounds

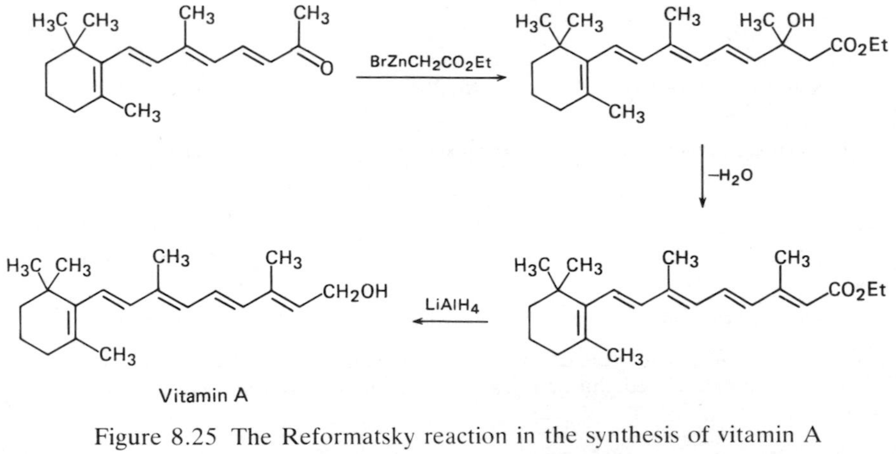

Figure 8.25 The Reformatsky reaction in the synthesis of vitamin A

Question 8.6 Devise three routes to 2-phenylbutan-2-ol using reactions of Grignard reagents with ketones.

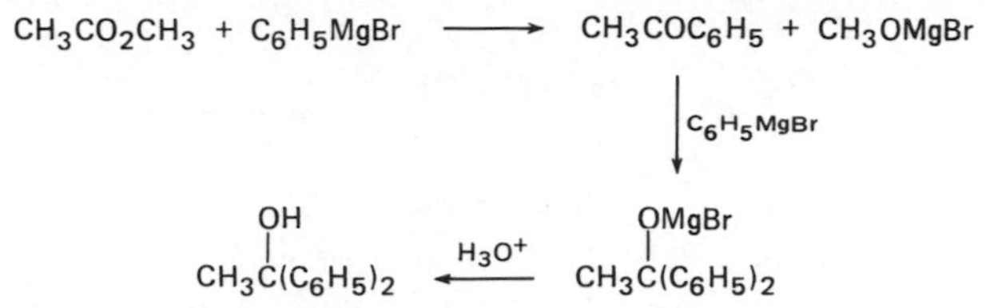

$$CH_3CO_2CH_3 + C_6H_5MgBr \longrightarrow CH_3COC_6H_5 + CH_3OMgBr$$

$$\downarrow C_6H_5MgBr$$

$$\underset{\underset{CH_3\overset{|}{C}(C_6H_5)_2}{\overset{|}{\underset{}{OH}}}}{} \xleftarrow{H_3O^+} \underset{\underset{CH_3\overset{|}{C}(C_6H_5)_2}{\overset{|}{\underset{}{OMgBr}}}}{}$$

Figure 8.26 The reaction of methyl acetate with phenyl-magnesium bromide

$$CH_3CO_2H + C_6H_5MgBr \longrightarrow CH_3CO_2MgBr + C_6H_6$$

$$C_2H_5CONH_2 + CH_3MgBr \longrightarrow C_2H_5CONHMgBr + CH_4$$

$$C_6H_5CONHCH_3 + CH_3MgBr \longrightarrow \underset{\underset{MgBr}{|}}{C_6H_5CONCH_3} + CH_4$$

Figure 8.27 Reactions of carboxylic acids and amides with Grignard reagents

The initial products of the reactions of esters with Grignard reagents or metal alkynes are ketones, which then react further with excess reagent to give tertiary alcohols (Figure 8.26). Reformatsky reagents do not react with esters.

Carboxylic acids are deprotonated by organometallic reagents (which are strong bases), and then do not generally react further. Indeed, even the hydrogens attached to nitrogen in primary and secondary amides are sufficiently acidic to be removed by these reagents (Figure 8.27).

(e) Reactions with hydrogen cyanide

Reactions of carboxylic acids and their derivatives with hydrogen cyanide, or indeed with sodium cyanide, are beyond the scope of this book. However, the reactions of aldehydes and ketones with hydrogen cyanide, which require a small quantity of cyanide anion as catalyst, are extremely important. The products are *cyanohydrins* (Figure 8.28(a)), which are used in the homologation of sugar molecules (Chapter 11) and in the synthesis

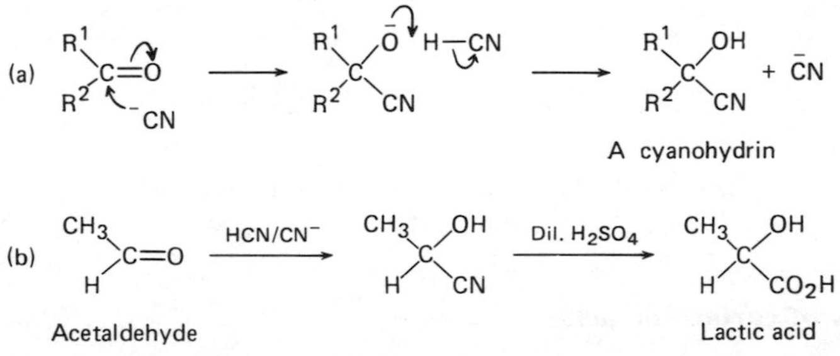

Figure 8.28 Reactions of aldehydes and ketones with hydrogen cyanide

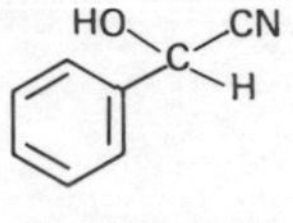

Mandelonitrile

Figure 8.29 The structure of mandelonitrile

of hydroxyacids such as lactic acid (Figure 8.28(b)). When ammonia is included in the reaction mixture, the products are amino acids (Strecker synthesis, see Chapter 12).

Cyanohydrins present a convenient, non-toxic way for an organism to store hydrogen cyanide. For example, mandelonitrile (Figure 8.29), the cyanohydrin of benzaldehyde, is found, either free or combined with a sugar molecule (Chapter 11), in the defensive glands of certain millipedes and in bitter almonds. On disturbing the insect or crushing the bitter almonds the cyanohydrin is mixed with an enzyme which promotes its decomposition, liberating highly toxic hydrogen cyanide.

(f) Reactions of carboxylic acids with thionyl chloride (sulphur dichloride oxide)

Thionyl chloride (sulphur dichloride oxide) reacts with carboxylic acids to give acyl chlorides. The initial intermediate, formed by elimination of hydrogen chloride, is unstable and eliminates sulphur dioxide (Figure 8.30). Since both by-products are gaseous, the acyl chloride is easy to purify, and the reaction is therefore widely used.

Acyl chlorides are also produced by reactions of carboxylic acids with chlorides of phosphorus, but the reactions are less widely used because of the greater difficulty of purifying the products.

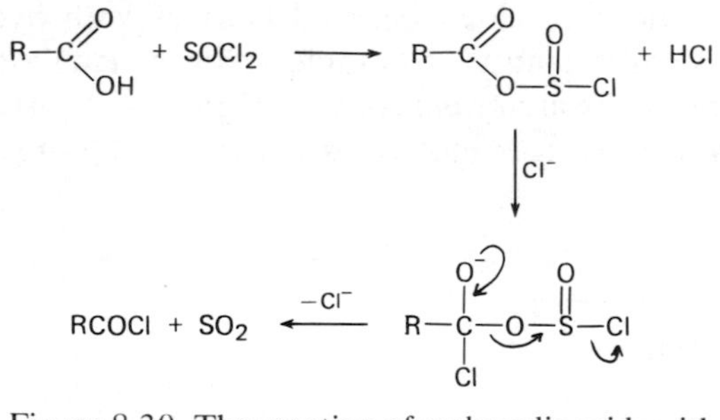

Figure 8.30 The reaction of carboxylic acids with
thionyl chloride

8.5 Acidity of carboxylic acids

As the name suggests, carboxylic acids are acidic compounds, readily forming salts on reaction with basic substances such as sodium hydrogen

$$RCO_2H + NaHCO_3 \longrightarrow RCO_2^- \, Na^+ + H_2O + CO_2$$

$$RCO_2H + NaOH \longrightarrow RCO_2^- \, Na^+ + H_2O$$

$$RCO_2H + NH_3 \longrightarrow RCO_2^- \, NH_4^+$$

$$R^1CO_2H + NR_3^2 \longrightarrow R^1CO_2^- \, R_3^2NH^+$$

$$RCO_2H + CH_3MgBr \longrightarrow RCO_2^- \, {}^+MgBr + CH_4$$

Figure 8.31 Reactions of carboxylic acids with bases

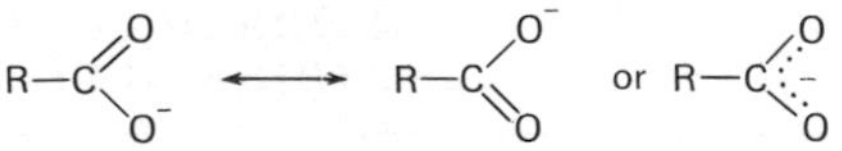

Figure 8.32 Stabilization of the carboxylate anion

carbonate, sodium hydroxide, ammonia, amines, or Grignard reagents (Figure 8.31).

The acidity is a result of stabilization of the carboxylate anion by resonance, the negative charge thereby being spread over all three atoms of the group (Figure 8.32).

Despite this stabilization by resonance, most carboxylate anions are much less stable than anions such as Cl^-, HSO_4^-, or NO_3^-, and hence carboxylic acids are much less dissociated than mineral acids in aqueous solution. For example, a one tenth molar aqueous solution of acetic (ethanoic) acid is only about 1% dissociated, whereas a comparable solution of hydrochloric acid is completely dissociated.

The strength of an acid can be expressed quantitatively by means of the equilibrium constant for its dissociation, known as the *dissociation constant*, K_a (Figure 8.33). The larger the value of K_a, the stronger is the acid. Thus, acetic acid ($K_a = 1.8 \times 10^{-5}$) is clearly a weak acid (compare HCl, $K_a \simeq \infty$) though nevertheless considerably stronger than an alcohol ($K_a \simeq 10^{-16}$). Acid strengths are sometimes also expressed on a logarithmic scale as pK_a values, where $pK_a = -\log_{10}K_a$.

Within the series of carboxylic acids there is considerable variation in acid strength (Table 8.4). Thus, acetic and butanoic acids have very comparable K_a values, but formic acid is considerably stronger. This observation

$$RCO_2H + H_2O \rightleftharpoons RCO_2^- + H_3O^+$$

$$\text{Dissociation constant, } K_a = \frac{[RCO_2^-]\,[H_3O^+]}{[RCO_2H]}$$

Figure 8.33 Dissociation of carboxylic acids in water

Table 8.4 Strengths of carboxylic acids

Acid	Structure	$K_a \times 10^5$
Formic (methanoic)	HCO_2H	20
Acetic (ethanoic)	CH_3CO_2H	1.8
Butanoic	$CH_3(CH_2)_2CO_2H$	1.5
Chloroacetic (chloroethanoic)	$ClCH_2CO_2H$	155
Trichloroacetic (trichloroethanoic)	Cl_3CCO_2H	90 000
2-Chlorobutanoic	$CH_3CH_2CHClCO_2H$	139
3-Chlorobutonoic	$CH_3CHClCH_2CO_2H$	8.9
4-Chlorobutanoic	$CH_2ClCH_2CH_2CO_2H$	3.0
Benzoic	$C_6H_5CO_2H$	6.3
4-Nitrobenzoic	$4\text{-}O_2NC_6H_4CO_2H$	40
4-Methoxybenzoic	$4\text{-}CH_3OC_6H_4CO_2H$	3.3

may be rationalized in terms of destabilization of the acetate and butanoate anions by electron-donating alkyl groups. In contrast, replacement of the hydrogen atoms in the alkyl chains by halogen atoms increases the acidity, because the carboxylate anion is stabilized by inductive electron withdrawal. This effect is greatest when the halogen atoms are attached at the α-positions.

Question 8.7 Rationalize the difference in K_a values of benzoic, 4-nitrobenzoic, and 4-methoxybenzoic acids (Table 8.4).

8.6 Acidity of α-hydrogens in carbonyl compounds

Removal of a proton from the α-position of an aldehyde, ketone, ester, or thioester produces an anion which is stabilized by resonance in the same way as the carboxylate anion (Figure 8.34(a)). However, the degree of stabilization is less because one of the canonical forms has the negative charge localized on carbon, instead of the more electronegative oxygen, and thus makes a smaller contribution to the resonance hybrid. Thus, while aldehydes, ketones, esters, and thioesters are not usually described as acids, their α-hydrogen atoms are indeed acidic, and they undergo a number of reactions involving enolate anions.

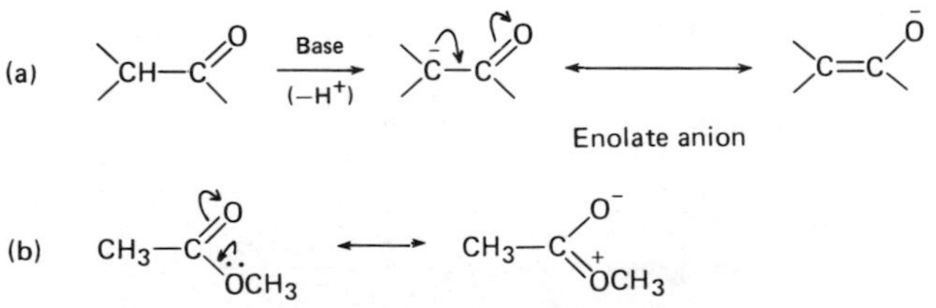

Figure 8.34 (a) α-Deprotonation of aldehydes, ketones, or esters, and (b) resonance in esters

Esters are less acidic than aldehydes or ketones because the ability of the carbonyl group to accept electron density is reduced by its interaction with the electron pairs already present on oxygen (Figure 8.34(b)). Stronger bases are thus required for removal of a proton from esters than from aldehydes or ketones. The electron pairs on a sulphur atom interact less strongly with a carbonyl group due to the larger size of the sulphur atom, and thioesters are almost as acidic as aldehydes and ketones.

(a) Aldol and Claisen condensations

Enolate anions, like most anionic species, are nucleophilic. In the presence of their parent carbonyl compounds they attack the carbonyl carbon atom of the latter to give nucleophilic addition or substitution products. These reactions are known as *aldol* and *Claisen* condensations when undergone by aldehydes/ketones and esters respectively (Figure 8.35). The reactions are carried out by treating the appropriate carbonyl compound with a base such as sodium hydroxide or sodium ethoxide. Other derivatives of carboxylic acids can undergo similar condensation reactions.

These reaction types are very important, both in nature and in the laboratory. For example, the biosynthesis of many natural products occurs through the intermediacy of a polyketide (polyketone), formed by multiple

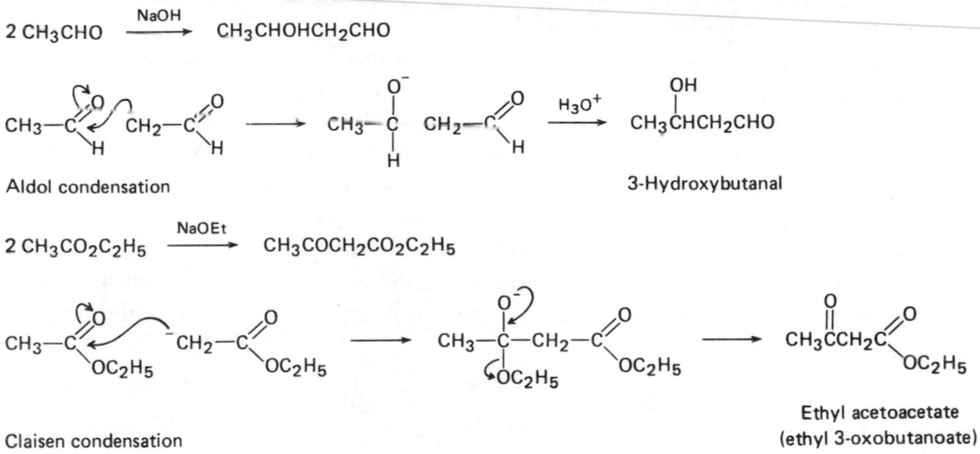
2 CH3CHO
NaOH
CH3CHOHCH2CHO
H3O+
Aldol condensation
3-Hydroxybutanal
2 CH3CO2C2H5
NaOEt
CH3COCH2CO2C2H5
Claisen condensation
Ethyl acetoacetate
(ethyl 3-oxobutanoate)

Figure 8.35 Aldol and Claisen condensations

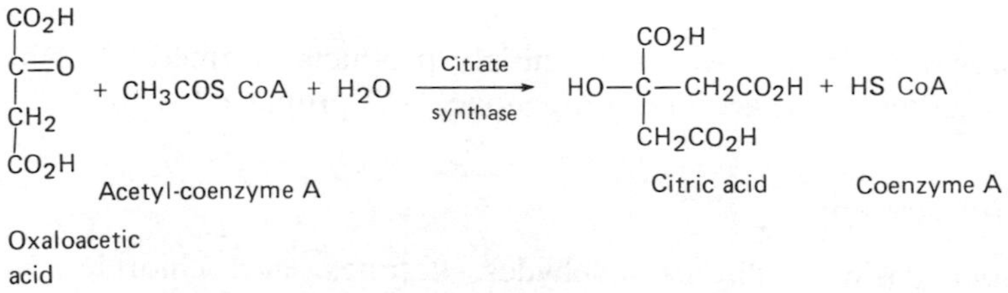
Citrate
synthase
+ CH3COS CoA + H2O
HO—C—CH2CO2H + HS CoA
Acetyl-coenzyme A
Citric acid
Coenzyme A
Oxaloacetic
acid

Figure 8.36 Biosynthesis of citric acid by an aldol-type reaction

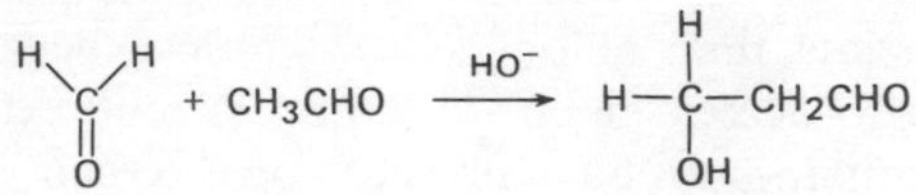

Figure 8.37 A mixed aldol reaction in the presence of excess formaldehyde

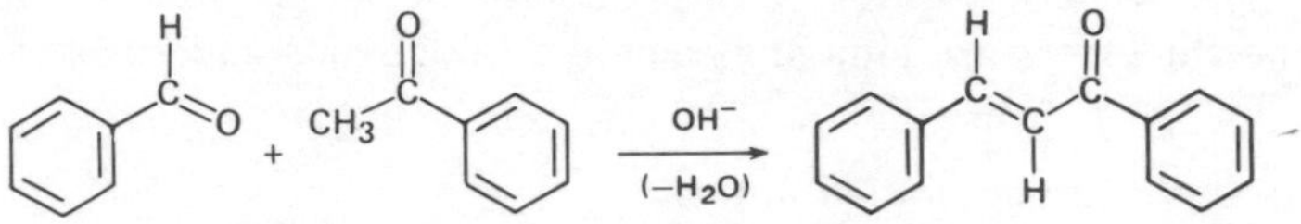

Figure 8.38 Formation of an $\alpha\beta$-unsaturated ketone by an aldol reaction

biochemical equivalents of the Claisen condensation (Chapter 15). Similarly, citric acid is biosynthesized in the tricarboxylic acid cycle (Chapter 15) by a reaction which is of the aldol type (Figure 8.36).

In the laboratory aldol and Claisen condensations cannot usually be carried out satisfactorily between two different carbonyl compounds. The problems arise because four possible products can be obtained. 'Mixed' aldol reactions can, however, be achieved when one of the carbonyl compounds is devoid of α-hydrogen atoms (e.g. formaldehyde, benzaldehyde) and is used in excess (Figure 8.37). Similarly, mixed Claisen reactions can sometimes be carried out.

The initial product of the aldol condensation is a β-hydroxycarbonyl compound, which may lose water under the conditions of the reaction to give an $\alpha\beta$-unsaturated aldehyde or ketone. This occurs particularly easily when the double bond is further conjugated (Figure 8.38). Also, since the products of aldol condensations are themselves aldehydes or ketones, further condensations may occur. Indeed, under the influences of concentrated alkali a polymeric resin may be obtained.

Aldol condensations also occur under acidic conditions, the reaction then occurring between a molecule of the enol and a protonated carbonyl molecule (Figure 8.39).

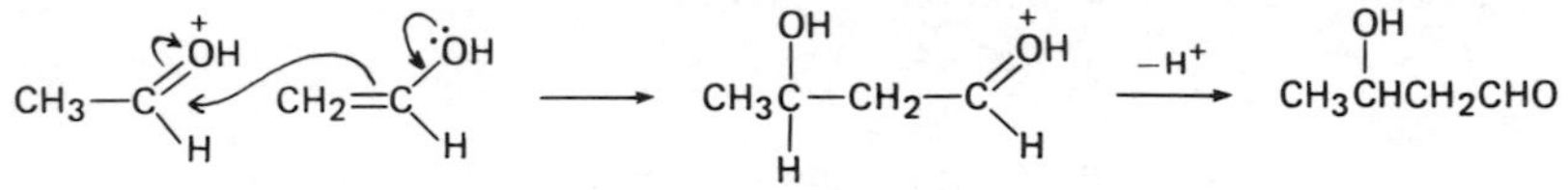

Figure 8.39 An aldol condensation under acid conditions

Question 8.8 What are the initial products formed in the aldol condensations of (i) acetone (propanone), (ii) propanal?

(b) Halogenation

Under basic conditions aldehydes, ketones, acyl chlorides, and acid anhydrides react with halogens to give α-halogeno products (Figure 8.40).

In the presence of excess halogen, further substitution may occur. The second halogen atom becomes attached preferentially to the same site as the first, because the electron-withdrawing halogen atom renders the adjacent hydrogen atoms even more acidic.

Methyl ketones, $RCOCH_3$, react with iodine to give the 1,1,1-triiodo product, which breaks down under the alkaline conditions to give the carboxylate anion and iodoform (triiodomethane). The latter precipitates as a yellow solid and this provides a visual test for methyl ketones, or compounds which are readily converted into methyl ketones under the reaction conditions. It is known as the *iodoform test* (Figure 8.41).

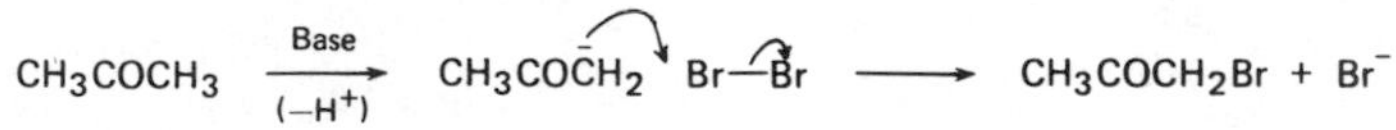

Figure 8.40 Bromination of a ketone under basic conditions

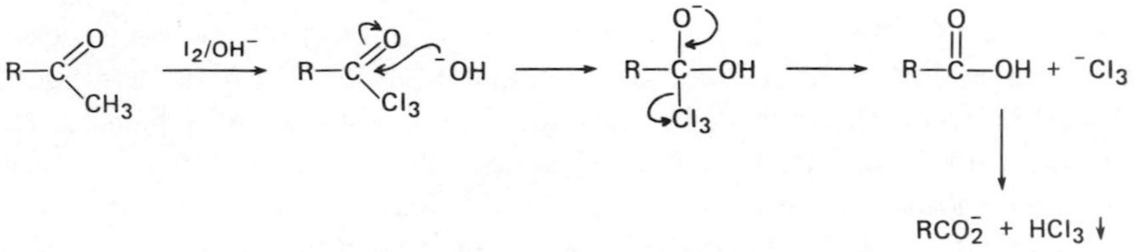

Figure 8.41 The iodoform reaction

Question 8.9 Replacement of α-hydrogens by halogens also occurs under acidic conditions. Assuming that the reaction involves the enol form, suggest a mechanism.

8.7 Oxidation and reduction of carbonyl compounds

(a) Oxidation

Aldehydes are exceptional amongst carbonyl compounds in that they are readily oxidized, giving carboxylic acids (Figure 8.42). Indeed, many aldehydes are oxidized on prolonged contact with air. On a preparative scale it may be convenient to use powerful oxidizing agents such as $KMnO_4$ or $K_2Cr_2O_7$, provided that the rest of the molecule is stable towards such reagents. However, much weaker oxidizing agents are also effective, and some of these provide simple visual tests for distinguishing between aldehydes and ketones. For example, Tollens' reagent, $Ag(NH_3)_2^+$ in alkaline solution, is reduced by an aldehyde to silver metal, which forms a mirror on the walls of the glass vessel. Alternatively, Fehling's or Benedict's reagents (blue solutions of Cu^{2+} complexed with tartrate or citrate ions respectively) react with aldehydes to give a red precipitate of Cu_2O. This latter method is commonly used to detect the presence of D-glucose in urine (Chapter 11).

$$RCHO \xrightarrow{[O]} RCO_2H$$

Figure 8.42 Oxidation of an
aldehyde

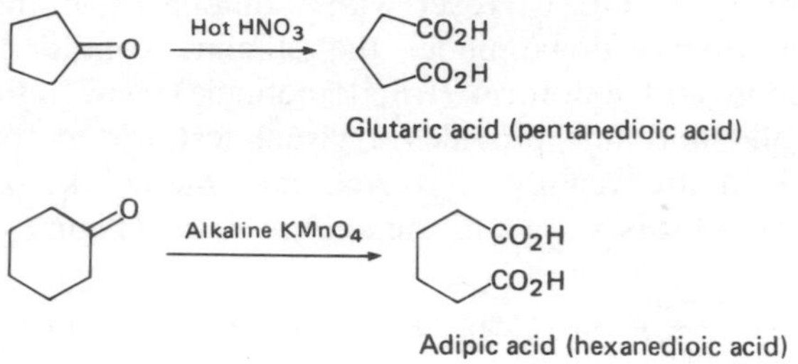

Glutaric acid (pentanedioic acid)

Adipic acid (hexanedioic acid)

Figure 8.43 Formation of dicarboxylic acids by oxidation of cyclic ketones

Oxidation of other carbonyl compounds requires very vigorous conditions, such as hot nitric acid or hot alkaline $KMnO_4$. With these reagents
the bond between the carbonyl group and an α-carbon atom is broken, and
each fragment is oxidized to a carboxylic acid. The reaction is of some
importance industrially for the production of dicarboxylic acids from cyclic
ketones (Figure 8.43). Dicarboxylic acids such as adipic acid are used in
the production of some types of nylon.

Question 8.10 Whereas cyclopentanone and cyclohexanone give single
products on vigorous oxidation, pentan-2-one gives a mixture of four different carboxylic acids. Why is this?

(b) Reduction

Reduction of carbonyl compounds is an extremely important process in
nature and in the laboratory. Complex metal hydrides such as $LiAlH_4$
(lithium tetrahydroaluminate or lithium aluminium hydride) or $NaBH_4$
(sodium tetrahydroborate or sodium borohydride) are convenient reagents
for such reductions (Figure 8.44). The former is very reactive and reduces
all types of carbonyl compounds, but since it also reacts rapidly with water,
the reductions must be carried out in an anhydrous solvent such as diethyl
ether. $NaBH_4$, on the other hand, is much less reactive, and can be used in
aqueous or alcoholic solutions. This reagent reduces aldehydes and ketones
readily, esters only slowly, and acids or amides not at all. It is therefore
especially useful for carrying out selective reductions of aldehydes or
ketones in the presence of other carbonyl groups.

These reductions involve nucleophilic attack by the MH_4^- ion (M=B,Al)
on the carbonyl group, and in this respect are similar to the reactions of
carbonyl compounds with organometallic reagents (Section 8.4(d)). When
reactions are carried out in anhydrous solvents, water must be added

$$RCHO \xrightarrow{\text{LiAlH}_4 \text{ or NaBH}_4} RCH_2OH$$

$$R^1COR^2 \xrightarrow{\text{LiAlH}_4 \text{ or NaBH}_4} R^1CHOHR^2$$

$$R^1CO_2R^2 \xrightarrow{\text{LiAlH}_4} R^1CH_2OH + R^2OH$$

$$R^1CO_2H \xrightarrow{\text{LiAlH}_4} R^1CH_2OH$$

$$R^1CONR_2^2 \xrightarrow{\text{LiAlH}_4} R^1CH_2NR_2^2$$

Figure 8.44 Reduction of carbonyl
compounds with complex hydrides

as a final step in order to decompose the metal complexes which are the
initial products of the reaction.

Similar reductions can be effected by dissolving metals such as sodium in
ethanol or zinc in dilute hydrochloric acid, or by catalytic hydrogenation
over a nickel catalyst. Complex hydride and dissolving metal reductions are
usually selective for carbonyl groups in the presence of carbon–carbon
double bonds, but catalytic hydrogenation preferentially reduces carbon–carbon double bonds.

Catalytic reduction of acid chlorides over a partially poisoned catalyst
consisting of palladium deposited on barium sulphate is a useful method
for the synthesis of aldehydes (Rosenmund reduction; Figure 8.45).

There are three ways of reducing the carbonyl group of an aldehyde or
ketone directly to a CH_2 group. The Clemmensen reaction is carried out
under strongly acidic conditions, using zinc and concentrated hydrochloric
acid, whereas the Wolff–Kishner reaction involves strongly basic conditions, viz. hydrazine and concentrated alkali (Figure 8.46(a)). The method
chosen depends upon the stability of the rest of the molecule to the
appropriate conditions. A milder procedure involves conversion to a
thioacetal or thioketal, followed by a desulphurization with Raney nickel, a
finely divided Ni powder with hydrogen adsorbed on its surface (Figure
8.46(b)).

$$RCOCl \xrightarrow[\text{sulphur poison}]{\text{H}_2/\text{Pd–BaSO}_4} RCHO$$

Figure 8.45 Catalytic reduction of
an acyl halide (Rosenmund reaction)

(a) $PhCOCH_2CH_2CO_2H \xrightarrow[\text{or NH}_2\text{NH}_2/\text{KOH}]{\text{Zn–Hg/conc. HCl}} PhCH_2CH_2CH_2CO_2H$

(b)
$$\begin{matrix} R^1 \\ R^2 \end{matrix} C=O + \begin{matrix} HS-CH_2 \\ HS-CH_2 \end{matrix} \xrightarrow{H^+} \begin{matrix} R^1 \\ R^2 \end{matrix} C \begin{matrix} S-CH_2 \\ S-CH_2 \end{matrix} \xrightarrow[\text{Ni/H}_2]{\text{Raney}} \begin{matrix} R^1 \\ R^2 \end{matrix} CH_2$$

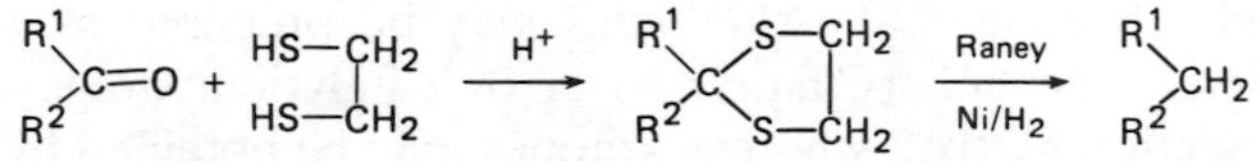

Figure 8.46 Three methods of converting the carbonyl group of
an aldehyde or ketone into a –CH_2 group

(c) Redox reactions in nature

Redox reactions are widespread in nature, but for obvious reasons the oxidizing and reducing systems do not include reagents such as $KMnO_4$, or zinc amalgam. One of the most common oxidizing agents in natural systems is NAD^+ (nicotinamide adenine dinucleotide; see Chapters 13 and 15). An example of its application to the oxidation of carbonyl compounds is the conversion of glyceraldehyde-3-phosphate to glyceric acid-1,3-diphosphate in the presence of inorganic phosphate (Figure 8.47 and Chapter 15).

In other circumstances the reduced form, NADH, behaves as a reducing agent, itself becoming oxidized to NAD^+. For example, pyruvic acid is reduced to lactic acid in this way (Figure 8.48 and Chapter 15).

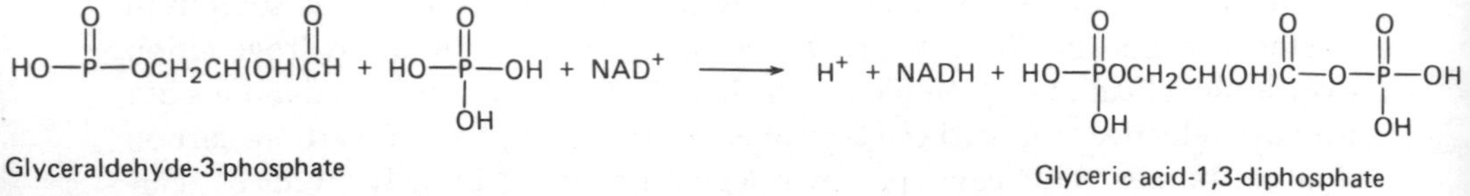

$$HO-\overset{\overset{O}{\|}}{P}-OCH_2CH(OH)CH + HO-\overset{\overset{O}{\|}}{\underset{\underset{OH}{|}}{P}}-OH + NAD^+ \longrightarrow H^+ + NADH + HO-\overset{\overset{O}{\|}}{\underset{\underset{OH}{|}}{P}}OCH_2CH(OH)\overset{\overset{O}{\|}}{C}-O-\overset{\overset{O}{\|}}{\underset{\underset{OH}{|}}{P}}-OH$$

Glyceraldehyde-3-phosphate   Glyceric acid-1,3-diphosphate

Figure 8.47   An example of the biochemical oxidation of an aldehyde

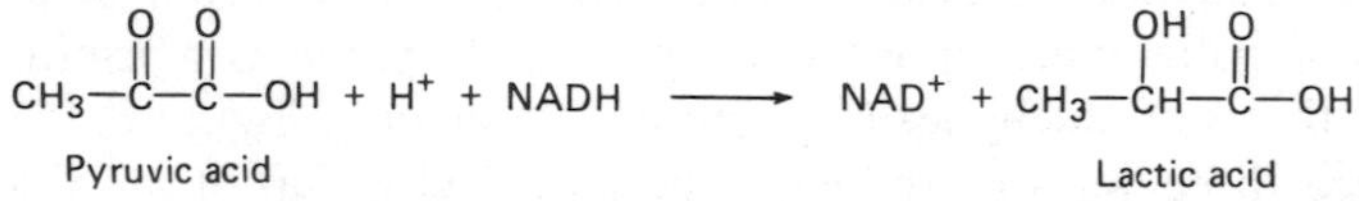

$$CH_3-\overset{\overset{O}{\|}}{C}-\overset{\overset{O}{\|}}{C}-OH + H^+ + NADH \longrightarrow NAD^+ + CH_3-\overset{\overset{OH}{|}}{CH}-\overset{\overset{O}{\|}}{C}-OH$$

Pyruvic acid   Lactic acid

Figure 8.48 An example of the biochemical reduction of a carbonyl compound

An interesting example of aldehyde reduction occurs in the retina, and plays an important part in the detection of light by the eye (Figure 8.49). Rhodopsin, a red complex of 11-*cis*-retinal with the protein opsin gives *trans*-retinal on absorption of light. Reduction of this aldehyde to *trans*-vitamin A is followed by isomerization, in the dark, to 11-*cis*-vitamin A, and reoxidation to 11-*cis*-retinal

8.8 Preparation of carbonyl compounds

In the laboratory, carbonyl compounds are usually synthesized by the most simple and direct routes. Thus, carboxylic acids are obtained by oxidation of primary alcohols (Chapter 6) or of aldehydes (Section 8.7(a)), hydrolysis of nitriles (Chapter 7), reactions of Grignard reagents with carbon dioxide (Chapter 6), or, in the case of aromatic acids, by oxidation of methylbenzenes (Chapter 5). Derivatives of carboxylic acids are usually synthesized either directly from the acid, or via the intermediacy of the acyl chloride (Section 8.4). Aldehydes may be prepared by *partial* oxidation of primary alcohols (Chapter 6) or by catalytic hydrogenation of acyl chlorides (Section 8.7(b)), whereas ketones may be obtained by oxidation of secondary alcohols (Chapter 6). Aromatic aldehydes and ketones may also be contained by Friedel Crafts and related reactions (Chapter 5).

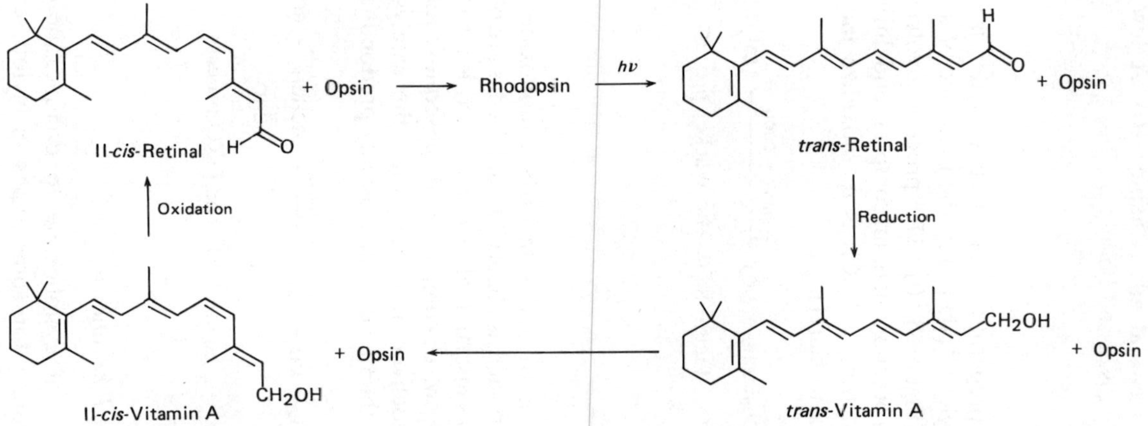

Figure 8.49 The oxidation–reduction reactions important in vision

In industry cost is normally the limiting factor, so processes involving oxidation of hydrocarbons (available cheaply from oil) with air, although requiring careful control and sophisticated apparatus, are often the routes of choice. In addition many carboxylic acids are cheaply obtained by hydrolysis of natural fats and oils (Chapter 10). A few important industrial processes are briefly discussed below.

(a) Wacker process

The industrial preparation of acetaldehyde is often achieved by autoxidation of ethylene (ethene) in the presence of a catalyst composed of $PdCl_2$ and $CuCl_2$ (Figure 8.50). The process involves the formation and decomposition of an organopalladium complex, and during the reaction the Pd is reduced. The function of $CuCl_2$ is to reoxidize the Pd(0) to Pd(II), air oxidizing Cu(I) back to Cu(II).

$$CH_2{=}CH_2 + \tfrac{1}{2}O_2 \xrightarrow{\ PdCl_2\ +\ CuCl_2\ } CH_3CHO$$

Figure 8.50 The Wacker process

(b) OXO process

When an alkene is treated with carbon monoxide and hydrogen in the presence of a catalyst such as $HCo(CO)_4$ aldehydes are obtained (Figure 8.51). The reaction involves insertion of carbon monoxide into the C—Co bond of an intermediate organocobalt complex, and subsequent reaction with hydrogen gives rise to the aldehyde and regenerates the catalyst. Of the two major aldehyde products, the linear isomer usually predominates.

$$RCH{=}CH_2 + CO + H_2 \xrightarrow{\ HCo(CO)_4\ } \underset{\underset{CHO}{|}}{RCHCH_3} + RCH_2CH_2CHO$$

Figure 8.51 The OXO process

(c) Oxidation of cyclohexane

The autoxidation of cyclohexane to cyclohexanone (Figure 8.52), an important intermediate for some types of nylon, exemplifies the direct

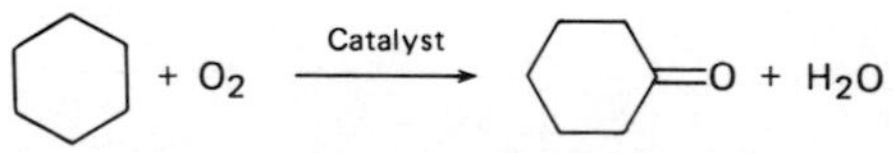

Figure 8.52 Controlled oxidation of cyclohexane

oxidation of an alkane. Careful control of conditions is necessary in order to get just the right degree of oxidation.

8.9 Sulphonic acids and their derivatives

Compounds of the type RSO_3H are called sulphonic acids. Like carboxylic acids they can lose a proton to give a resonance-stabilized anion (Figure 8.53), but the stabilization is more efficient, so that sulphonic acids are much stronger, approaching mineral acids in their strength.

Aromatic sulphonic acids are readily available by direct sulphonation of the parent hydrocarbons (Chapter 5), but aliphatic sulphonic acids are encountered much less frequently.

Derivatives of sulphonic acids, such as the acid chlorides (RSO_2Cl), esters (RSO_3R^1), and amides (RSO_2NH_2, RSO_2NHR^1, $RSO_2NR_2^1$), fulfil similar roles to their counterparts in the carboxylic acid series. Indeed, the similarity between carbonyl compounds and sulphonyl compounds (i.e. compounds containing the $—SO_2—$ group) enables the latter to substitute for the corresponding carbonyl compounds in some living systems. Once incorporated, however, they interfere with the normal metabolic pathways and may thus bring about the destruction of the cell. This explains the bacteriostatic effect of the sulphonamide drugs which revolutionized medicine after their introduction by Gerhard Domagk in 1935. Domagk administered a sulphonamide dyestuff to his own daughter in a desperate, but successful, attempt to save her from dying of a streptococcal infection. Subsequently, scientists experimented with modified structures and obtained many useful drugs, of which a few of the better known examples are illustrated in Figure 8.54. Domagk was later awarded a Nobel Prize.

Because of the stability of the sulphonate anion it is readily displaced from sulphonate esters, which behave as alkylating agents in the same way as alkyl halides (Chapter 6).

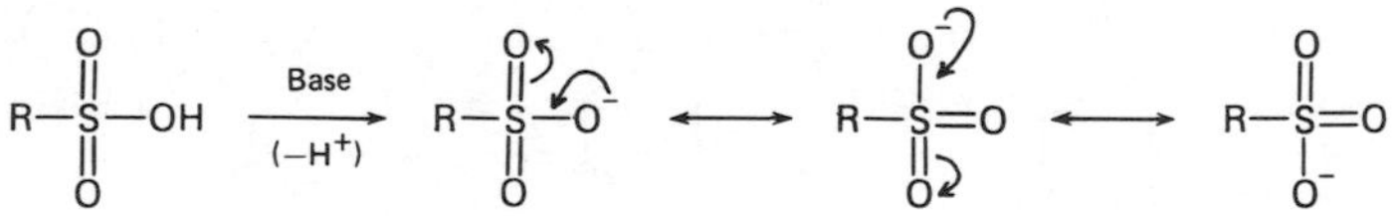

Figure 8.53 Ionization of sulphonic acids

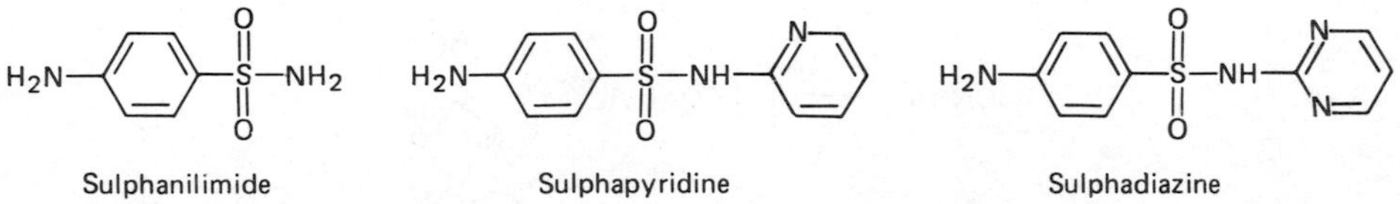

Sulphanilimide   Sulphapyridine   Sulphadiazine

Figure 8.54 Some sulphonamide drugs

Answers to questions

8.1

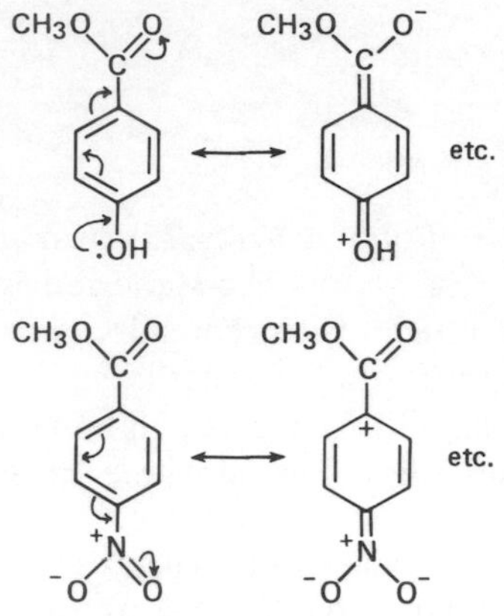

8.2

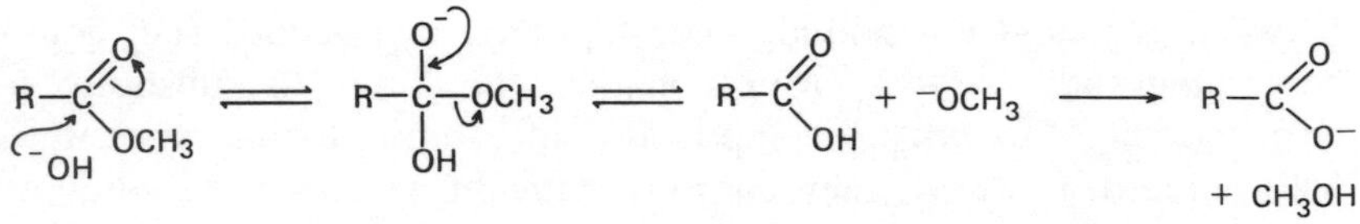

8.3 The mechanism involves the same equilibrium as that depicted in Figure 8.12; the conditions are arranged so that the reverse direction is favoured (e.g. use of excess CH_3OH).

8.4

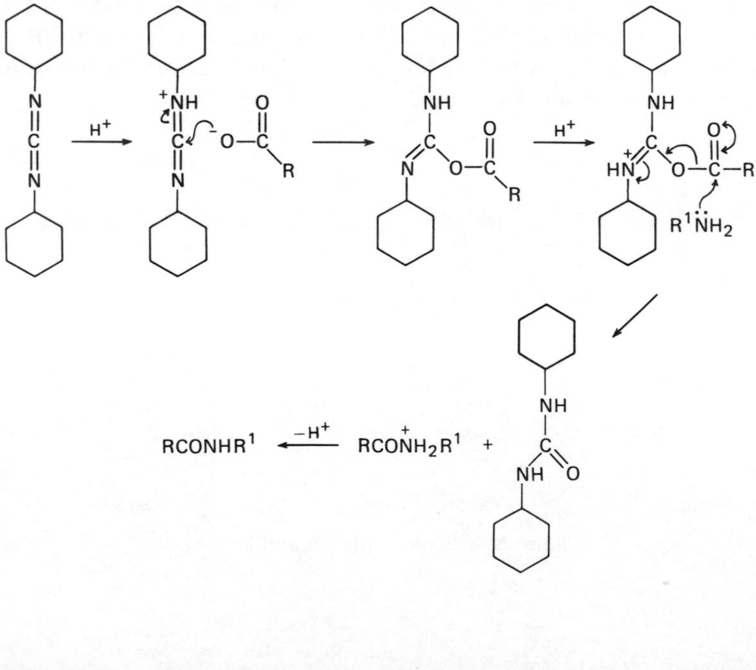

$RCONHR^1 \xleftarrow{-H^+} RC\overset{+}{O}NH_2R^1 +$

8.5

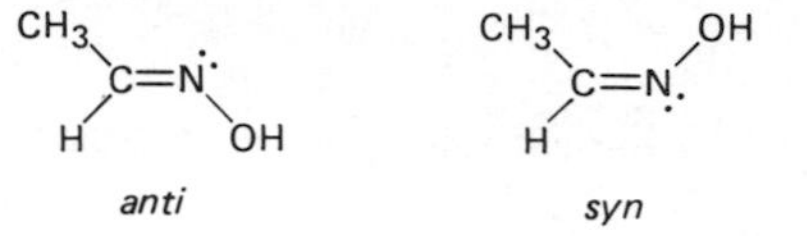

anti   syn

8.6

$$Ph-\underset{\underset{CH_3}{|}}{\overset{\overset{OH}{|}}{C}}-C_2H_5$$

O
‖
PhCCH₃ + C₂H₅MgBr   CH₃CC₂H₅ + PhMgBr

O
‖
PhCC₂H₅ + CH₃MgBr

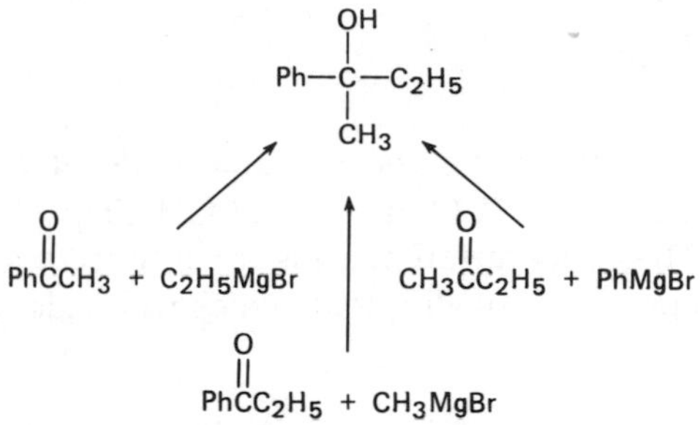

8.7   The negative charge on the carboxylate anion is stabilized when the benzene ring has electron-withdrawing groups (e.g. NO_2) attached, especially when they are in the 2- or 4- positions, because then a canonical form can be drawn in which there is a positive charge on the ring atom next to the carboxylate group:

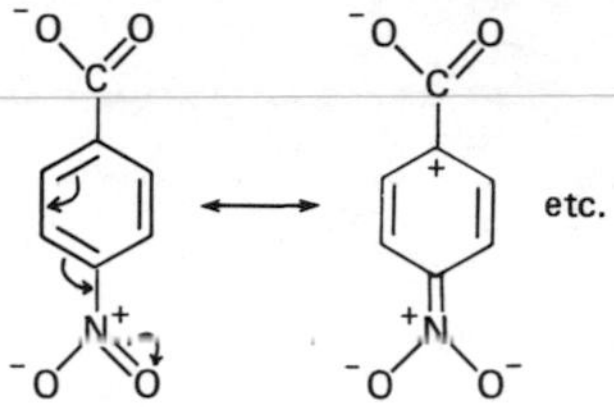

etc.

Electron-donating substituents destabilize the anion by the reverse process, so 4-methoxybenzoic acid is weaker than benzoic acid:

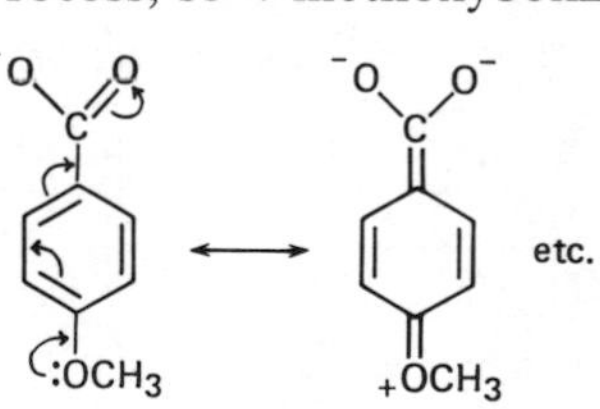

etc.

8.8   (i)

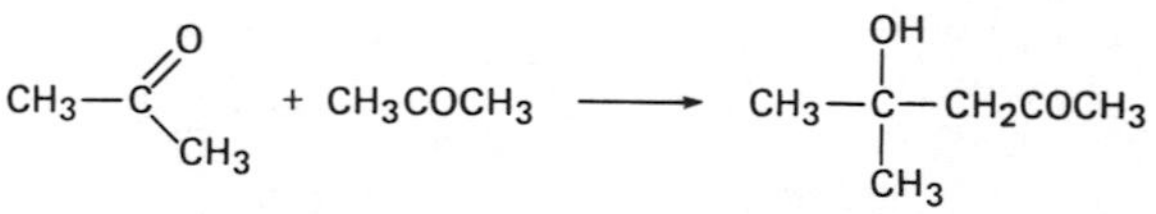

$$CH_3-\overset{\overset{O}{\|}}{C}-CH_3 \; + \; CH_3COCH_3 \longrightarrow CH_3-\underset{\underset{CH_3}{|}}{\overset{\overset{OH}{|}}{C}}-CH_2COCH_3$$

(ii)

$$CH_3CH_2-\overset{\overset{O}{\|}}{C}\diagdown_H \; + \; \underset{\underset{CH_3}{|}}{CH_2CHO} \longrightarrow CH_3CH_2-\underset{\underset{H}{|}}{\overset{\overset{OH}{|}}{C}}-\underset{\underset{CH_3}{|}}{CHCHO}$$

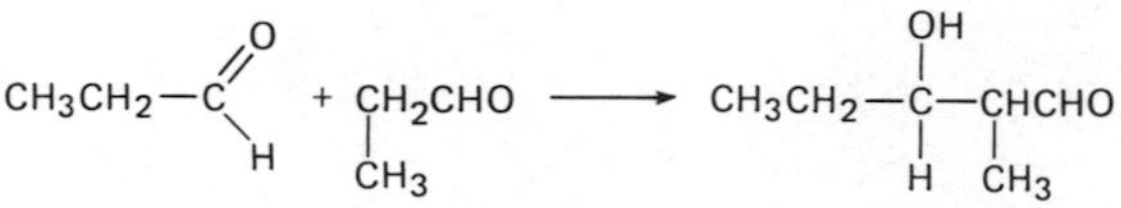

8.9

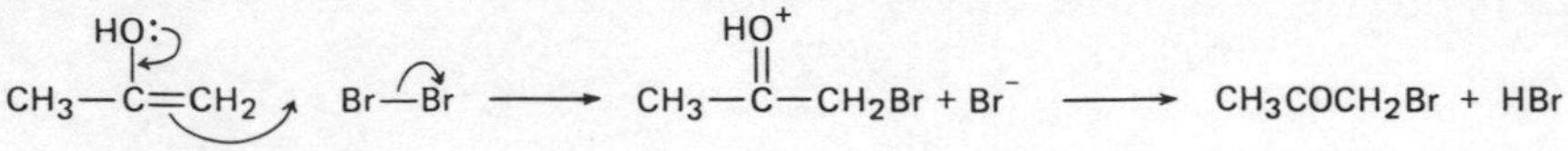

$$CH_3-\overset{\overset{\displaystyle HO:}{|}}{C}=CH_2 \quad Br-Br \longrightarrow CH_3-\overset{\overset{\displaystyle HO^+}{||}}{C}-CH_2Br + Br^- \longrightarrow CH_3COCH_2Br + HBr$$

8.10

$$CH_3\overset{\overset{\displaystyle O}{||}}{\underset{\underset{\displaystyle a \quad\ b}{}}{C}}CH_2CH_2CH_3$$

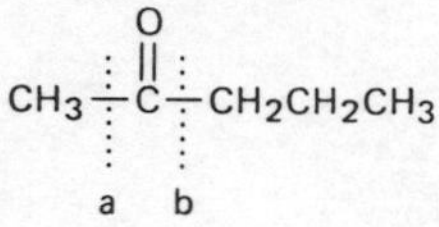

There are two possible sites of cleavage, a and b. Each mode of cleavage gives a mixture of two acids: a, HCO_2H + $CH_3CH_2CH_2CO_2H$; b, CH_3CO_2H + $CH_3CH_2CO_2H$. Thus, four different acids are produced, whereas cyclopentanone gives the same dicarboxylic acid whichever side of the carbonyl group is cleaved.

CHAPTER 9

Bifunctional Molecules

Previous chapters have been primarily concerned with the chemistry of compounds containing a single functional group. However a large number of important organic compounds contain two or more functional groups. In many cases the chemistry of such compounds is very similar to that of the corresponding monofunctional compounds, but in some cases the presence of two groups in a molecule, particularly when they are in close proximity, gives rise to unique chemical and physical properties. For example, the carbon–carbon double bond of an $\alpha\beta$-unsaturated carbonyl compound readily undergoes nucleophilic addition (Figure 9.1(a)), in contrast to the more usual situation in which most alkenes react only with electrophiles (Chapter 4). Conversely the carbon–carbon double bond of an enamine is so strongly nucleophilic that it will even react with alkyl halides (Figure 9.1(b)).

Another aspect of the chemistry of bifunctional compounds is that even when the two functional groups are widely separated they may interact with each other. For example, diketones and diesters undergo *intramolecular* aldol and Claisen condensations (Section 8.6(a)) under basic conditions to give cyclic products (e.g. Figure 9.2). Alternatively, bifunctional molecules may link together in a non-cyclic manner (i.e. *intermolecularly*) to give polymeric products (Section 9.7).

Some of the main types of bifunctional compounds are listed in Table 9.1. This chapter is concerned with bifunctional compounds in which interaction

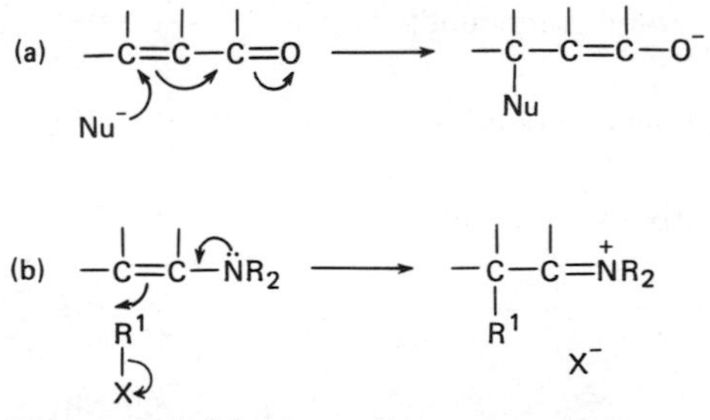

Figure 9.1 (a) An $\alpha\beta$-unsaturated carbonyl compound undergoing nucleophilic attack, and (b) an enamine acting as a nucleophile

173

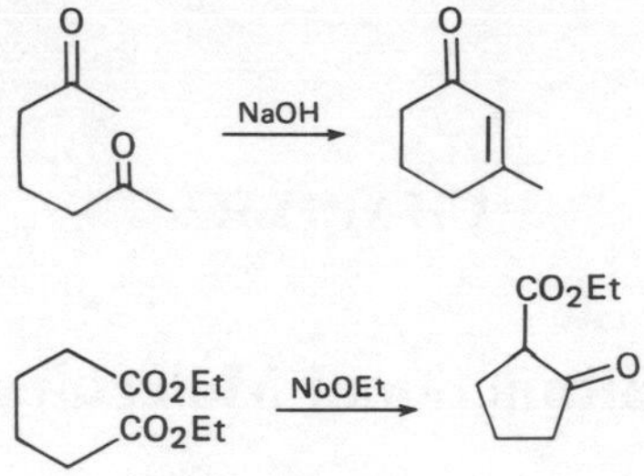

Figure 9.2 Intramolecular aldo- and
Claisen-type condensation of ketones
and esters

of the two functional groups gives rise to properties different from those of
the corresponding monofunctional compounds. Compounds containing
more than two groups are also mentioned where appropriate. Some
multifunctional compounds such as amino acids (examples of aminocarbonyl
compounds) and carbohydrates (derivatives of polyhydroxycarbonyl com-

Table 9   Bifunctional compounds

Compound type	Functional groups present	
Dienes	C=C	C=C
Unsaturated carbonyl compounds	C=C	C=O
Dicarbonyl compounds	C=O	C=O
Enols	C=C	$-\text{C—OH}$
Enamines	C=C	$-\text{C—NR}_2$
Halogenoalkenes	$-\text{C—Cl}$	C=C
Hydroxycarbonyl compounds	$-\text{C—OH}$	C=O
Aminocarbonyl compounds	$-\text{C—NR}_2$	C=O
Halogenocarbonyl compounds	$-\text{C—Cl}$	C=O
Diols	$-\text{C—OH}$	$-\text{C—OH}$
Diamines	C—NR_2	$-\text{C—NR}_2$
Dihalides	$-\text{C—Cl}$	$-\text{C—Cl}$

(R=H, alkyl or aryl)

pounds) are of such considerable chemical and biochemical importance that they are dealt with separately (Chapters 12 and 11 respectively).

9.1 Compounds containing cumulated double bonds

Two π bonds which share a common carbon atom are said to be *cumulated*. Examples of such systems which are encountered in organic chemistry are shown in Figure 9.3.

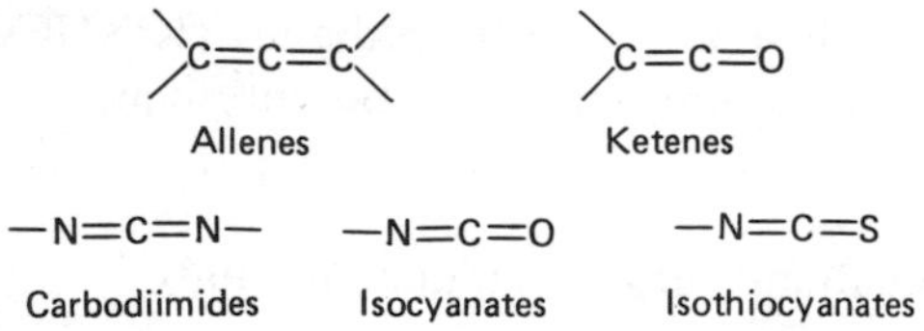

Figure 9.3 Examples of cumulated π bond systems

The central carbon atom in a cumulated system is sp^1 hybridized and the two double bonds are therefore at right angles to one another. This is responsible for the fact that some *allenes* are chiral (Section 2.8). Allenes are isomeric with alkynes (Chapter 4) and the two classes of compounds can often be interconverted. The reactions of allenes are similar to those of alkenes and alkynes.

Question 9.1   Which of the following allenes are chiral:(a) buta-1,2-diene; (b) penta-2,3-diene; (c) penta-2,3-dienoic acid?

Ketenes are usually very reactive compounds. They react with water, alcohols, thiols, and primary and secondary amines to give carboxylic acids and their derivatives (Figure 9.4). In the absence of such reagents many ketenes dimerize and must therefore be prepared immediately prior to use. One method by which they can be prepared is by base catalysed elimination of hydrogen chloride from an acyl halide (Figure 9.4).

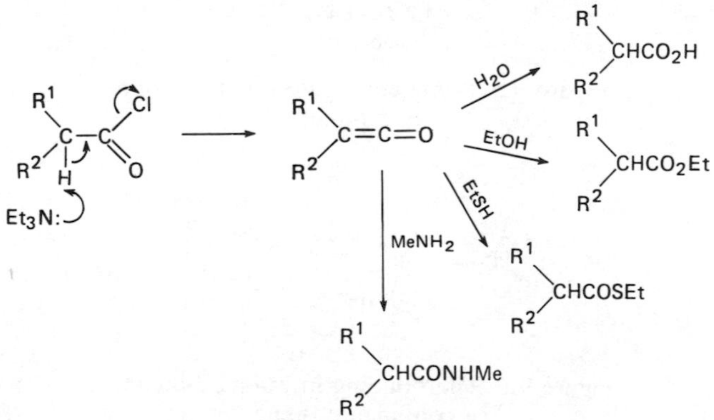

Figure 9.4 Formation and reactions of ketenes

176

Question 9.2 Propose a mechanism for the reaction of dimethylketene with water.

The chemistry of carbodiimides, isocyanates, and isothiocyanates is beyond the scope of this book, but it is interesting to note that carbodiimides are used in peptide synthesis and that phenyl isothiocyanate is used in protein sequencing (Chapter 12). Isocyanates are formed in the Hofmann rearrangement of amides (Section 7.6). They are readily hydrolysed and react with alcohols to form urethanes ($R^1NHCO_2R^2$), a reaction which is important for the preparation of polyurethanes.

9.2 Compounds containing conjugated double bonds

One of the most important categories of bifunctional compounds is that in which two double bonded carbon atoms are separated by a single bond. Some examples of groupings of this type are shown in Figure 9.5.

In order to see why conjugated dienes and $\alpha\beta$-unsaturated carbonyl compounds exhibit modified chemical and physical properties, it is necessary to consider the bonding and hybridization in such molecules. In conjugated dienes, for example, there is interaction between the unhybridized 2p atomic orbitals on all four carbon atoms (Figure 9.6) and the π electrons are delocalized over the entire system. This picture is supported in the case of

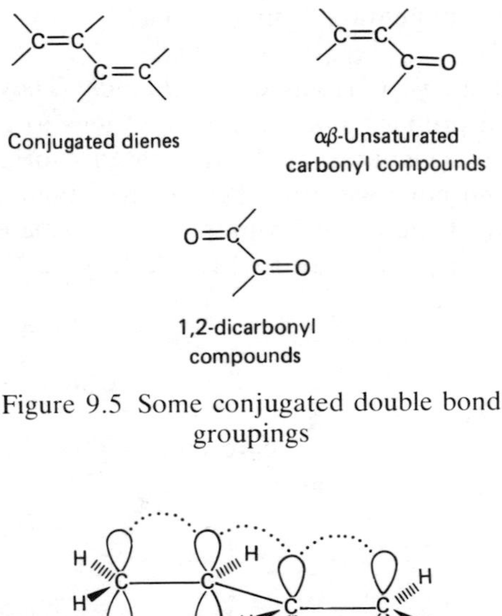

Conjugated dienes

$\alpha\beta$-Unsaturated carbonyl compounds

1,2-dicarbonyl compounds

Figure 9.5 Some conjugated double bond groupings

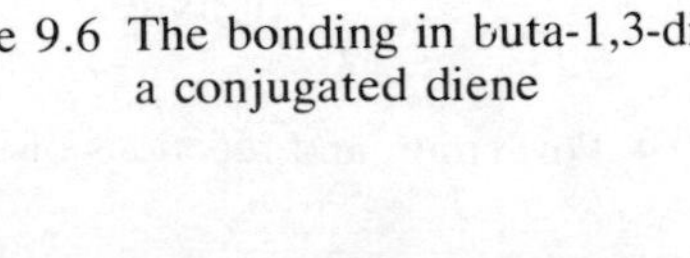

Figure 9.6 The bonding in buta-1,3-diene, a conjugated diene

buta-1,3-diene by physical measurements which show that the carbon–carbon double bonds are slightly longer than those of ethene, whereas the carbon–carbon single bond is slightly shorter than that of ethene. This suggests that there is some partial double bond character between carbon atoms 2 and 3.

However it should be clear from Figure 9.6 that the interaction between the unhybridized 2p orbitals on carbon atoms 2 and 3 can only occur when the molecule is planar. This situation imposes a slight restriction to rotation about the central C—C bond, so that the preferred conformations of the molecule are those in which all four carbon atoms are coplanar. Nevertheless, at room temperature rotation about the central C—C bond takes place easily.

Question 9.3 Draw the structures of the two preferred conformations of buta-1,3-diene.

Conjugated compounds exhibit characteristic ultraviolet spectra due to the presence of delocalized molecular orbitals (Chapter 3).

(a) Conjugated dienes

Conjugated dienes react with the same kinds of reagent (i.e. electrophiles) as do alkenes. However in some cases different products are obtained. Consider for example the reaction of one mole of buta-1,3-diene with one mole of bromine. Two products are obtained (Figure 9.7(a)). The minor product corresponds to the usual 1,2-addition of bromine across a double bond. The major product, however, is formed by 1,4-addition (Figure 9.7(b)).

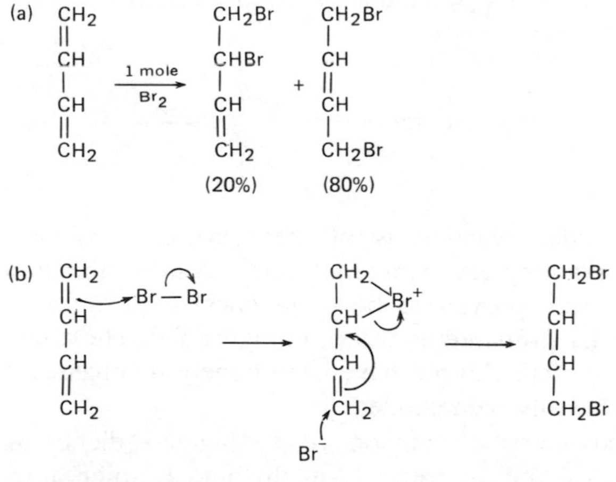

Figure 9.7 The reaction of buta-1,3-diene with bromine

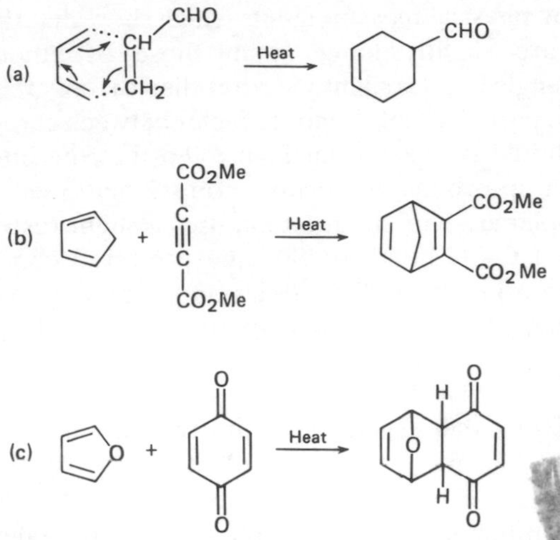

Figure 9.8 Examples of the Diels Alder reaction

Conjugated dienes also undergo one reaction which simple alkenes do not undergo at all. This is the *Diels Alder reaction* of a diene with a *dienophile*, usually an alkene or alkyne bearing one or more electron-withdrawing groups (Figure 9.8). The Diels Alder reaction is in fact only one example of a general class of concerted (one-step) reactions, sometimes referred to as pericyclic reactions. The mechanism can be regarded as a redistribution of six electrons in a six-membered cyclic transition state, hence the term *pericyclic* reaction.

Question 9.4 Draw the structure of the product formed in the following reaction:

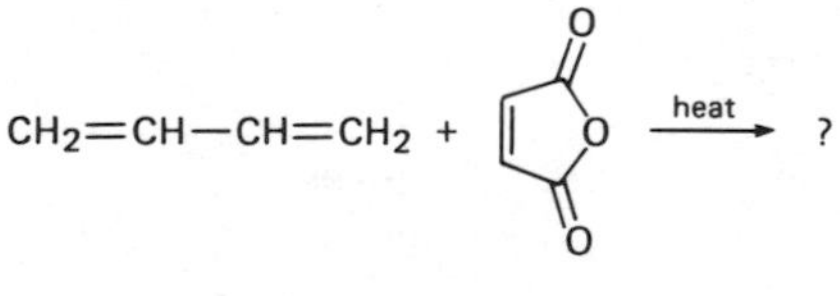

The Diels Alder reaction is of great synthetic value for preparing compounds containing six-membered rings. Among the products of such reactions that have proved to be of practical importance are *aldrin* and *dieldrin* which are prepared as shown in Figure 9.9. These compounds were widely used as insecticides but have been largely superceded by compounds which are more easily biodegraded.

Another characteristic reaction of conjugated dienes is their ready polymerization. When isoprene (2-methylbuta-1,3-diene) is treated with conventional free radical initiators it readily forms a polymer (Figure 9.10(a)). This synthetic rubber contains mainly *trans* (E) double bonds.

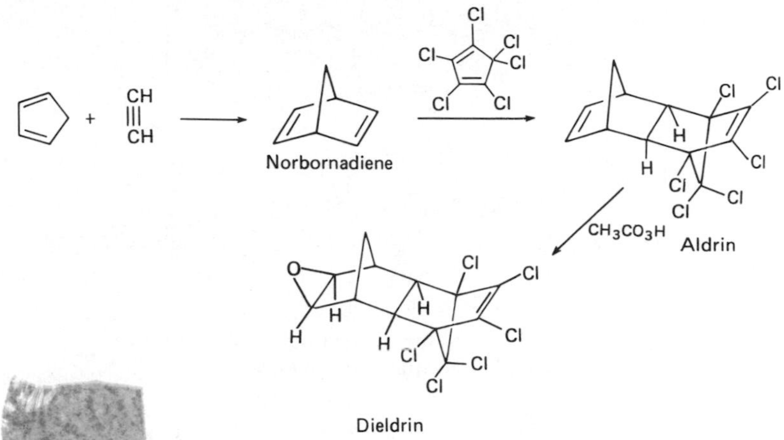

Figure 9.9 Preparation of aldrin and dieldrin via the Diels Alder reaction

Polymerization using organometallic catalysts on the other hand gives a product with *cis* (Z) double bonds, as found in natural rubber (Figure 9.10(b)).

The ____ bonds of crude rubber are capable of undergoing many of the r____ o b____ lkenes. Thus rubber is readily attacked by ozone and is subj____ ve deterioration when exposed to oxygen or air for long peric____ However the useful properties of rubber can be enhanced and p____ if it is subjected to the process of *vulcanization*, in which about 3% of sulphur is incorporated into the crude rubber. Carbon–sulphur bonds are formed during vulcanization and the improved properties are due to cross-linking of the polymer chains and the disappearance of some of the carbon–carbon double bonds. Other conjugated dienes, e.g. 2-chlorobuta-1,3-diene, also give rise to polymers having useful commercial properties.

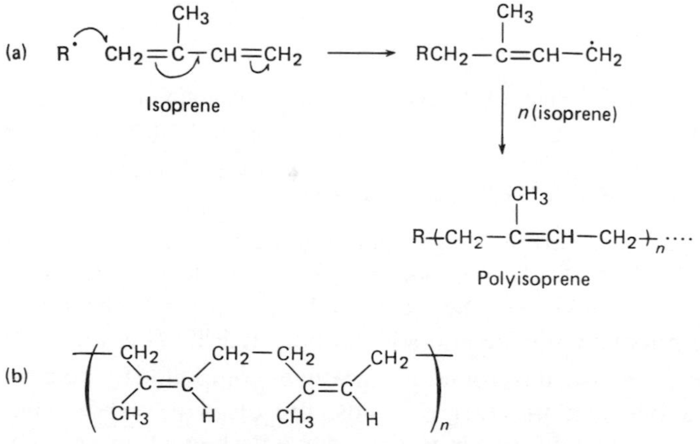

Figure 9.10 Free radical polymerization of isoprene

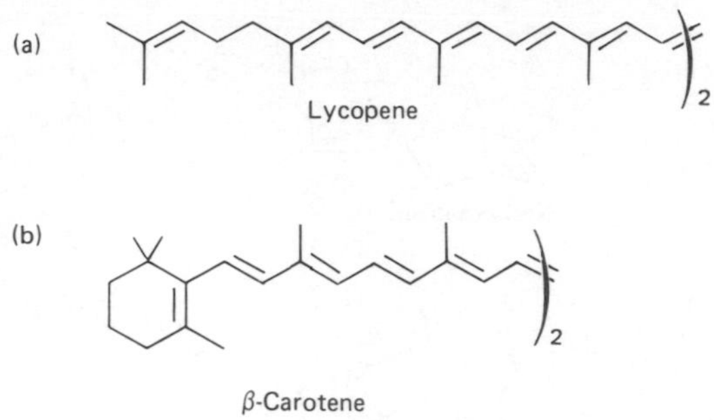

(a)

Lycopene

(b)

β-Carotene

Figure 9.11 Two natural products containing extended conjugated systems

Many natural products contain extended conjugated systems and some are highly coloured. Examples include the carotenoids lycopene (Figure 9.11(a)), a red pigment in ripe tomatoes and watermelons, and β-carotene (Figure 9.11(b)), an orange pigment present in many plants and first isolated from carrots. Notice that these compounds contain a similar arrangement of carbon atoms to that in rubber. Indeed their biosynthesis involves the linking together of five carbon units related to isoprene (Chapter 15).

β-Carotene and similar cyclic carotenoids are important because they can be cleaved in the liver to give vitamin A (Figure 9.12 and Chapter 14). Vitamin A is also closely related to retinene, a component of the photosensitive substance 'rhodopsin' that is found in the retina of the eye (see Section 8.7(c)).

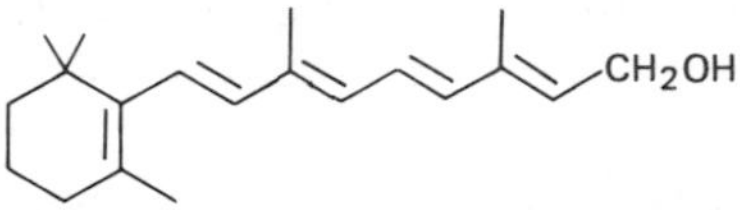

CH₂OH

Vitamin A

Figure 9.12 The structure of vitamin A

(b) αβ-Unsaturated carbonyl compounds

αβ-Unsaturated carbonyl compounds undergo many reactions similar to those of simple carbonyl compounds. For example they are susceptible to nucleophilic attack at the carbonyl group by reagents such as 2,4-dinitrophenylhydrazine and sodium borohydride (Figure 9.13).

However, in αβ-unsaturated carbonyl compounds, because of the interaction between the two π bonds, the electron-withdrawing effect of the carbonyl group is transmitted to the β-carbon atom. As a result such compounds are also susceptible to nucleophilic attack at this position

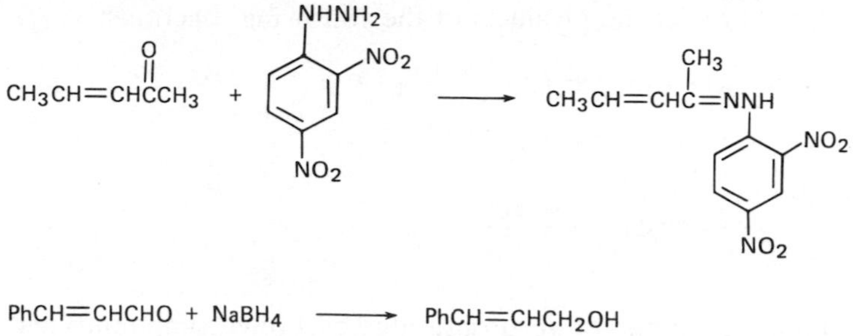

Figure 9.13 Examples of nucleophilic attack on the carbonyl group of $\alpha\beta$-unsaturated carbonyl compounds

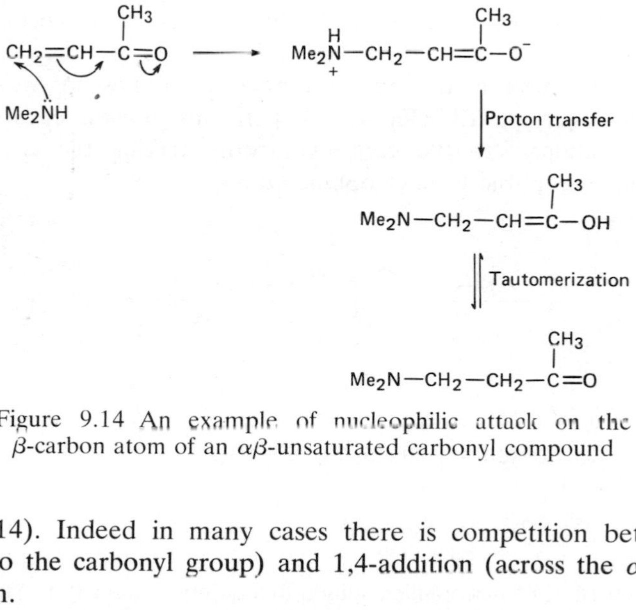

Figure 9.14 An example of nucleophilic attack on the β-carbon atom of an $\alpha\beta$-unsaturated carbonyl compound

(Figure 9.14). Indeed in many cases there is competition between 1,2-addition (to the carbonyl group) and 1,4-addition (across the $\alpha\beta$-unsaturated system.

Reactions involving β-attack by a carbanion are called *Michael reactions*. An example is the reaction of diethyl malonate with acrolein in the presence of sodium ethoxide (Figure 9.15). The carbanion is formed by removal of a proton from the active methylene group of diethyl malonate (Section 9.4).

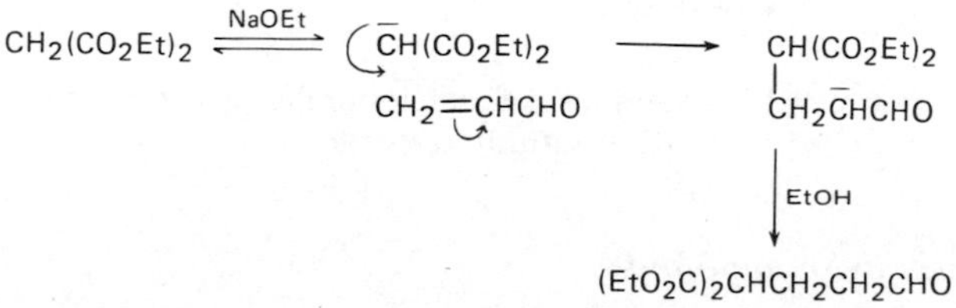

Figure 9.15 An example of the Michael reaction

Question 9.5   Predict the products of the following reactions:

(a) $CH_3CH{=}CHCO_2Et + CH_2(CO_2Et)_2 \xrightarrow{\text{NaOEt}} ?$

(b) $CH_2{=}CHCO_2Et + CH_3COCH_2CO_2Et \xrightarrow{\text{NaOEt}} ?$

(c) 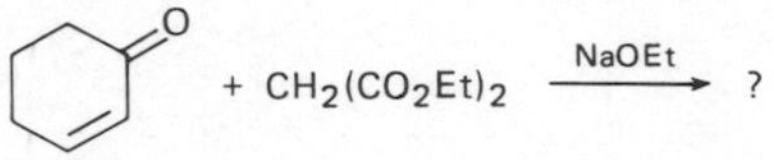 $+ CH_2(CO_2Et)_2 \xrightarrow{\text{NaOEt}} ?$

Quinones are examples of $\alpha\beta$-unsaturated carbonyl compounds and as such undergo nucleophilic addition (Figure 9.16(a)). Quinones are important because of their widespread occurrence in nature as products of plant and animal metabolism. Thus vitamin K_1 (Figure 9.16(b)) is found in green plants and plays a vital role in maintaining the coagulant properties of blood (Chapter 14).

A related reaction is the addition of nucleophiles to $\alpha\beta$-unsaturated nitriles such as acrylonitrile (Figure 9.17). In this case the cyano group acts in a similar manner to the carbonyl group making the carbon–carbon double bond susceptible to nucleophilic attack.

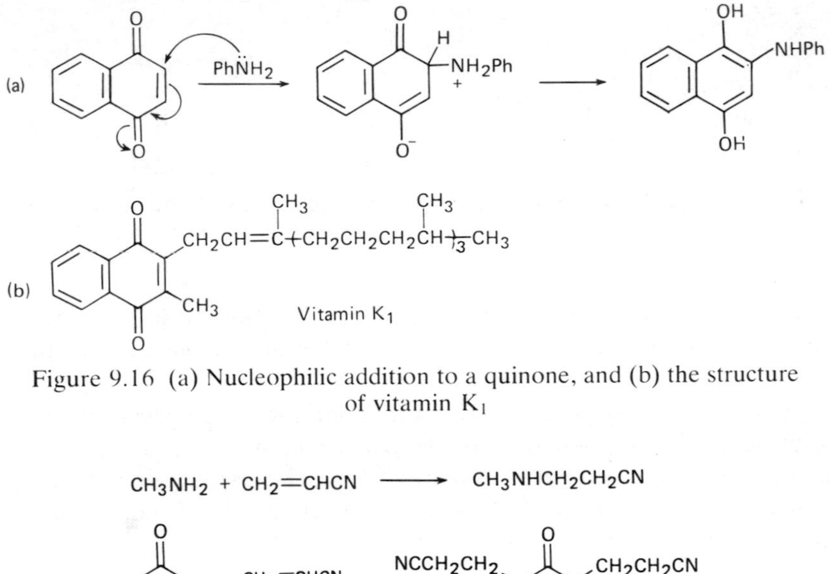

Figure 9.16 (a) Nucleophilic addition to a quinone, and (b) the structure of vitamin K_1

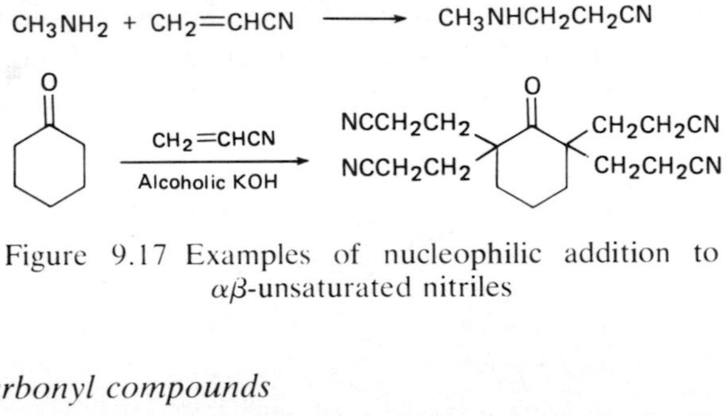

Figure 9.17 Examples of nucleophilic addition to $\alpha\beta$-unsaturated nitriles

(c) 1,2-Dicarbonyl compounds

1,2-Dicarbonyl compounds are usually very reactive due to the close

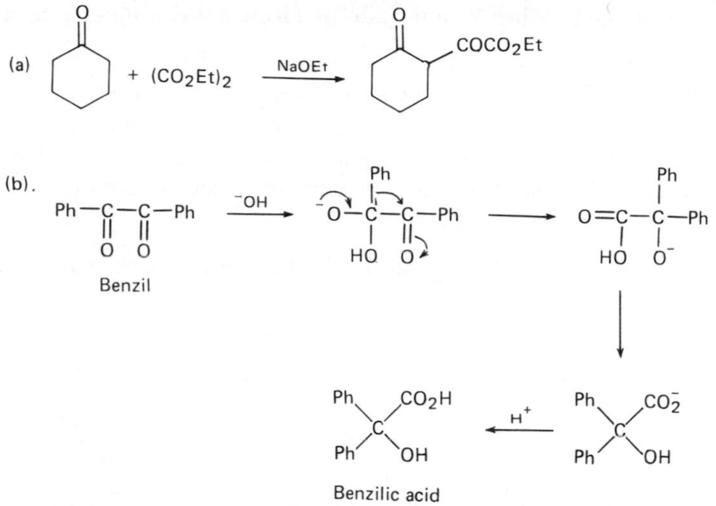

Figure 9.18 (a) The mixed Claisen ester condensation of diethyl oxalate, and (b) the rearrangement of benzil to benzilic acid

proximity of the two electrophilic carbonyl groups. Two of the more interesting reactions of 1,2-dicarbonyl compounds are illustrated by the mixed Claisen ester condensation of diethyl oxalate (Figure 9.18(a)) and the rearrangement of benzil under alkaline conditions to give benzilic acid (Figure 9.18(b)).

Question 9.6 Suggest a mechanism for the reaction shown in Figure 9.18(a) (see Section 8.6(a)).

Pyruvic acid (CH_3COCO_2H) occupies a central role in glucose metabolism (Chapter 15). In some organisms (e.g. yeast) under anaerobic conditions, pyruvic acid is decarboxylated to acetaldehyde and the latter reduced to ethanol by NADH (Figure 9.19). In muscle cells under normal conditions pyruvic acid is oxidized to three molecules of CO_2 (Chapter 15). When the supply of oxygen is inadequate, for example during vigorous exercise, some pyruvic acid is reduced to lactic acid.

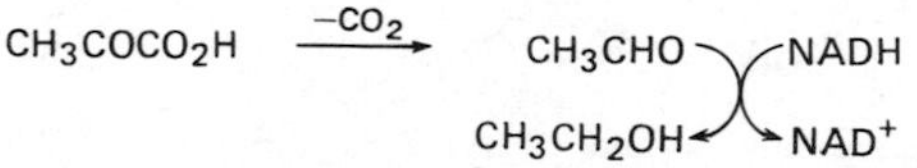

Figure 9.19 Biochemcial decarboxylation of pyruvic acid to acetaldehyde and subsequent reduction to ethanol

9.3 Compounds containing a lone pair of electrons conjugated to a double bond

When an atom bearing a lone pair of electrons is directly attached to a carbon–carbon double bond, the lone pair and π orbitals interact (Figure 9.20). In principle carboxylic acids and their derivatives are examples of very similar systems, but these are usually regarded as containing a single

functional group (i.e. $-C\overset{O}{\underset{\ddot{X}}{\diagup}}$) and are therefore dealt with separately (Chapter 8).

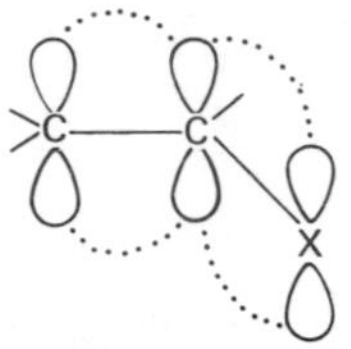

Figure 9.20 The bonding structure of a compound containing a lone pair of electrons conjugated to a double bond

Some examples of systems containing a lone pair of electrons conjugated with a carbon–carbon double bond are shown in Figure 9.21. In all of these systems interaction between the lone pair of electrons and the double bond introduces some double bond character into the carbon–heteroatom bond and increases the electron density at carbon atom 2.

Enols are simply tautomers of carbonyl compounds and are covered in Section 8.3. The chemistry of enol ethers is beyond the scope of this book. Alkenyl halides exhibit properties analogous to those of alkyl halides except that they are much less susceptible to nucleophilic substitution due to the partial double bond character between carbon and chlorine (Section 6.3).

Enamines are formed by reacting aldehydes or ketones with secondary amines (Figure 9.22(a)). Like amines, enamines are nucleophiles, but due to

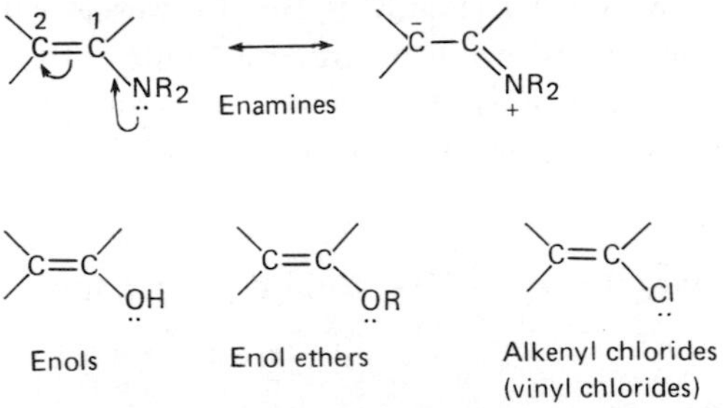

Enamines

Enols   Enol ethers   Alkenyl chlorides (vinyl chlorides)

Figure 9.21 Some examples of compounds containing a lone pair of electrons conjugated to a double bond

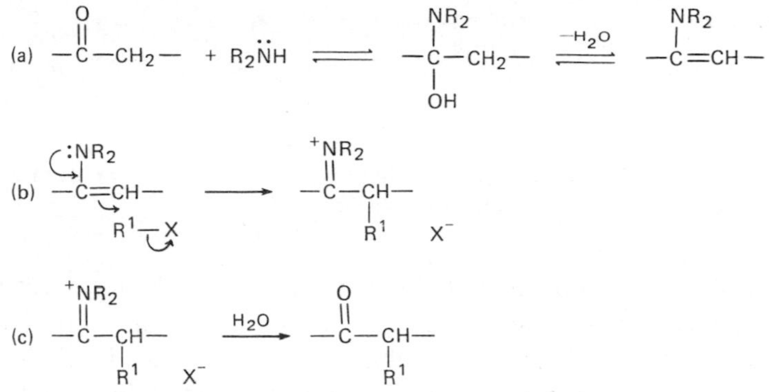

$$\text{(a)} \quad -\overset{\overset{\displaystyle O}{\|}}{C}-CH_2- \;+\; R_2\overset{\cdot\cdot}{N}H \;\rightleftharpoons\; -\overset{\overset{\displaystyle NR_2}{|}}{\underset{\underset{\displaystyle OH}{|}}{C}}-CH_2- \;\overset{-H_2O}{\rightleftharpoons}\; -\overset{\overset{\displaystyle NR_2}{|}}{C}=CH-$$

$$\text{(b)} \quad -\overset{\overset{\displaystyle :NR_2}{|}}{C}=CH- \quad\longrightarrow\quad -\overset{\overset{\displaystyle {}^{+}NR_2}{\|}}{C}-\overset{\underset{\displaystyle R^1}{|}}{C}H- \quad X^-$$
$$R^1{-}X$$

$$\text{(c)} \quad -\overset{\overset{\displaystyle {}^{+}NR_2}{\|}}{C}-\overset{\underset{\displaystyle R^1}{|}}{C}H- \quad X^- \quad\overset{H_2O}{\longrightarrow}\quad -\overset{\overset{\displaystyle O}{\|}}{C}-\overset{\underset{\displaystyle R^1}{|}}{C}H-$$

Figure 9.22 Formation and some reactions of enamines

the conjugation within the molecule they can react with electrophiles at either the nitrogen atom or carbon atom 2. In reactions with alkyl halides alkylation at carbon atom 2 usually predominates (Figure 9.22(b)). Alkylated enamines can be readily hydrolysed (Figure 9.22(c)) and enamines are therefore valuable synthetic reagents because they afford an indirect method for alkylating the α-carbon atoms of aldehydes and ketones. The advantage of alkylation through the enamine, as compared to direct alkylation of the carbonyl compound, is that polyalkylation does not usually occur and the monoalkylated aldehyde or ketone is obtained in good yield.

Question 9.7 Draw the structure of the enamine formed by reacting cyclohexanone with a secondary amine (R_2NH) and give the structures of the products that would be formed by reacting the enamine with (a) CH_3I, (b) $ClCH_2OCH_3$, (c) $CH_2{=}CHCH_2Br$.

9.4 Compounds containing non-conjugated double bonds

When two double bonds are separated by two or more saturated carbon atoms they behave in most respects like the corresponding monofunctional compounds. Thus, for example, dicarbonyl compounds of this type undergo aldol and Claisen condensations as do simple carbonyl compounds. Sometimes cyclic products result (Section 9.6).

When two double bonds are separated by just one saturated carbon atom the hydrogen atoms attached to that carbon atom show enhanced reactivity. This is true in 1,4-dienes and is responsible for the hardening of highly unsaturated vegetable oils (e.g. linseed oil) into polymeric materials (Chapter 10). However the effect is much more marked in 1,3-dicarbonyl compounds, which are considerably more acidic than simple carbonyl compounds (Figure 9.23). The hydrogen atoms attached to the carbon atom between the two carbonyl groups are easily lost because a resonance-stabilized anion is thereby formed.

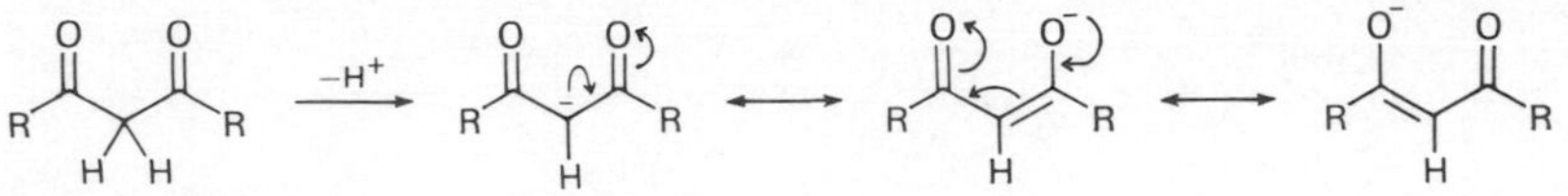

(R = Alkyl, aryl or alkoxyl)

Figure 9.23 The enhanced reactivity of hydrogen atoms at carbon atom 2 in 1,3-dicarbonyl compounds

Thus ethyl acetoacetate and diethyl malonate readily form carbanions when treated with a base (Figure 9.24). Such carbanions react as nucleophiles, for example with alkyl halides, giving rise to substituted 1,3-dicarbonyl compounds. Since these products still contain one acidic hydrogen atom, a second alkyl group can be introduced in a similar manner, leading to disubstituted compounds.

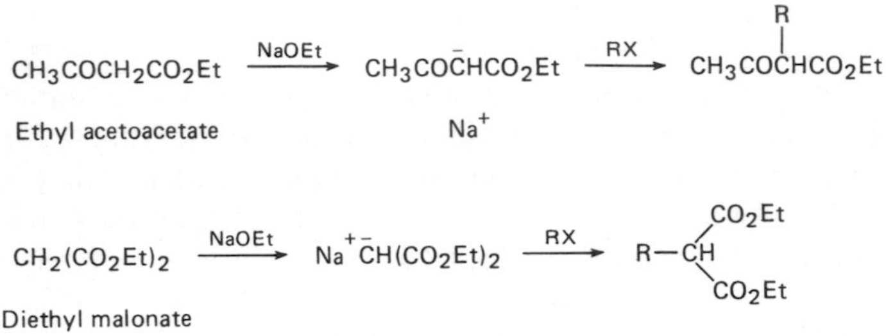

Ethyl acetoacetate

Diethyl malonate

Figure 9.24 The formation and reactions of carbanions derived from ethyl acetoacetate and diethyl malonate

Question 9.8 How could each of the following compounds be prepared starting from ethyl acetoacetate or diethyl malonate:

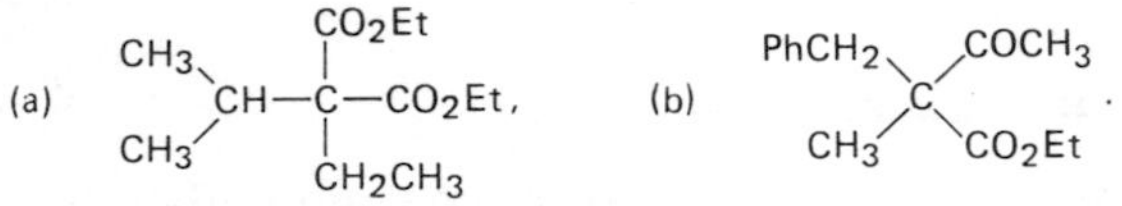

The carbanions generated from 1,3-dicarbonyl compounds also react with other electrophiles such as acyl chlorides, and $\alpha\beta$-unsaturated carbonyl compounds (Section 9.2(b)).

Substituted β-keto acids and malonic acids readily decarboxylate (i.e. lose CO_2) on heating because of the interaction of the two groups (Figure 9.25(a)). It is therefore a very simple matter to convert substituted β-keto esters and malonic esters into methyl ketones and monocarboxylic acids (Figure 9.25(b)).

A similar series of reactions is involved in the biosynthesis of polyketides and fatty acids (Figure 9.26). Condensation of malonyl coenzyme A with the carbonyl group of acetyl coenzyme A is the biochemical equivalent of acylation of a malonic ester. The product readily loses CO_2 to give acetoacetyl coenzyme A which can be reduced to a saturated carboxylic

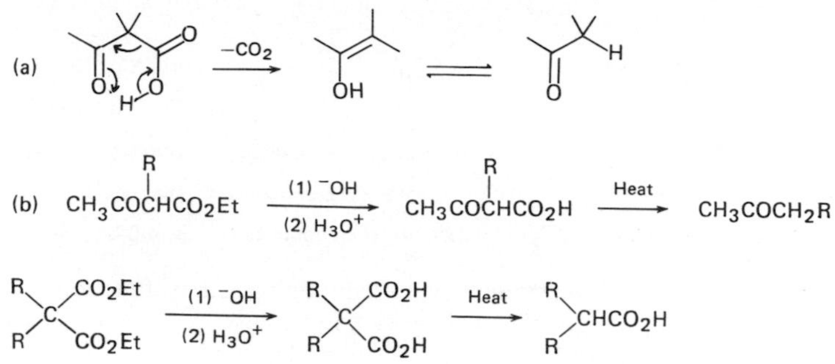

(a)

(b) $CH_3COCHCO_2Et$ (with R above) $\xrightarrow[(2) H_3O^+]{(1)\ ^-OH}$ $CH_3COCHCO_2H$ (with R above) $\xrightarrow{Heat}$ CH_3COCH_2R

Figure 9.25 (a) Decarboxylation reactions of β-keto acids and (b) the conversion of substituted β-keto esters and malonic esters into methyl ketones and monocarboxylic acids

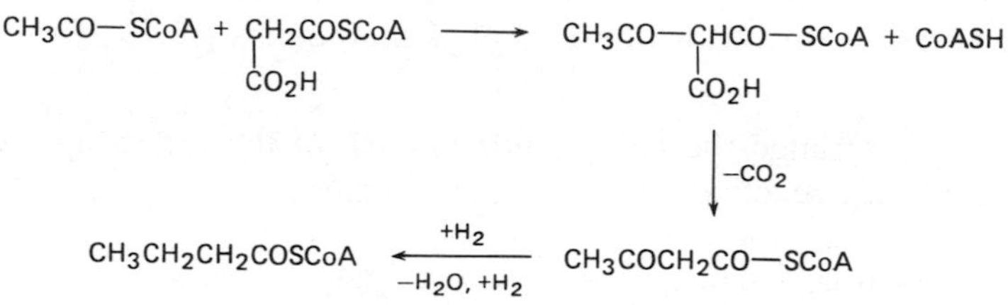

$CH_3CO-SCoA + CH_2COSCoA$ (with CO_2H below) $\longrightarrow$ $CH_3CO-CHCO-SCoA + CoASH$ (with CO_2H below)

$\downarrow -CO_2$

$CH_3CH_2CH_2COSCoA$ $\xleftarrow[-H_2O, +H_2]{+H_2}$ $CH_3COCH_2CO-SCoA$

Figure 9.26 The biosynthesis of polyketides and fatty acids

acid derivative or can condense with further malonyl coenzyme A units to give a polyketide chain (Chapter 15).

The decarboxylation of a β-keto acid also occurs in the citric acid cycle when oxalosuccinic acid is converted into α-ketoglutaric acid (Chapter 15).

It should be noted that 1,3-dicarbonyl compounds are extensively enolized (Section 8.3) and show many reactions characteristic of the enol tautomer. Thus ethyl acetoacetate not only forms derivatives with hydroxylamine and phenylhydrazine, reactions which are characteristic of a ketone, but also reacts rapidly with bromine, gives an intense colour with $FeCl_3$, and readily forms a methyl ether, reactions which are characteristic of enols. Under favourable circumstances the keto and enol tautomers of ethyl acetoacetate can actually be separated, though they rapidly revert to the equilibrium mixture.

9.5 Compounds containing other combinations of functional groups

Compounds containing isolated functional groups generally undergo reactions which are typical of each of the functional groups, although in some cases cyclic or polymeric products may result (Section 9.6 and 9.7). Some compound types containing functional groups in close proximity are shown in Figure 9.27.

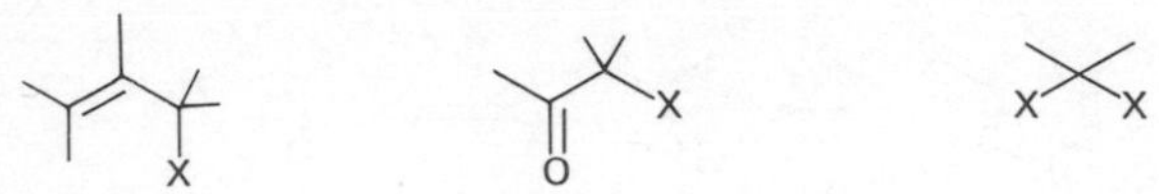

X = Cl	Allyl chlorides	α-Chlorocarbonyl compounds	gem-Dichlorides
X = OH	Allyl alcohols	α-Hydroxycarbonyl compounds	gem-Diols
X = NR$_2$	Allyl amines	α-Aminocarbonyl compounds	gem-Diamines

Figure 9.27 Some compound types containing other functional groups in close proximity

Allyl compounds are usually very reactive by comparison with the corresponding saturated compounds. Thus allyl chloride (3-chloroprop-1-ene), for example, is much more susceptible to nucleophilic substitution than 1-chloropropane. Benzyl compounds (e.g. PhCH$_2$Cl) also show many properties analogous to those of allyl compounds (see Section 6.3).

Question 9.9. Arrange the compounds in each of the following groups in order of increasing reactivity towards nucleophiles:

(a) CH_2=$CHCH_2CH_2Cl$, CH_2=$CHCHCH_3$, CH_2=CCH_2CH_3;
      (with Cl substituents shown below the CH groups)

(b) $PhCH_2CH_2Cl$, $PhCHCH_3$ (Cl below), and the o-(CH₂CH₃)(Cl) substituted benzene ring.

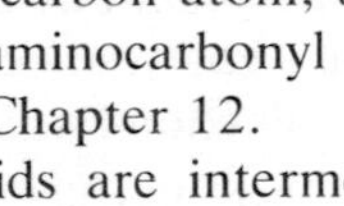

α-Halogenocarbonyl compounds also readily undergo nucleophilic substitution reactions. This is because the carbonyl group increases the electrophilicity of the α-carbon atom, thus promoting S$_N$2 attack. The only important group of α-aminocarbonyl compounds are the α-amino acids which are dealt with in Chapter 12.

Several α-hydroxy acids are intermediates in carbohydrate metabolism (Chapter 15) and carbohydrates themselves are derivatives of α-hydroxyaldehydes and ketones (Chapter 11). One characteristic property of α-hydroxyaldehydes and ketones is their ability to undergo isomerization, especially under basic conditions (Figure 9.28). Simple enolization of the keto tautomer gives an enediol which can retautomerize to give either of the two isomeric carbonyl compounds.

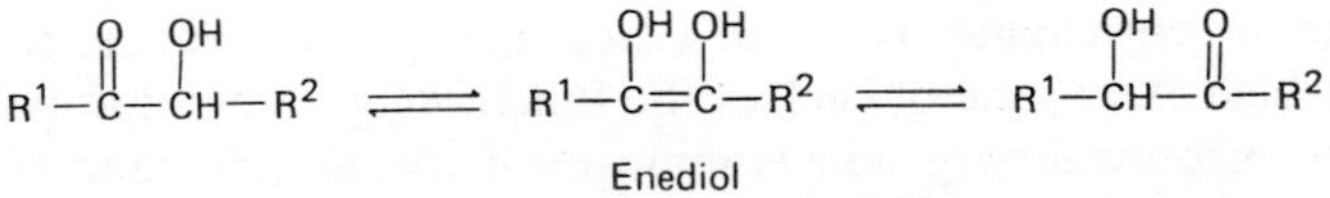

Figure 9.28 The isomerization of α-hydroxyaldehydes and ketones

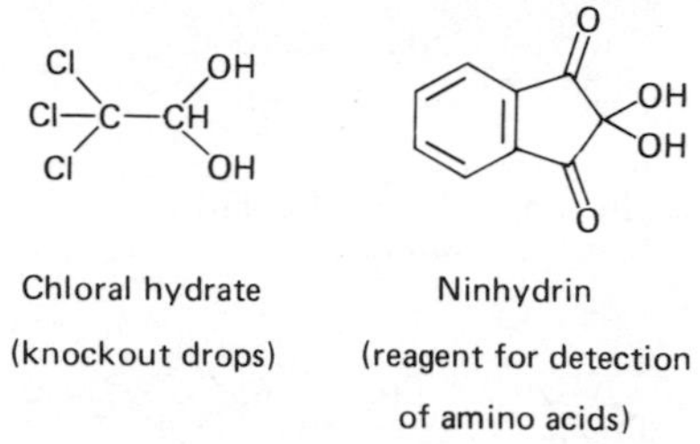
Cl
Cl—C—CH
Cl
OH
OH
O
OH
OH
O
Chloral hydrate
(knockout drops)
Ninhydrin
(reagent for detection
of amino acids)

Figure 9.29 Two *gem*-diols

Most *gem*-diols only exist in equilibrium in aqueous solutions of the corresponding carbonyl compounds (Section 8.4(a)). However, when the carbonyl group is attached to strongly electron-withdrawing groups the 'hydrate' is much more stable (Figure 9.29).

9.6 Cyclization of bifunctional molecules

Many reactions are known in which bifunctional molecules react intramolecularly to give cyclic products. Two widely used methods for ring formation involve the aldol and Claisen condensations, which are particularly effective for the synthesis of five-, six-, and seven-membered rings (e.g. Figure 9.30).

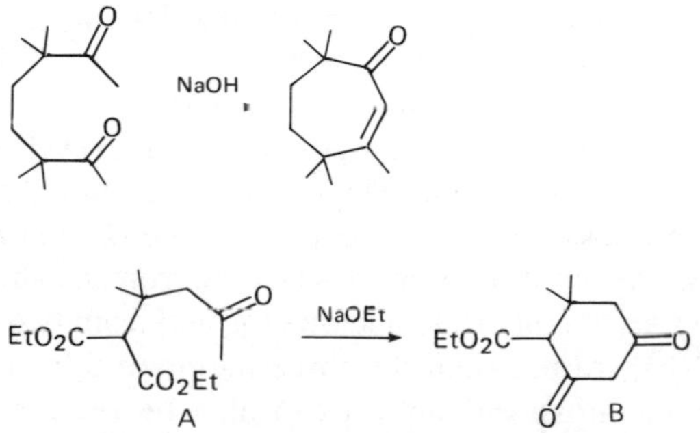
NaOH
NaOEt
EtO_2C
CO_2Et
A
EtO_2C
B

Figure 9.30 Cyclization reactions of bifunctional molecules

Question 9.10 (*a*) Propose a synthesis of compound A in Figure 9.30 (see Section 9.2(b)). (*b*) Give the structure of the product obtained by hydrolysis of the ester group in compound B, followed by heating (see Section 9.4).

Another important ring forming reaction involves the intramolecular alkylation of a 1,3-dicarbonyl compound. One example of such a reaction is shown in Figure 9.31.

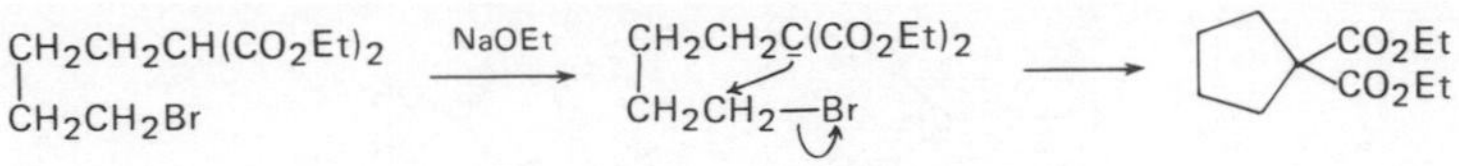

$$CH_2CH_2CH(CO_2Et)_2 \atop CH_2CH_2Br \quad \xrightarrow{NaOEt} \quad CH_2CH_2C(CO_2Et)_2 \atop CH_2CH_2-Br$$

Figure 9.31 The intramolecular alkylation of a 1,3-dicarbonyl compound

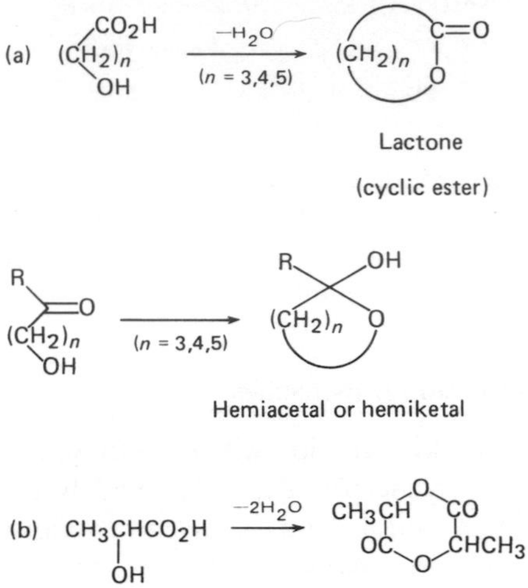

(a) $(CH_2)_n{\nearrow^{CO_2H}_{\searrow OH}}$ $\xrightarrow[(n = 3,4,5)]{-H_2O}$ $(CH_2)_n$ Lactone

Lactone

(cyclic ester)

$R{\searrow}_{(CH_2)_n{\searrow OH}}{=}O$ $\xrightarrow{(n = 3,4,5)}$ $R{\searrow OH \atop (CH_2)_n}$

Hemiacetal or hemiketal

(b) $CH_3CHCO_2H \atop OH$ $\xrightarrow{-2H_2O}$ $CH_3CH{\nwarrow O \atop OC}CO \atop \searrow O \atop CHCH_3$

Figure 9.32 (a) Cyclization of hydroxycarbonyl compounds, and (b) dehydration
of lactic acid

4-, 5-, and 6-hydroxycarbonyl compounds readily cyclize to give cyclic hemiacetals, hemiketals, or lactones (Figure 9.32(a)). Small ring hemiacetals, hemiketals, and lactones do not form very easily due to the ring strain present in such compounds. Attempts to prepare a lactone fron an α-hydroxy acid, such as lactic acid, give a six-membered dilactone formed from two molecules of the acid (Figure 9.32(b)), rather than the three-membered lactone.

Cyclic amides (lactams) and imines can also be prepared. Again three- and four-membered rings are strained and therefore difficult to prepare. Nevertheless the widely used penicillin and cephalosporin antibiotics contain a four-membered β-lactam ring (e.g. Figure 9.33).

The formation of a cyclic imine is involved in the biosynthesis of the amino acid proline from ornithine (Figure 9.34).

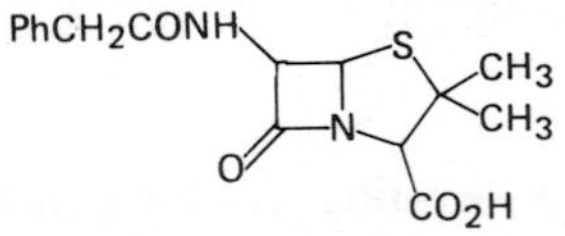

Figure 9.33 The structure of
penicillin G

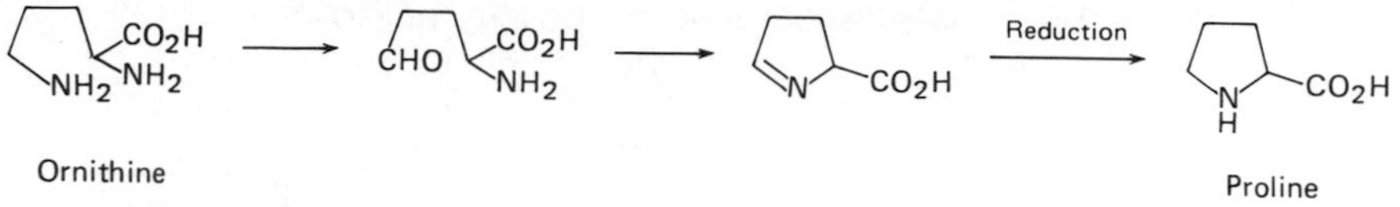

Figure 9.34 Biochemical cyclization of ornithine to form proline

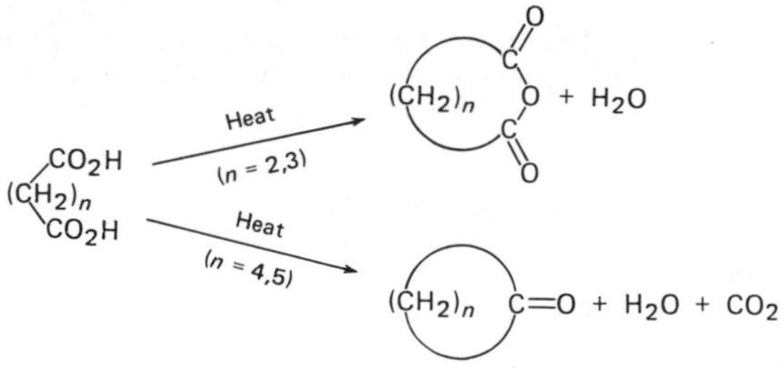

Figure 9.35 The formation of cyclic anhydrides or ketones from dicarboxylic acids

Dicarboxylic acids on heating yield cyclic anhydrides or ketones depending upon the size of the ring so formed (Figure 9.35). Once again the formation of five- and six-membered rings is favoured. The anhydrides are formed by elimination of water while an initial Claisen-type condensation, followed by elimination of CO_2 from the resulting β-keto acid, gives ketones.

Bifunctional molecules are widely used for the synthesis of heterocyclic compounds (Section 7.10(g)).

9.7 Polymers derived from bifunctional molecules

Polyesters are formed by condensation of dihydroxy compounds with dicarboxylic acids or esters. The most important polyester is Terylene® (Dacron®) which is a condensation product of dimethyl terephthalate and ethylene glycol (Figure 9.36).

If glycerol (propane-1,2,3-triol) is used instead of ethylene glycol, along with a dibasic acid such as phthalic acid, the thermosetting alkyd resins are produced. When heated, crosslinking occurs between residual carboxyl and hydroxyl groups producing a hard durable product (Figure 9.37).

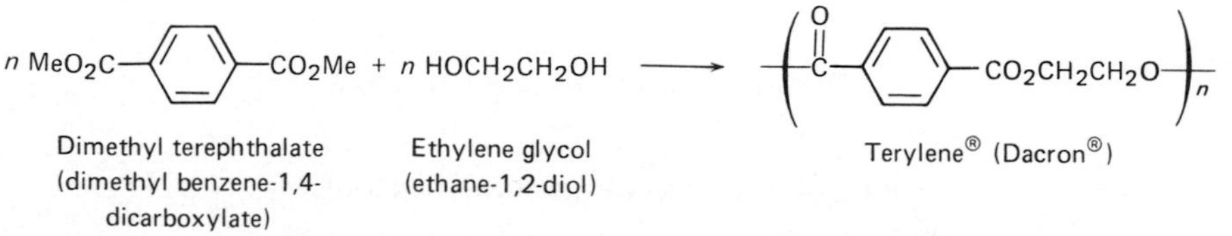

Dimethyl terephthalate (dimethyl benzene-1,4-dicarboxylate)   Ethylene glycol (ethane-1,2-diol)   Terylene® (Dacron®)

Figure 9.36 The synthesis of Terylene® (Dacron®) by condensation polymerization

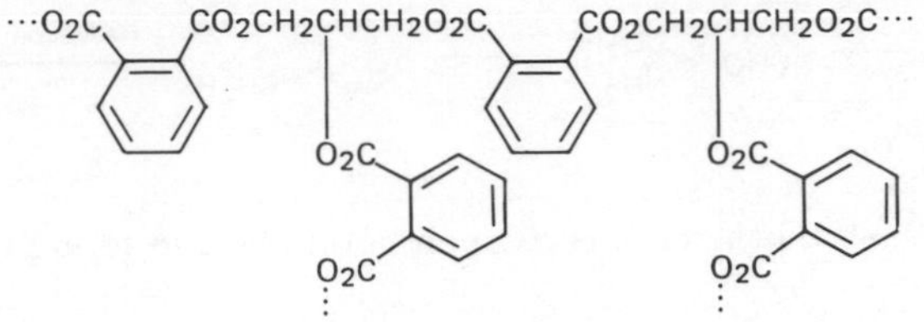

···O$_2$C   CO$_2$CH$_2$CHCH$_2$O$_2$C   CO$_2$CH$_2$CHCH$_2$O$_2$C···

O$_2$C   O$_2$C

O$_2$C   O$_2$C

Figure 9.37 The structure of a polymer formed by condensation polymerization of glycerol and phthalic acid

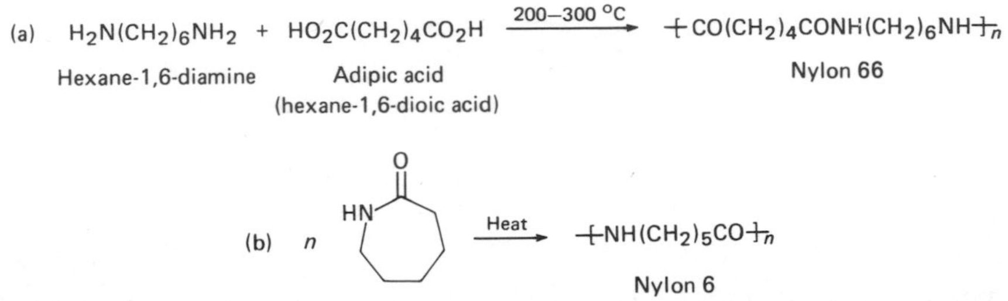

(a)   H$_2$N(CH$_2$)$_6$NH$_2$ + HO$_2$C(CH$_2$)$_4$CO$_2$H $\xrightarrow{200-300\ ^\circ C}$ $+$CO(CH$_2$)$_4$CONH(CH$_2$)$_6$NH$+_n$

Hexane-1,6-diamine   Adipic acid   Nylon 66
                  (hexane-1,6-dioic acid)

(b) n $\xrightarrow{\text{Heat}}$ $+$NH(CH$_2$)$_5$CO$+_n$

Nylon 6

Figure 9.38 The formation of Nylon 66 and Nylon 6

Polyurethanes are produced by reacting diols with bis-isocyanates (Section 9.1), in a process similar to the production of polyesters.

Nylon 66 is a polyamide formed by condensation of hexane-1,6-diamine and adipic acid (Figure 9.38(a)). A second polyamide, Nylon 6, is produced by polymerizing caprolactam (Figure 9.38(b)). Nylon is used in the manufacture of textiles, carpets, ropes, and moulded objects.

Another important group of polymers is derived from formaldehyde. The condensation of urea and formaldehyde gives a partially polymerized material which, when subjected to heat and pressure, crosslinks to give a hard infusible product (Figure 9.39).

In phenol–formaldehyde polymers (e.g. bakelite) phenol units are linked by CH$_2$ groups derived from formaldehyde (Figure 9.40).

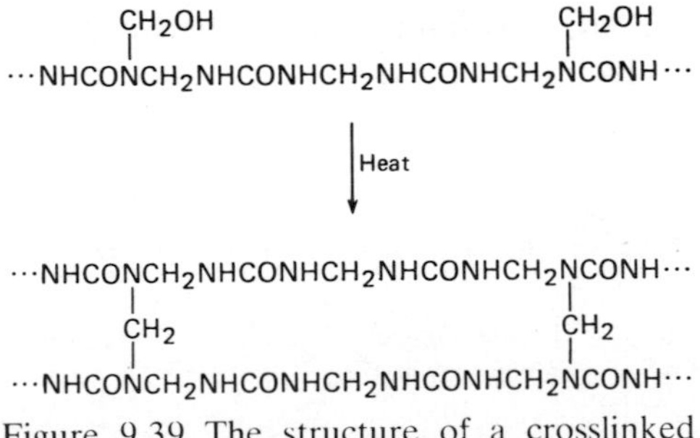

CH$_2$OH   CH$_2$OH
···NHCONCH$_2$NHCONHCH$_2$NHCONHCH$_2$NCONH···

Heat

···NHCONCH$_2$NHCONHCH$_2$NHCONHCH$_2$NCONH···
        CH$_2$ CH$_2$
···NHCONCH$_2$NHCONHCH$_2$NHCONHCH$_2$NCONH···

Figure 9.39 The structure of a crosslinked urea–formaldehyde condensation polymer

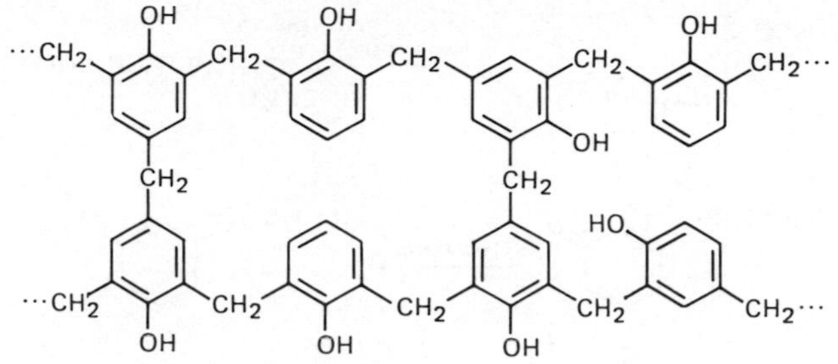

Figure 9.40 The structure of bakelite

Answers to questions

9.1

(a)

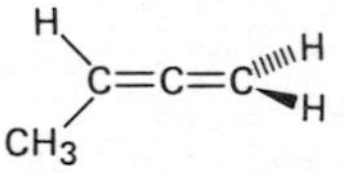

is superimposable on its mirror image and hence *achiral*

(b)

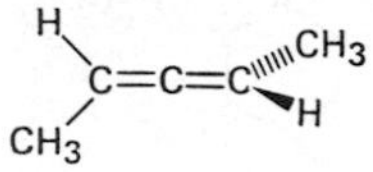

is not superimposable on its mirror image and is therefore *chiral*

(c)

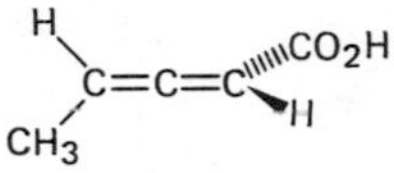

chiral

9.2

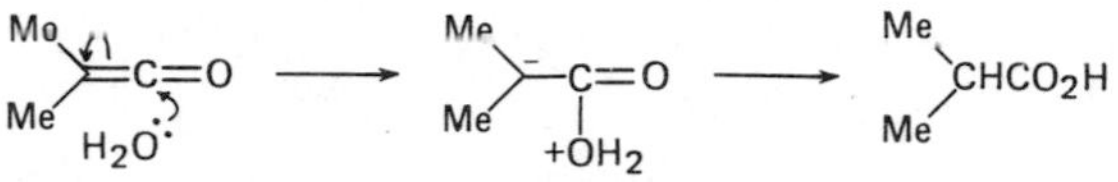

9.3

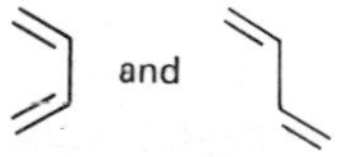

and

9.4

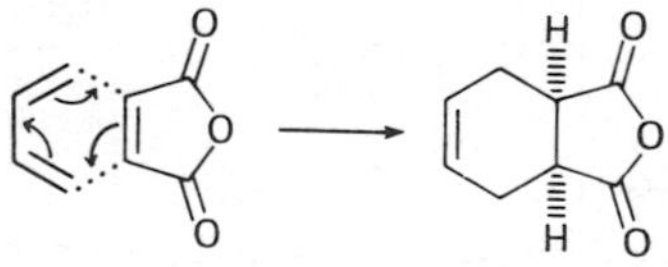

9.5 (a)

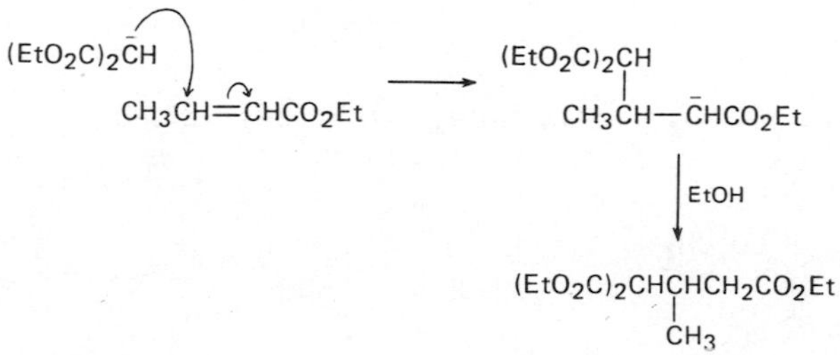

(b)

$$CH_3CO\!-\!CH(EtO_2C)\quad CH_2\!=\!CHCO_2Et \xrightarrow{\text{(Two steps)}} CH_3CO\!-\!C(EtO_2C)\!-\!CHCH_2CH_2CO_2Et$$

(c)

$$(EtO_2C)_2\bar{C}H \quad \xrightarrow{\text{(Two steps)}} (EtO_2C)_2CH\text{—}$$

9.6

$$\xrightarrow{\text{NaOEt}} \quad \xleftarrow{\begin{array}{c}CO_2Et\\CO_2Et\end{array}} \quad \xrightarrow{\begin{array}{c}O^-\ CO_2Et\\OEt\end{array}} \quad \longrightarrow \quad COCO_2Et$$

9.7

$$\xrightarrow{R_2NH} \quad \overset{R_2N\ \ OH}{\rule{0pt}{0pt}} \quad \rightleftharpoons \quad \overset{NR_2}{\rule{0pt}{0pt}} \quad \xrightarrow{CH_3I} \quad \overset{{}^+NR_2\ \ I^-}{\underset{CH_3}{\rule{0pt}{0pt}}}$$

$$\xrightarrow{ClCH_2OCH_3} \quad \overset{{}^+NR_2\ \ Cl^-}{\underset{CH_2OCH_3}{\rule{0pt}{0pt}}}$$

$$\xrightarrow{CH_2=CHCH_2Br} \quad \overset{{}^+NR_2}{\underset{CH_2CH=CH_2\ \ Br^-}{\rule{0pt}{0pt}}}$$

9.8 (a)

$$CH_2(CO_2Et)_2 \xrightarrow{\text{NaOEt}} \bar{C}H(CO_2Et)_2 \xrightarrow{\text{EtI}} EtCH(CO_2Et)_2$$

$$\downarrow \text{NaOEt}$$

$$EtC(CO_2Et)_2 \xleftarrow{\text{Me}_2CHI} Et\bar{C}(CO_2Et)_2$$
$$|$$
$$\dot{C}HMe_2$$

(b)

$$CH_3COCH_2CO_2Et \xrightarrow{\text{NaOEt}} CH_3CO\bar{C}HCO_2Et \xrightarrow{\text{PhCH}_2Br} CH_3COCHCO_2Et$$
$$|$$
$$CH_2Ph$$

$$\downarrow \text{NaOEt}$$

$$\overset{CH_3}{\underset{CH_2Ph}{CH_3CO\!-\!\overset{|}{C}\!-\!CO_2Et}} \xleftarrow{\text{CH}_3I} CH_3CO\bar{C}CO_2Et$$
$$|$$
$$CH_2Ph$$

9.9 (a)

$$CH_2\!=\!\underset{Cl}{CCH_2CH_3} < CH_2\!=\!CHCH_2CH_2Cl < CH_2\!=\!CH\underset{Cl}{CHCH_3}$$

(b)

$$\underset{Cl}{\overset{CH_2CH_3}{\rule{0pt}{0pt}}} < PhCH_2CH_2Cl < Ph\underset{Cl}{CHCH_3}$$

195

9.10

(a)

(b)

CHAPTER 10

Fats, Oils, Waxes, and Detergents

Lipids are compounds which are extracted from plant or animal tissues with relatively non-polar solvents. Some of the compounds of lipid mixtures are hydrolysed by aqueous alkalis, a process known as saponification. These are referred to as *saponifiable* lipids and are the subject of this chapter. *Non-saponifiable* components of lipid mixtures include steroids, a class to which several sex hormones belong (Chapter 14), and terpenes (Chapter 15).

The most important categories of saponifiable lipids are the simple fats and oils, which are triesters of glycerol (triglycerides), and are hydrolysed by aqueous alkali to give glycerol and sodium salts of long-chain carboxylic acids (fatty acids, Figure 10.1).

Question 10.1 Glycerol also forms triesters with inorganic acids. The triester of nitric acid is known by the trivial name 'nitroglycerine'. Suggest a structure for this compound.

Animal fats and *vegetable oils* are foodstuffs of high calorific value. Excess food intake is largely converted into fat for storage. In addition to their role as foodstuffs, fats and oils also find application in a variety of important industries such as the paint, margarine, and soap industries.

Other classes of saponifiable lipids include *waxes,* which are esters of long-chain acids with long-chain alcohols (Section 10.6), *sphingolipids,* which are found in brain tissue, and *phospholipids,* which are found in nerve tissue. Representative structures are shown in Figure 10.2.

$$
\underset{\text{A triglyceride}}{
\begin{array}{c}
\overset{\displaystyle O}{\underset{\displaystyle \|}{}}\\
CH_2OCR^1\\
|\quad \overset{\displaystyle O}{\underset{\displaystyle \|}{}}\\
CHOCR^2\\
|\quad \overset{\displaystyle O}{\underset{\displaystyle \|}{}}\\
CH_2OCR^3
\end{array}
}
\xrightarrow{\text{NaOH}}
\underset{\text{Glycerol}}{
\begin{array}{c}
CH_2OH\\
|\\
CHOH\\
|\\
CH_2OH
\end{array}
}
+
\underset{\text{Fatty acid salts}}{
\begin{array}{c}
R^1CO_2^-\ Na^+\\
R^2CO_2^-\ Na^+\\
R^3CO_2^-\ Na^+
\end{array}
}
$$

Figure 10.1 Hydrolysis of a triglyceride

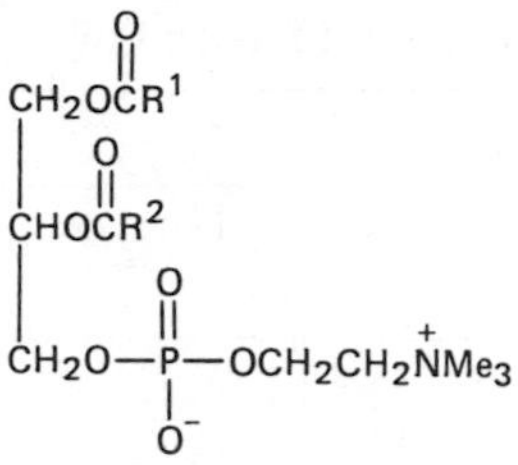

$$CH_2OCR^1$$

$$CHOCR^2$$

$$CH_2O-P-OCH_2CH_2\overset{+}{N}Me_3$$

$$CH_3(CH_2)_{12}CH=CHCHOH$$

$$CHNHCR$$

$$CH_2O-P-OCH_2CH_2\overset{+}{N}Me_3$$

A lecithin
(one type of phospholipid)

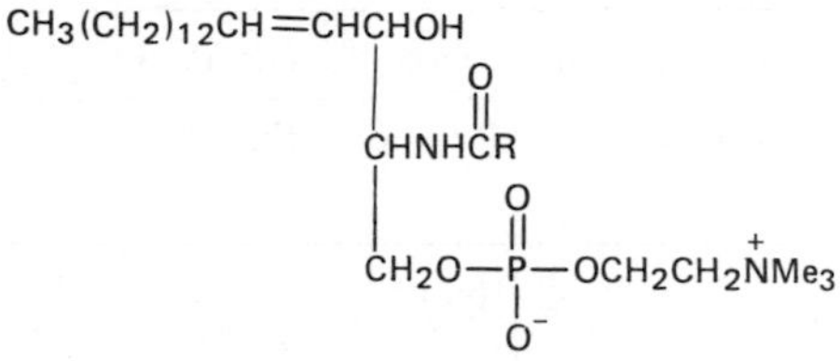

A sphingomyelin
(one type of sphingolipid)

Figure 10.2 Examples of phospholipids and sphingolipids

10.1 Occurrence and composition of fats and oils

Fats and oils are widely distributed in nature. The vegetable products are generally liquids (i.e. oils) and can usually be obtained by pressing the seeds or fruits where they occur in greatest abundance. Examples include olive, coconut, linseed, sunflower, corn, cottonseed, and rapeseed oils. Some useful oils can also be obtained from marine animals (e.g. cod liver oil, whale oil). Animal fats are generally solids at ordinary temperatures and are therefore obtained by heating the meat to liberate the molten fat. Examples include bacon fat, lard, beef tallow, and butter, although the last named is obtained by churning milk rather then by heating. Two of the major functions of animal fats are to protect vital organs and to insulate the body against the loss of heat.

The carboxylic acids obtained by hydrolysis of triglycerides typically contain 14, 16, or 18 carbon atoms, and the more common examples are shown in Table 10.1. The double bonds all have the *cis* configuration. The absence of carboxylic acids containing odd numbers of carbon atoms is a consequence of their biosynthesis (Chapter 15).

The relative proportions of the various fatty acid residues in a mixture of triglycerides has a significant effect on the physical and chemical properties of the mixture. The fatty acid compositions of some typical natural fats and oils are indicated in Table 10.2. Such compositions are determined by

Table 10.1 Common fatty acids

Usual name	Systematic name	Formula
Myristic	Tetradecanoic	$CH_3(CH_2)_{12}CO_2H$
Palmitic	Hexadecanoic	$CH_3(CH_2)_{14}CO_2H$
Stearic	Octadecanoic	$CH_3(CH_2)_{16}CO_2H$
Oleic	Octadec-9-enoic	$CH_3(CH_2)_7CH=CH(CH_2)_7CO_2H$
Linoeic	Octadeca-9,12-dienoic	$CH_3(CH_2)_3(CH_2CH=CH)_2(CH_2)_7CO_2H$
Linolenic	Octadeca-9,12,15-trienoic	$CH_3(CH_2CH=CH)_3(CH_2)_7CO_2H$

Table 10.2 Typical fatty acid compositions of some natural fats and oils

Fat or oil	Constituent carboxylic acid (% by weight)						
	Lower than C_{14}	Myristic	Palmitic	Stearic	Oleic	Linoleic	Linolenic
Butter	13	10	25	12	35	5	0
Lard	1	2	30	15	45	7	0
Human fat	8	3	25	8	46	10	0
Olive oil	0	0	10	2	84	4	0
Corn oil	0	2	10	3	34	51	0
Linseed oil	0	0	5	3	5	62	25
Cod liver oil*	0	4	11	1	28		

*Also contains large quantities of other unsaturated acids (C_{20} and above) not listed in the table.

hydrolysis of the fat or oil, conversion of the acids into their methyl esters, and gas chromatography (Chapter 3) of this mixture.

It can be seen from Table 10.2 that butter and human fat contain substantial quantities of short-chain fatty acids, containing as few as four carbon atoms, whereas most vegetable oils contain few short-chain acids. Since the very short-chain acids, especially butanoic (C_4), have extremely unpleasant odours, it is understandable that any hydrolysis of fats or oils containing these acid residues leads to an unpleasant smell. This accounts, at least in part, for the unpleasant odours associated with rancid butter and human perspiration.

Table 10.2 also shows that vegetable oils contain a higher proportion of unsaturated acid residues than do animal fats. Indeed, this is the very reason that vegetable oils are liquids and animal fats solids at ordinary temperatures. Increasing unsaturation causes a lowering of melting point, because the *cis* double bonds create 'kinks' in the hydrocarbon chains which disturb the packing arrangements of the molecules.

The presence of double bonds in the molecules also renders them liable to light-initiated oxidation (by air), producing shorter-chain aldehydes and carboxylic acids. Since these again are odorous compounds, such oxidation is another cause of rancidity in fats and oils. Oxidation can be minimized by storing the products in a cool, dark place, by tight wrapping, and by addition of chemical antioxidants.

10.2 Analysis of fats and oils

Rather than do a complete fatty acid analysis by gas chromatography (Section 10.1), it is often more convenient to get information about the degree of unsaturation and average chain length by titration methods.

Iodine adds to carbon–carbon double bonds in the same way as bromine (Chapter 4). The weight in grams of iodine taken up by 100 g of a fat is described as its *iodine value* or *iodine number*. The higher the iodine value, the greater the degree of unsaturation. Experimentally, a known weight of fat is allowed to react with a known weight (excess) of iodine, and unreacted iodine is then titrated against standard sodium thiosulphate solution.

The weight in milligrams of KOH necessary to neutralize 1 g of fat after complete hydrolysis is described as its *saponification value* or *saponification number*. A high proportion of short-chain acid residues results in a lower average molecular mass, and therefore more acid residues per unit mass. Thus, the higher the saponification value, the greater the proportion of short-chain acid residues.

Some typical iodine values and saponification values are recorded in Table 10.3.

Table 10.3

Fat or oil	Iodine value	Saponification value
Butter	27	220
Lard	53	200
Olive oil	85	192
Corn oil	121	192
Linseed oil	190	192

Question 10.2 Glyceryl monostearate (GMS, formula below) is a synthetic food additive useful for its emulsification properties. Calculate the theoretical iodine and saponification values for pure GMS. (Atomic masses: H=1, C=12, O=16, K=39).

$$CH_2OH$$
$$|$$
$$CHOH$$
$$|$$
$$CH_2OCO(CH_2)_{16}CH_3$$

Glyceryl monostearate

10.3 Hardening of vegetable oils; margarine

The double bonds in the carboxylic acid residues of a vegetable oil can be hydrogenated over a catalyst as can double bonds in simple alkenes

(Chapter 4). If this is done, the melting point of the oil increases and the oil eventually becomes a fat. This is the basis of the margarine industry, and the process is referred to as *hardening* of vegetable oils. Hardening can be carried out to differing extents. If relatively high proportions of unsaturated acid residues remain a 'soft' margarine is obtained. Further hydrogenation leads to 'hard' margarines, though the process is never carried to completion otherwise the product would be hard and brittle.

From a dietary viewpoint margarine differs from butter in several respects, although there is no substantial difference in calorific value. Since margarine is made from vegetable oils containing negligible proportions of short-chain carboxylic acids, little odour develops on hydrolysis. The storage properties are further improved by addition of antioxidants to reduce oxidation by air. Furthermore, whereas butter contains cholesterol (a steroid alcohol) and a whole range of volatile flavour-giving substances, margarine contains very little of these compounds. Since cholesterol is implicated in obstructing the flow of blood in the arteries, people suffering from arterial diseases are recommended to eat margarine.

Another difference between some types of margarine and butter is the proportion of unsaturated compounds present. In recent years, a number of soft margarines have been developed. These can be produced in two ways. One is simply to emulsify a hard margarine with water. Alternatively, a partially hardened margarine is produced. In the latter case, the proportion of unsaturated and 'polyunsaturated' acid residues is inevitably higher. This may be advantageous since polyunsaturated residues are believed to help the breakdown of fatty arterial deposits and to hinder their formation.

10.4 Drying oils; paints and lacquers

In polyunsaturated fatty acids such as linoleic and linolenic acids, CH_2 groups situated between two double bonds are very reactive because they can produce stabilized intermediates by loss of a hydrogen atom (Section 9.4). Oils (e.g. linseed oil) which contain a high proportion of such acid residues react with air to produce hydroperoxides and subsequently crosslinked peroxides (Figure 10.3). If spread thinly on a surface, such *drying oils* set to a hard film.

Question 10.3 Draw out resonance structures to show the stabilization of the radical formed by removal of a hydrogen atom ($H^{\cdot}$) from the doubly allylic CH_2 group of methyl linoleate.

When linseed oil is mixed with a pigment a paint is obtained. Usually a free radical catalyst is added to accelerate the 'drying' process. At one time paints of this type formed the basis of almost the entire surface coatings industry. Nowadays, however, they have been largely superseded by other

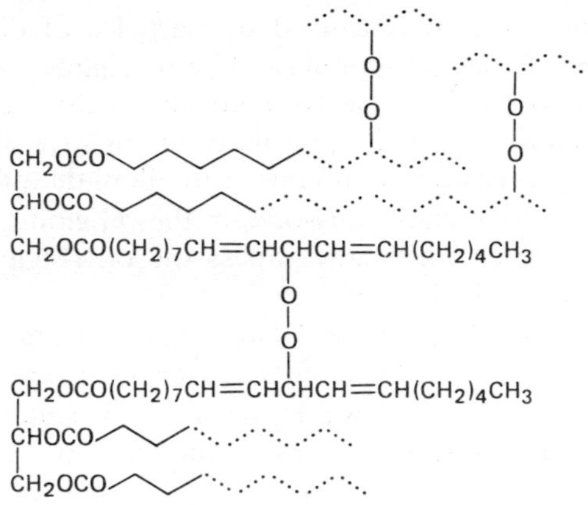

CH$_2$OCO

CHOCO

CH$_2$OCO(CH$_2$)$_7$CH=CHCHCH=CH(CH$_2$)$_4$CH$_3$

CH$_2$OCO(CH$_2$)$_7$CH=CHCHCH=CH(CH$_2$)$_4$CH$_3$

CHOCO

CH$_2$OCO

Figure 10.3 Crosslinking of a polyunsaturated triglyceride

materials such as polyurethanes (Section 9.7), alkyds (Section 9.7), silicones, and 'vinyl' polymers (Chapter 4).

10.5 Soaps and detergents

The products of hydrolysis of fats and oils under alkaline conditions are alkali-metal salts of long-chain carboxylic acids, which are called soaps (Figure 10.1). Commercial soaps are mixed with perfumes and other additives to enhance their attractiveness.

Soap molecules contain a very polar carboxylate group which has a tendency to dissolve in water (i.e. which is *hydrophilic*), and a very non-polar, hydrocarbon chain which has no affinity for an aqueous environment (i.e. which is *hydrophobic*). The outcome of this is that soap molecules tend to group together in water to form aggregates (micelles) which have all of the hydrocarbon chains in the centre of the aggregate, and the hydrophilic carboxylate groups on the outside, forming the interface with water (Figure 10.4(a)).

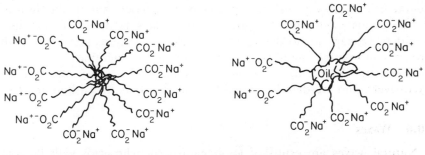

(a) Soap micelle

(b) Oil – soap complex micelle

Figure 10.4 Symbolic representations of soap and oil–soap micelles

Soiling of clothing or skin is caused by particles of dirt which adhere through the agency of tiny oil globules. Any mechanism of dissolving the oil loosens the dirt so that it is easily removed. In dry cleaning processes, solvents such as halogenoalkanes are used to dissolve the oil. In more traditional washing procedures soap is used to allow the oil to be dispersed in water. The hydrocarbon chains dissolve in the oil globules to form water soluble, modified micelles which encompass the oil (Figure 10.4(b)), thus allowing the oil to be washed away.

Soaps are not very effective in 'hard' water, from which they are precipitated as their Mg^{2+} or Ca^{2+} salts, which form an unsightly 'scum'. Also, since soaps are salts of weak acids, their solutions are slightly alkaline. These disadvantages can be overcome by use of synthetic detergents, which are also cheaper to produce. A few examples are listed in Table 10.4. All have a common feature—a long hydrocarbon 'tail' and a polar 'head'. Detergents may be 'anionic', 'cationic', or 'non-ionic' according to the formal charge on the head group. Anionic detergents (including soap) are by far the most important category, although non-ionic detergents now account for a substantial part of the detergent market.

Table 10.4 Some typical detergents

Type	Structure
Sodium alkylsulphate	$CH_3(CH_2)_nCH_2OSO_3^-\ Na^+$
Sodium alkylbenzenesulphonate	$CH_3(CH_2)_nCH_2$—⟨benzene⟩—$SO_3^-\ Na^+$
Alkyltrimethylammonium bromide	$CH_3(CH_2)_nCH_2N^+Me_3\ Br^-$
Alkylpolyethyleneglycol	$CH_3(CH_2)_nCH_2O(CH_2CH_2O)_nCH_2CH_2OH$

Some of the early synthetic detergents contained branched-chain alkyl residues and were not easily biodegradable. These accumulated and presented a considerable threat to marine life. Most of the detergents used nowadays are biodegradable, although alkylbenzenesulphonates are degraded only slowly and will probably be phased out in the future. A more serious environmental problem nowadays is posed by the use of 'builders' (cheap aids to detergent performance) such as phosphates, which are present in large amounts in many detergent formulations.

Non-ionic detergents have the advantage that they do not foam excessively, and are therefore ideal for use in automatic washing machines.

10.6 Waxes

Natural waxes are esters of long-chain carboxylic acids with long-chain alcohols. Like fats and oils they are often mixtures. They are found in such diverse places as the head of the sperm whale (spermaceti wax), bees' honeycombs (beeswax), the coatings of certain palm leaves (carnauba wax), and the coating of wool (lanolin).

In general, waxes contain largely saturated alkyl residues. The uses to which they are put depend largely upon their melting points. For example, carnauba wax is predominantly $CH_3(CH_2)_{24}CO_2(CH_2)_{29}CH_3$ and has a high melting point, 80–90 °C. It is used where a rigid, hard coating is required, as in car and floor polishes. Beeswax, mainly $CH_3(CH_2)_{12}CO_2(CH_2)_{25}CH_3$, melts at 60–65 °C and is used where a rather less brittle but nevertheless permanent wax is required, such as in high-grade candles and in shoe polishes. Spermaceti wax is low melting (40–50 °C) and contains largely $CH_3(CH_2)_{14}CO_2(CH_2)_{15}CH_3$. It is used as a base for cosmetics, since it is supple at body temperature. Lanolin is another soft wax and is used in cosmetics and in ointments and lotions.

Notice that both the acid and alcohol residues in waxes contain even numbers of carbon atoms, and that the melting points of the waxes increase with increasing chain length.

Alternatives to natural waxes are now used for many purposes. Thus, ordinary candle wax is a hydrocarbon fraction from crude oil, while silicone waxes are often used in polishes.

Answers to Questions

10.1 CH_2ONO_2
$|$
$CHONO_2$
$|$
CH_2ONO_2

10.2   Iodine value = 0, because no double bonds present; saponification value =156; the saponification value is found as follows: molecular mass of GMS=358, thus 358 mg GMS would require 56 mg KOH and hence 1000 mg GMS require $\dfrac{1000}{358} \times 56$ mg KOH $\simeq 156$ mg.

10.3
$CH_3(CH_2)_4CH{=}CH{-}CH{-}CH{=}CH(CH_2)_7CO_2Me$

$CH_3(CH_2)_4CH{-}CH{=}CH{-}CH{=}CH(CH_2)_7CO_2Me$

$CH_3(CH_2)_4CH{=}CH{-}CH{=}CH{-}CH(CH_2)_7CO_2Me$

CHAPTER 11

Carbohydrates

The carbohydrates constitute a major class of naturally occurring organic compounds. They are essential to the maintenance of plant and animal life, and also provide raw materials for many industries such as the textile, paper, fermentation, and food industries. Carbohydrates are also of considerable chemical interest and their study has aided our understanding of topics such as stereochemistry.

The term *carbohydrates* originated because many compounds in this class (e.g. glucose, $C_6H_{12}O_6$, sucrose, $C_{12}H_{22}O_{11}$, and starch, $(C_6H_{10}O_5)_n$) have the empirical formula $C_x(H_2O)_y$ and were formally regarded as 'hydrates of carbon'. This definition is, however, most inappropriate and would in any case exclude many important compounds (e.g. deoxyribose, $C_5H_{10}O_4$, and 2-amino-2-deoxyglucose, $C_6H_{13}NO_5$). Nevertheless the term carbohydrate remains in general use. It is applied to polyhydroxy compounds which reduce Fehling's solution, either before or after hydrolysis with mineral acids.

In green plants carbohydrates are formed by *photosynthesis* (Chapter 13). In this process the energy from sunlight is absorbed with the aid of the plant pigment chlorophyll and used to convert carbon dioxide and water into carbohydrates and oxygen (Figure 11.1). Carbohydrates are a great storehouse of energy and photosynthesis is the process by which all life on this planet is sustained.

$$x \, CO_2 + y \, H_2O \xrightarrow[\text{Chlorophyll}]{h\nu} C_x(H_2O)_y + x \, O_2$$

Figure 11.1 The overall photosynthetic reaction

Carbohydrates can be conveniently divided into three groups, the *monosaccharides* such as glucose, xylose, and fructose, the *oligosaccharides* such as maltose and lactose, formed by the union of a small number (2–10) of monosaccharide units, and the polysaccharides such as starch and cellulose, in which a large number of monosaccharide units are combined. The term *sugar* is sometimes used to describe carbohydrates containing a small number (1, 2, or 3) of monosaccharide units.

11.1 Monosaccharides

Monosaccharides are generally colourless crystalline solids with a sweet taste. They are soluble in water, giving optically active solutions which reduce Fehling's solution and Tollens' reagent (Section 11.3(a)). They caramelize on heating, char in the presence of concentrated sulphuric acid, and suffer extensive decomposition on heating with concentrated alkalis.

Monosaccharides may be regarded as polyhydroxy aldehydes or ketones and can be classified as *aldoses* or *ketoses* depending upon whether they contain an aldehyde or ketone group. They can also be classified according to the number of carbon atoms they contain, a *pentose,* for example, being a monosaccharide containing five carbon atoms, and a *hexose* one containing six carbon atoms. The two classifications are often combined. Thus glucose is an *aldohexose* (i.e. six carbon atoms and an aldehyde group) while fructose is a *ketohexose* (i.e. six carbon atoms and a ketone group). They can be represented as shown in Figure 11.2.

Similar representations of the D-enantiomers of the simple aldoses containing between three and six carbon atoms are shown in Figure 11.3 using Fischer projection formulas. The prefix D (Chapter 2) indicates that the configuration at the asymmetric centre furthest from the aldehyde group is the same as that in D-(+)-glyceraldehyde. The naturally occurring forms of all the common monosaccharides belong to the D- series.

Question 11.1 Draw Fischer projection formulas of L-arabinose and L-gulose.

Since aldehydes react with alcohols, aldoses do not in fact exist as polyhydroxyaldehydes but as cyclic hemiacetals. Thus D-glucose cyclizes to form two cyclic hemiacetals called α- and β-D-glucose (Figure 11.4(a)). In water 99.8% of glucose molecules are in one or other of the cyclic forms and only 0.2% in the open-chain form.

The two cyclic hemiacetals of D-glucose differ only in their configuration at carbon atom 1, the *anomeric carbon atom*, and are called anomers. The

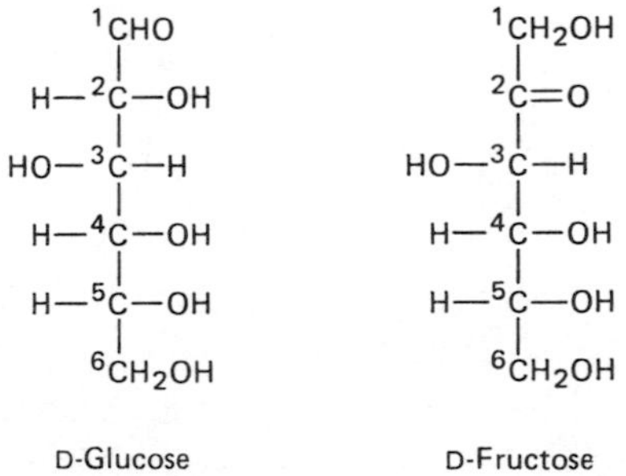

D-Glucose   D-Fructose

Figure 11.2 An aldohexose and a ketohexose

206

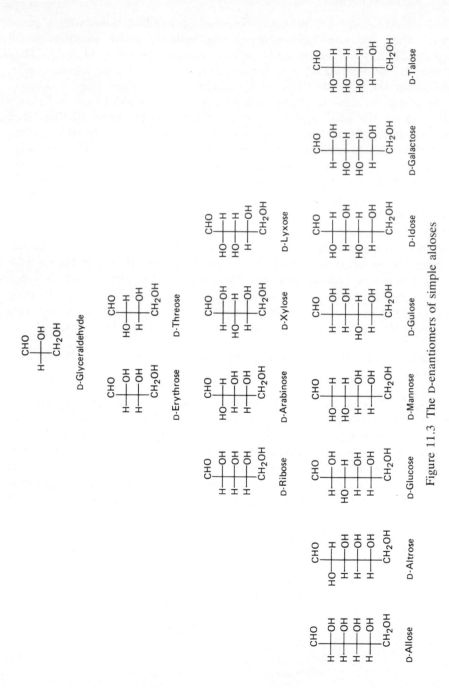

Figure 11.3 The D-enantiomers of simple aldoses

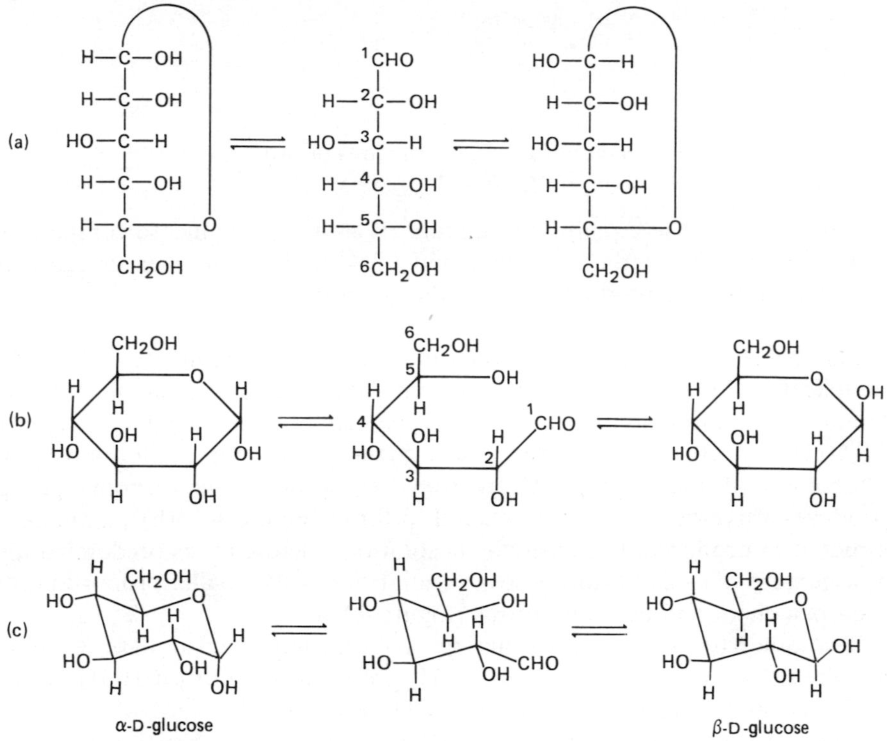

Figure 11.4 The isomeric structures of α- and β-D-glucose

various isomeric forms of D-glucose can be more conveniently represented by *Haworth formulas* (Figure 11.4(b)), although even these are only poor representations of the the actual shapes of the molecules since the six-membered rings are in fact puckered (Figure 11.4(c)), as in cyclohexane (Chapter 2).

Question 11.2 Using the Cahn–Ingold–Prelog rules (Section 2.9) work out the configuration (R or S) of the anomeric carbon atoms in α- and β-D-glucose.

In the solid state the two anomers of D-glucose are stable and they can be separated from each other. However, in aqueous solution each anomer is in equlilbrium with the open-chain form and hence the two anomers are in equilibrium with each other. α-D-Glucose has a specific rotation of $+113°$ while β-D-glucose has a specific rotation of $+19°$. In aqueous solution either pure anomer is slowly converted into an equilibrium mixture of the two, having a specific rotation of $+52.5°$. The conversion of pure α- or β-D-glucose into an equilibrium mixture may thus be followed by observing the change of specific rotation. The change of specific rotation with time which accompanies such a process is called *mutarotation*.

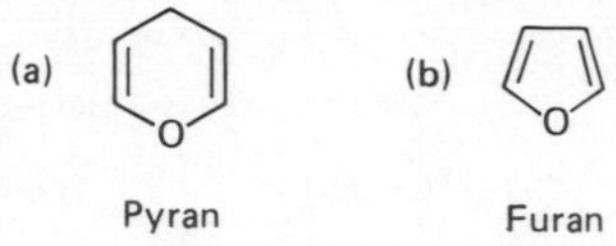

(a) Pyran (b) Furan

Figure 11.5 The structures of two
oxygen heterocycles

Question 11.3 The equilibrium mixture of α- and β-D-glucose contains, at
equilibrium, 37% of the α-anomer and 63% of the β-anomer. Suggest an
explanation for the predominance of the β-anomer.

As we have seen, monosaccharides exist predominantly as cyclic
hemiacetals. In glucose the oxygen-containing ring is six-membered and
bears a structural relationship to the oxygen heterocycle, pyran (Figure
11.5(a)). It therefore is referred to as a *pyranose* structure. Alternatively, if
interaction of the appropriate hydroxyl group with the carbonyl group
produces a five-membered ring related to furan (Figure 11.5(b)), a *furanose*
structure is produced. For example, in solution fructose exists predominantly
as a mixture of α- and β-furanoses in equilibrium with smaller proportions of
the open-chain and pyranose forms (Figure 11.6).

Two other types of monosaccharides are very important in nature. These
are the deoxy sugars, which have a CH_2 group in place of a CHOH group,
and the amino sugars, which have a $CHNH_2$ group in place of a CHOH
group. 2-Deoxyribose (Figure 11.7(a)) is an important constituent of the
nucleic acids (Chapter 12) and 2-amino-2-deoxyglucose (Figure 11.7(b)) is
the basic monosaccharide unit present in the polysaccharide chitin, of
which the shells of crabs and other crustacea are composed.

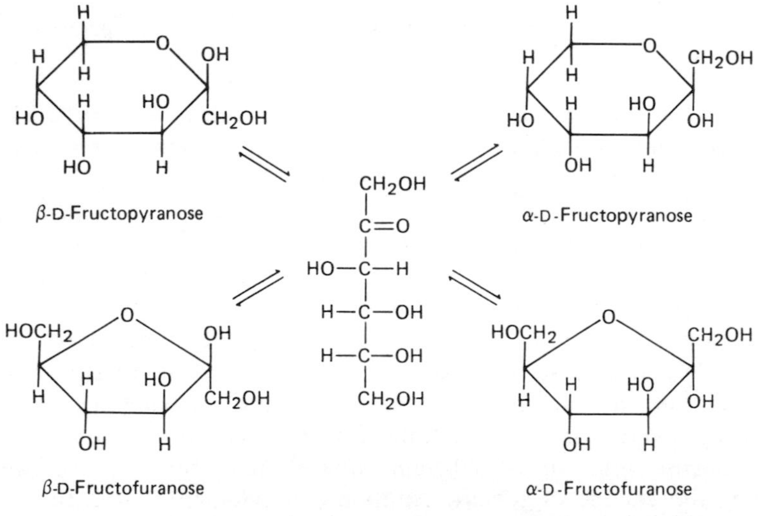

β-D-Fructopyranose

α-D-Fructopyranose

β-D-Fructofuranose

α-D-Fructofuranose

Figure 11.6 The equilibrium behaviour of fructose in solution

β-D-2-Deoxyribose   β-D-2-Amino-2-deoxyglucose

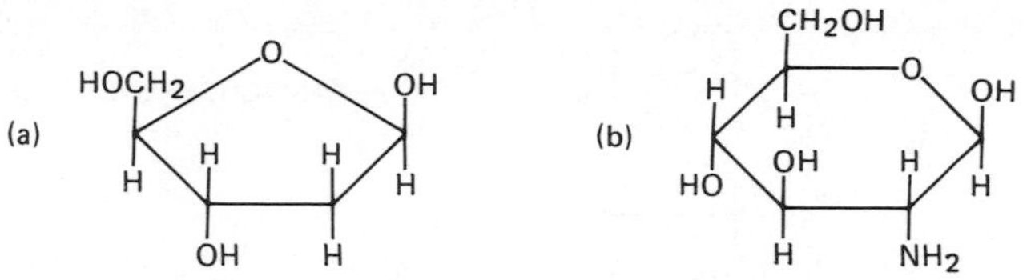

Figure 11.7 A deoxy sugar and an amino sugar

11.2 Glycosides

As mentioned, monosaccharides exist as cyclic hemiacetals and hemiketals. They will therefore react with a further molecule of alcohol to give acetals and ketals (Chapter 8). These are called *glycosides*. The specific name for an acetal of glucose is *glucoside*, whereas that for a ketal of fructose is *fructoside*. The α- and β-glucosides (Figure 11.8) can be separated and do not interconvert in aqueous solution, but being acetals they are easily hydrolysed by aqueous acids back to D-glucose. The noncarbohydrate portion of a glycoside is called the *aglycone*. Thus methanol is the aglycone of the methyl glucosides (Figure 11.8).

There are many naturally occurring glycosides. In the plant world they are found in the leaves and seeds of plants and in the bark of trees. Coniferin (Figure 11.9(a)), for example, is found in the sap of pine trees. In the animal world glycosides occur as cerebrosides (Figure 11.9(b)).

Plant glycosides are usually accompanied by an appropriate specific enzyme capable of hydrolysing the glycoside. The enzyme is not usually contained in the same cell as the glycoside but is present in a neighbouring cell and is brought into action when required. 3-Hydroxyindole occurs in nature as the glucoside indican (Figure 11.10(a)), formerly of great importance for the preparation of indigo. Hydrolysis of indican releases the 3-hydroxyindole and on atmospheric oxidation this is transformed into the dyestuff indigotin. Mandelonitrile (the cyanohydrin of benzaldehyde, Section 8.4(e)) occurs in nature as the glycoside amygdalin (Figure 11.10(b)) which is present in bitter almonds. When the almonds are crushed enzymes are released which hydrolyse the glycoside and liberate HCN.

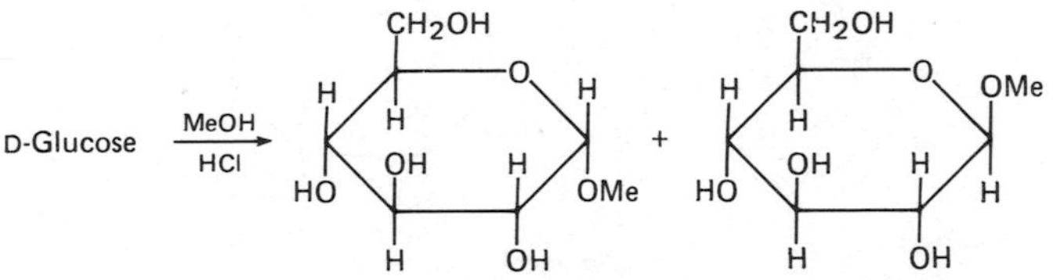

Methyl-α-D-glucopyranoside   Methyl-β-D-glucopyranoside

Figure 11.8 The formation of α- and β-glycosides

(a) Coniferin

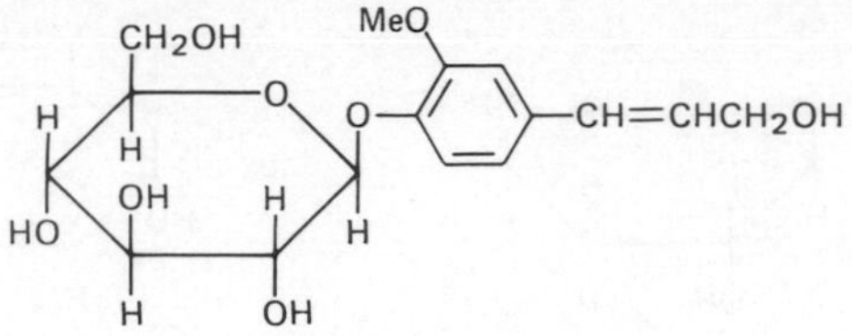

(b) A cerebroside

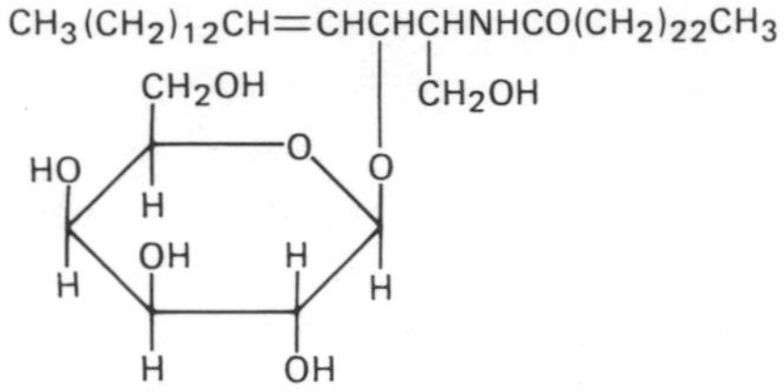

Figure 11.9 The structures of two naturally occurring glycosides

Among the medicinally important glycosides are those isolated from *Digitalis* species (e.g. foxglove) which are valuable in medicine for stimulating the heart muscles. Thus digitoxin is a glycoside of the steroid

(a) Indican

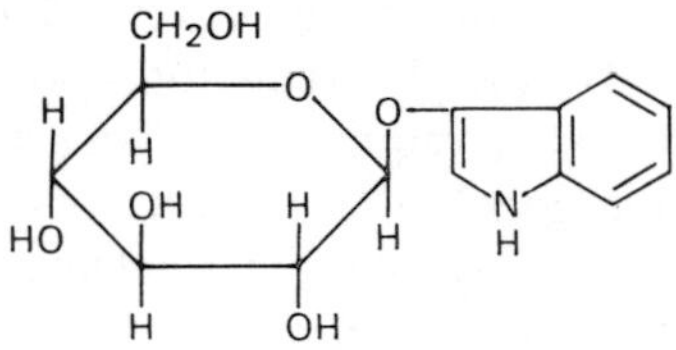

(b) Amygdalin

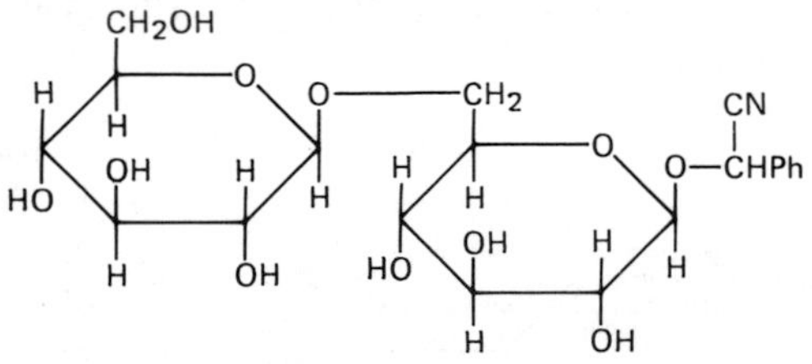

Figure 11.10 Two further naturally occurring glycosides

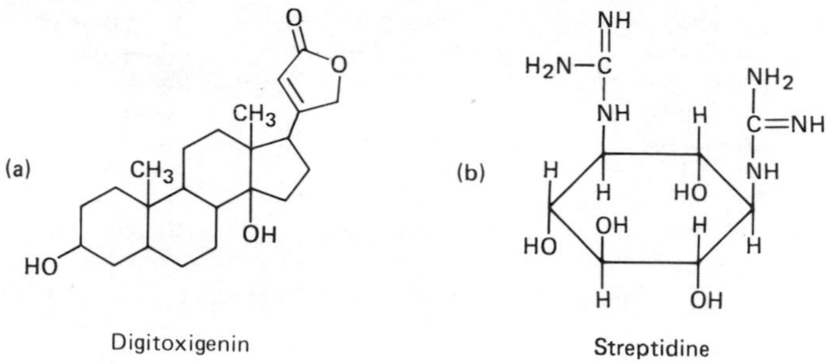

Figure 11.11 Two medically important glycosides

alcohol digitoxigenin (Figure 11.11(a)). Also important are the antibiotic glycosides, such as streptomycin, which inhibits protein synthesis in ribosomes. In this case the aglycone is streptidine (Figure 11.11(b)).

11.3 Reactions of monosaccharides

Monosaccharides show many of the characteristic chemical reactions of their component functional groups. As alcohols they can be converted into ethers and react with acids or their derivatives to form esters. As aldehydes or ketones they may be oxidized or reduced and undergo nucleophilic addition reactions.

(a) Oxidation

Simple aldoses and ketoses are *reducing sugars*, i.e. they reduce Tollens' reagent and other mild oxidizing agents such as Fehling's or Benedict's solution (Section 8.7(a)). The reaction with ketoses is somewhat surprising since ketones do not normally reduce these reagents. However ketoses are α-hydroxyketones and in basic solution rapidly interconvert with the isomeric aldoses (Figure 11.12, cf. Section 9.5).

Mild oxidation of an aldose using bromine water (Figure 11.13) affords a

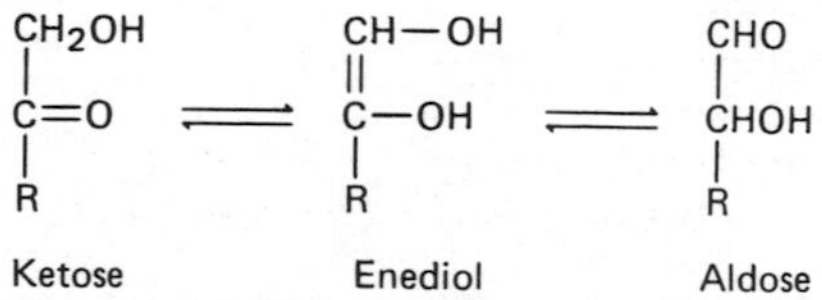

Figure 11.12 Ketone–aldose isomerization
in basic solution

H

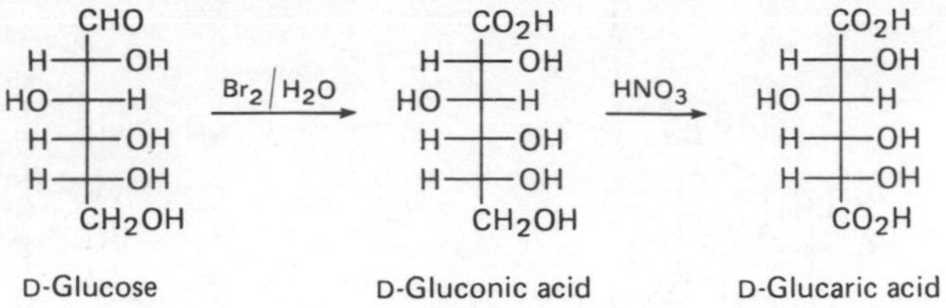

CHO   CO₂H   CO₂H

$$\begin{array}{c}\text{CHO}\\ \text{H}\!-\!\text{OH}\\ \text{HO}\!-\!\text{H}\\ \text{H}\!-\!\text{OH}\\ \text{H}\!-\!\text{OH}\\ \text{CH}_2\text{OH}\end{array} \xrightarrow{\text{Br}_2/\text{H}_2\text{O}} \begin{array}{c}\text{CO}_2\text{H}\\ \text{H}\!-\!\text{OH}\\ \text{HO}\!-\!\text{H}\\ \text{H}\!-\!\text{OH}\\ \text{H}\!-\!\text{OH}\\ \text{CH}_2\text{OH}\end{array} \xrightarrow{\text{HNO}_3} \begin{array}{c}\text{CO}_2\text{H}\\ \text{H}\!-\!\text{OH}\\ \text{HO}\!-\!\text{H}\\ \text{H}\!-\!\text{OH}\\ \text{H}\!-\!\text{OH}\\ \text{CO}_2\text{H}\end{array}$$

D-Glucose   D-Gluconic acid   D-Glucaric acid

Figure 11.13 Oxidation of D-glucose

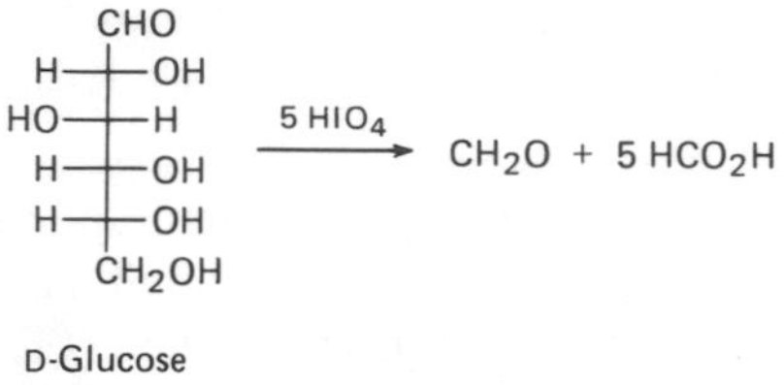

$$\begin{array}{c}\text{CHO}\\ \text{H}\!-\!\text{OH}\\ \text{HO}\!-\!\text{H}\\ \text{H}\!-\!\text{OH}\\ \text{H}\!-\!\text{OH}\\ \text{CH}_2\text{OH}\end{array} \xrightarrow{\text{5 HIO}_4} \text{CH}_2\text{O} + \text{5 HCO}_2\text{H}$$

D-Glucose

Figure 11.14 Periodic acid cleavage of
D-glucose

monocarboxylic acid. Ketoses react much less readily with bromine water and this reaction can therefore be used to distinguish between aldoses and ketoses. Further oxidation of the monocarboxylic acid (Figure 11.13), using for example nitric acid, yields a dicarboxylic acid.

1,2-Diols and α-hydroxycarbonyl compounds are cleaved by periodic acid (HIO_4). Terminal CH_2OH groups are oxidized to give formaldehyde (methanal) whereas CHOH or carbonyl groups are converted into formic acid (methanoic acid). Thus one mole of D-glucose consumes five moles of periodic acid and produces one mole of formaldehyde and five moles of formic acid (Figure 11.14). Reactions of this type have proved useful for the elucidation of carbohydrate structures.

Question 11.4 What would be the organic products formed when sorbitol (Figure 11.15) is oxidized with periodic acid?

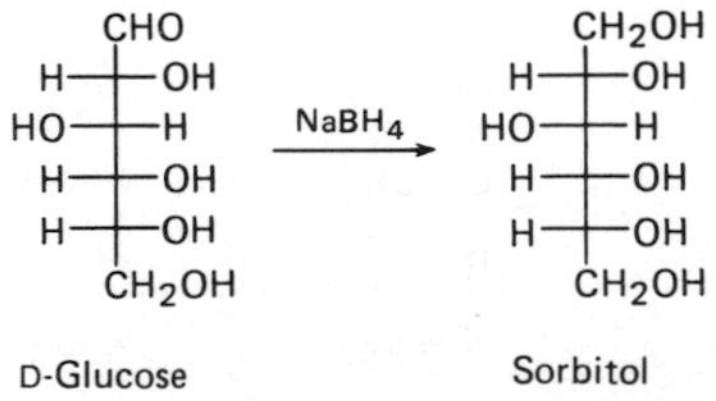

$$\begin{array}{c}\text{CHO}\\ \text{H}\!-\!\text{OH}\\ \text{HO}\!-\!\text{H}\\ \text{H}\!-\!\text{OH}\\ \text{H}\!-\!\text{OH}\\ \text{CH}_2\text{OH}\end{array} \xrightarrow{\text{NaBH}_4} \begin{array}{c}\text{CH}_2\text{OH}\\ \text{H}\!-\!\text{OH}\\ \text{HO}\!-\!\text{H}\\ \text{H}\!-\!\text{OH}\\ \text{H}\!-\!\text{OH}\\ \text{CH}_2\text{OH}\end{array}$$

D-Glucose   Sorbitol

Figure 11.15 Reduction of
D-glucose

(b) Reduction

Sodium amalgam or sodium borohydride reduce aldehyde or ketone groups of monosaccharides to the corresponding primary or secondary alcohols. For example, sodium borohydride reduction of glucose affords sorbitol (Figure 11.15).

This reaction is useful for distinguishing between an aldose and a ketose since an aldose gives only one product (Figure 11.15) whereas a ketose gives two diastereoisomers due to the formation of a new asymmetric centre.

Question 11.5 Reduction with sodium borohydride can also be used to distinguish between a sample of D-erythrose and D-threose (see Figure 11.3). Draw the product formed by reduction of each compound and indicate how they could be distinguished.

(c) Reaction with phenylhydrazine

It is sometimes possible to isolate simple phenylhydrazones (Section 8.4(c)) of aldoses and ketoses. However they usually react further with excess phenylhydrazine (three equivalents) to give bisphenylhydrazones called *osazones* (Figure 11.16).

On boiling with aqueous acid osazones are converted into the corresponding dicarbonyl compounds called osones (Figure 11.16). Mild reduction converts the osone into the ketose since the aldehyde group is reduced preferentially. Thus D-glucose for example can be converted into D-fructose by this method.

Question 11.6 Given that osazone formation involves only the hydroxyl group on carbon atom 1 of D-fructose what would be the differences between the osazones formed from D-glucose, D-fructose, and D-mannose?

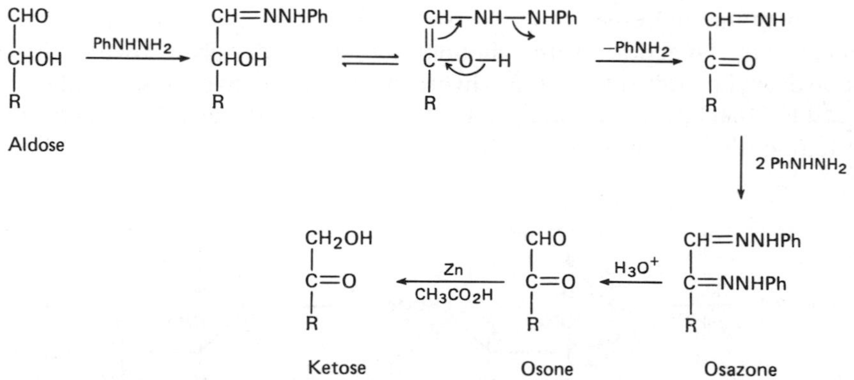

Figure 11.16 Conversion of an aldose into an osazone using phenylhydrazine

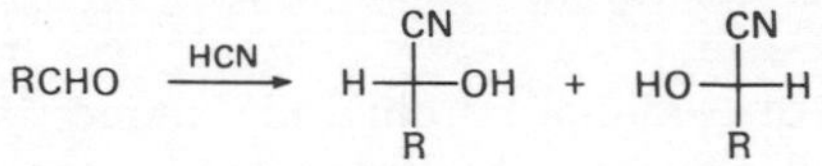

$$RCHO \xrightarrow{HCN} \begin{array}{c} CN \\ H-\!\!\!-OH \\ R \end{array} + \begin{array}{c} CN \\ HO-\!\!\!-H \\ R \end{array}$$

Figure 11.17 Formation of isomeric cyanohydrins

CHOH ... CHOH $\xrightarrow{(CH_3CO)_2O}$ CH$_2$OAc ... CHOAc

$$(Ac = CH_3 - \overset{\displaystyle O}{\overset{\|}{C}} -)$$

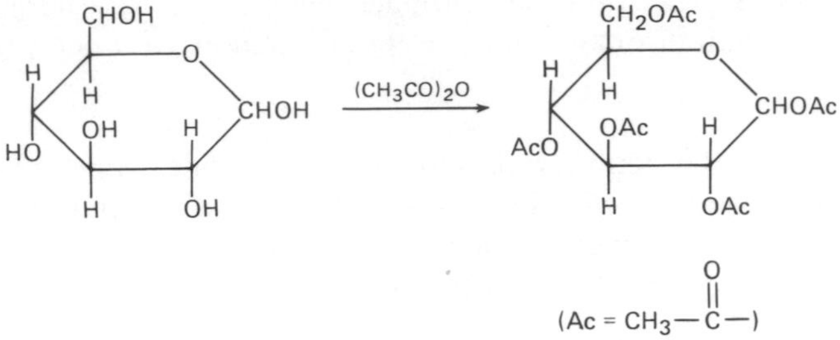

Figure 11.18 Formation of a carbohydrate ester

(d) Cyanohydrin formation

Monosaccharides react with hydrogen cyanide to form two diastereoisomeric cyanohydrins (Figure 11.17, cf. Section 8.4(e)). The two diastereoisomers are formed in unequal amounts due to the presence of an asymmetric centre in the starting material (Section 2.11). Hydrolysis gives the corresponding hydroxy acids, which can be separated as their crystalline lactones.

(e) Ester and ether formation

The hydroxyl groups of carbohydrates react with acids or acid derivatives to form esters. D-Glucose, for example, reacts with acetic anhydride (ethanoic anhydride) to form a pentaacetate (Figure 11.18).

The pentamethyl ether of D-glucose can be formed in a similar way, for example by reaction with methyl iodide and silver oxide in a non-hydroxylic solvent. Dilute hydrochloric acid hydrolyses only the glycosidic (acetal) group and thus converts the pentamethyl ether into the tetramethyl ether (Figure 11.19).

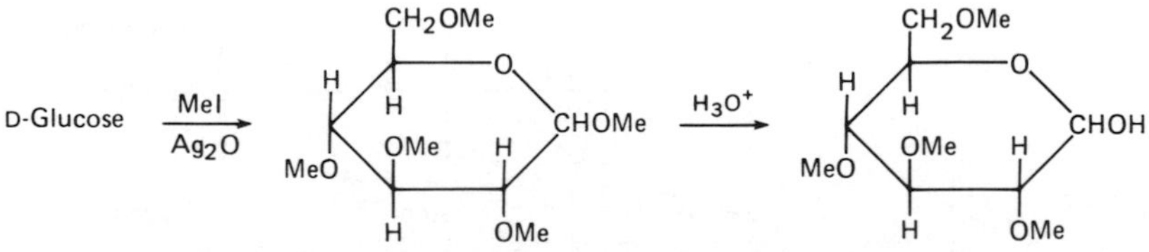

D-Glucose $\xrightarrow[Ag_2O]{MeI}$ CH$_2$OMe ... CHOMe $\xrightarrow{H_3O^+}$ CH$_2$OMe ... CHOH

Figure 11.19 Formation and acid hydrolysis of a carbohydrate ether

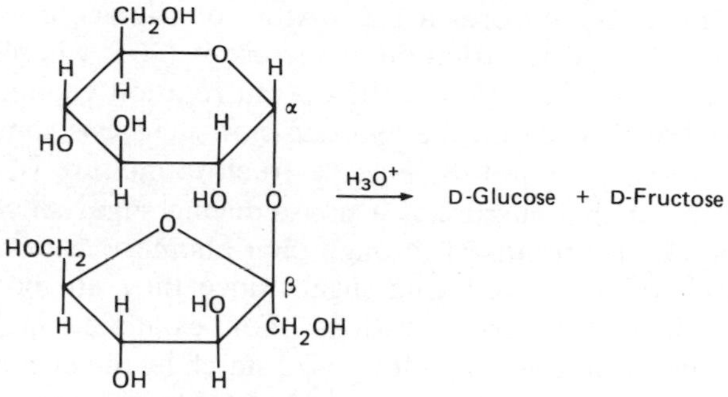

Figure 11.20 Hydrolysis of sucrose

11.4 Disaccharides

A disaccharide is a molecule composed of two monosaccharide units joined together by a *glycosidic linkage*. One of the most common disaccharides is sucrose (common sugar) which is found in all plants and is present in large amounts in sugar cane and sugar beet. Hydrolysis of

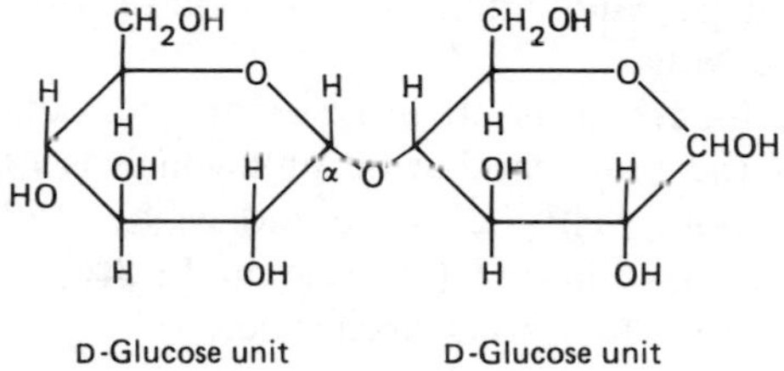

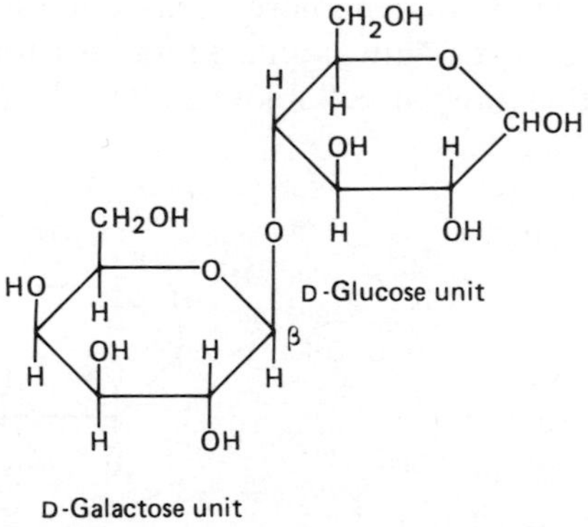

Figure 11.21 The structures of (a) maltose and (b) lactose

sucrose by aqueous acid affords a 1:1 mixture of D-glucose and D-fructose (Figure 11.20). The specific rotation of sucrose is $+66°$ whereas that of the hydrolysis mixture is $-20°$, so that the optical rotation changes sign during the hydrolysis. For this reason the hydrolysis of sucrose is sometimes called the 'inversion' of sucrose and the glucose–fructose mixture is referred to as 'invert' sugar. Note that sucrose is a non-reducing sugar since both of the monosaccharide units are linked through their anomeric carbon atoms.

Many disaccharides are reducing sugars since they are bonded through only one of their anomeric positions. For example, maltose (Figure 11.21(a)), the major product of hydrolysis of starch by the enzymes present in saliva (Section 11.5), and lactose (Figure 11.21(b)), the major disaccharide present in mammalian milk, are both reducing sugars.

In order to determine the structure of a disaccharide the following steps are carried out:

1. The disaccharide is hydrolysed and the component monosaccharides are identified.
2. The disaccharide is tested with Tollens' reagent to establish whether or not it is a reducing sugar, i.e. whether one or both of the anomeric positions are involved in glycosidic bonds.
3. Appropriate enzymes are used to determine whether the glycosidic bonds are α or β. Thus emulsin and maltase are specific for β and α linkages respectively.
4. For reducing sugars it is then necessary to determine the point of attachment of the monosaccharide unit which is not linked through its anomeric position. This can be done by fully methylating and hydrolysing the disaccharide (cf. Section 11.3(e)), and then identifying the partially methylated monosaccharide.

Question 11.7 Hydrolysis of cellobiose, a disaccharide obtained by partial hydrolysis of cellulose, gives only D-glucose and is catalysed by emulsin but not by maltase. Methylation of cellobiose followed by hydrolysis gives the following products:

Deduce the structure of cellobiose.

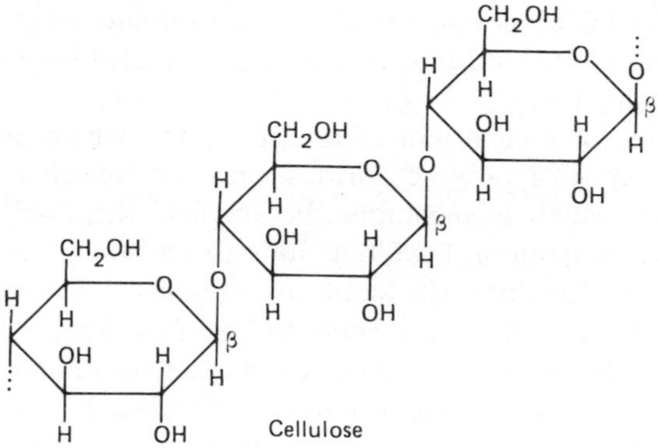

Figure 11.22 A part structure of cellulose

11.5 Polysaccharides

Polysaccharides occur widely in nature in such forms as pectin, agar, and alginates. Two of the most important are starch and cellulose which are both composed of repeating glucose units. Starch contains α-linked glucose units while cellulose is β-linked.

Cellulose is one of the main constituents of wood and plants. For example, wood contains approximately 50% cellulose, flax 80%, and jute 65%, whereas cotton is almost pure cellulose. The glucose units are linked from carbon atom 1 of one unit to carbon atom 4 of the next (Figure 11.22).

Higher animals do not have the enzymes necessary to break the β-glycosidic links of cellulose and therefore cannot utilize it as food. However some animals (e.g. deer and cows) can use cellulose as food because they have colonies of bacteria capable of hydrolysing cellulose in their digestive tracts.

Cellulose products are of considerable commercial importance. Thus cellulose acetate is used as a fibre and as a moulding plastic, whereas cellulose trinitrate is used as an explosive (gun cotton). Celluloid is made from cellulose nitrates, camphor, and alcohol. In the viscose process

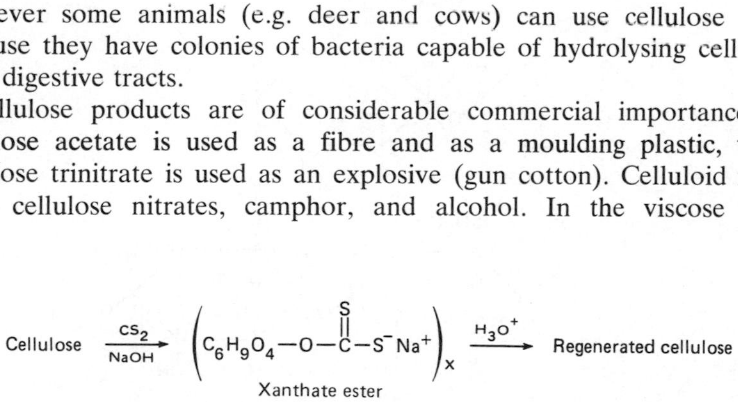

Figure 11.23 The cellulose viscose process

(Figure 11.23) cellulose is converted into a xanthate ester and a solution of the ester is then extruded into dilute acid which reliberates the cellulose as a filament or sheet.

Starch is the major constituent of rice, potatoes, wheat, and corn. It is a mixture of two types of polysaccharides, amylose, which is water soluble, and amylopectin, which is insoluble. In amylose the D-glucose units are α-linked from carbon atom 1 of one unit to carbon atom 4 of the next (Figure 11.24(a)). Amylopectin similarly contains 1,4-α-linked D-glucose units but also has crosslinks attached to the 6-position of some glucose units (Figure 11.24(b)). Animals do have the enzymes needed to cleave the α-glycosidic links of starch. When a starchy food such as corn or bread is chewed the enzyme amylase, present in saliva, breaks down the starch into maltose which has a sweet taste. Further breakdown to glucose occurs in the intestines. The glucose is then absorbed into the bloodstream to be used as an energy source (Chapter 15). Excess glucose is polymerized into glycogen, a polysaccharide resembling amylopectin, which is stored in the liver and muscles. Glycogen supplies the body with glucose during periods of exertion and between meals. The problem of fat build-up arises because

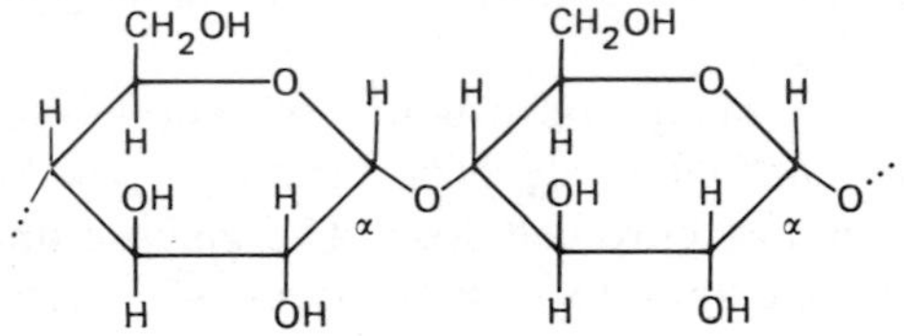

Amylose

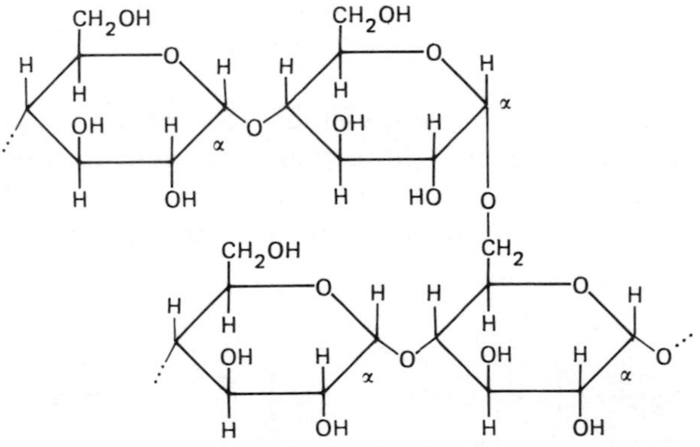

Amylopectin

Figure 11.24 The two polysaccharides found in starch

tissues are limited in the amount of glycogen they can store. After producing 50–60 g of glycogen per kilogram of tissue, glucose is diverted from glycogen synthesis to fat synthesis.

Question 11.8 The major product formed when amylose is methylated and then hydrolysed is 2,3,6-tri-*O*-methyl-D-glucose. However about 0.5% of 2,3,4,6-tetra-*O*-methyl-D-glucose is also obtained. What is the average chain length of the amylose molecules?

Answers to Questions

11.1

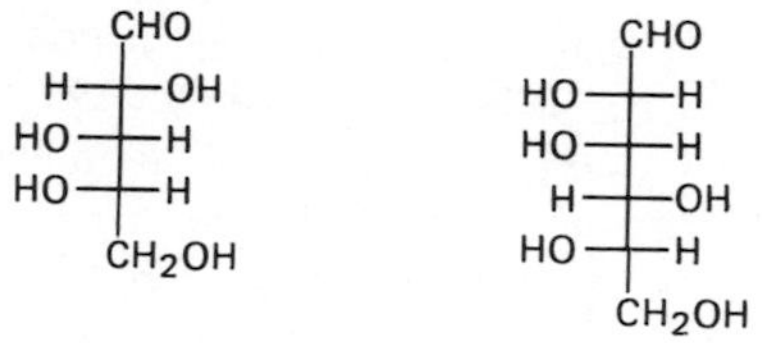

CHO
H——OH
HO——H
HO——H
CH$_2$OH

L-Arabinose

CHO
HO——H
HO——H
H——OH
HO——H
CH$_2$OH

L-Gulose

11.2 *S* and *R* respectively.

11.3 The β-anomer, predominates because the hydroxyl group at carbon atom 1 is equatorial, whereas in the α-anomer it is axial.

11.4

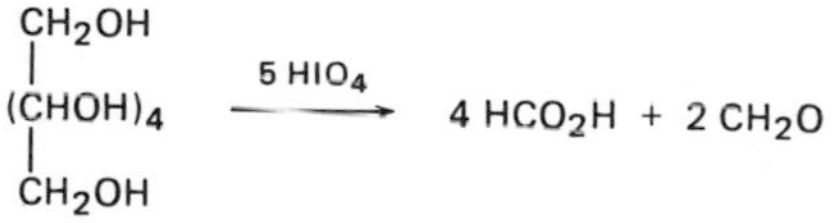

CH$_2$OH
|
(CHOH)$_4$ $\xrightarrow{\text{5 HIO}_4}$ 4 HCO$_2$H + 2 CH$_2$O
|
CH$_2$OH

11.5

D-Erythrose $\xrightarrow{\text{NaBH}_4}$

CH$_2$OH
H——OH
H——OH
CH$_2$OH

; D-Threose $\xrightarrow{\text{NaBH}_4}$

CH$_2$OH
HO——H
H——OH
CH$_2$OH

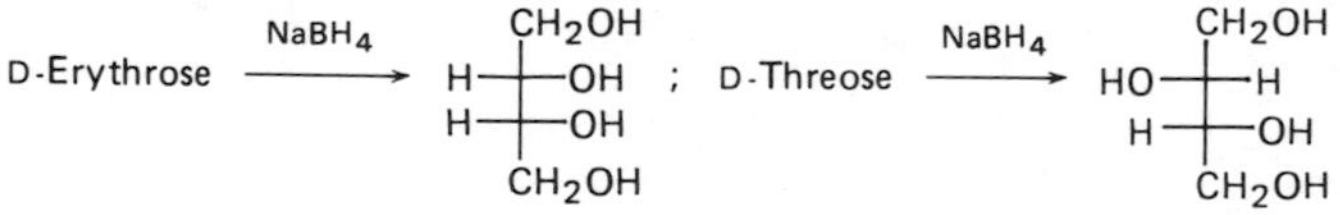

The reduction product from D-erythrose has a plane of symmetry and is therefore achiral whereas the product from D-threose is chiral. The two products can therefore be distinguished on the basis of their optical activity.

11.6 They all give the same osazone.

11.7

11.8 Since each amylose molecule gives rise to only one molecule of tetramethyl glucose the average number of glucose units per amylose molecule is 100÷0.5=200.

CHAPTER 12

Proteins and Nucleic Acids

It is difficult to overemphasize the importance of proteins and nucleic acids in the chemistry of life. Proteins are the principle ingredients of skin, muscle, fingernails, hair, blood, antibodies, and enzymes, as well as of many hormones and antibiotics. Indeed, a single living organism requires many thousands of different proteins in order to function, and each species requires its own unique set. Nucleic acids are the essential components of chromosomes, the carriers of genetic information within the nucleus of a cell. They control heredity and direct the synthesis of proteins.

Proteins and nucleic acids are high molecular mass compounds consisting of many smaller subunits. This chapter deals with their structures and properties, the methods by which they may be synthesized and a little about their biochemical function.

12.1 Structure of amino acids

Hot hydrochloric acid hydrolyses proteins to give a mixture of compounds called *α-amino acids*. As the name implies, these are carboxylic acids substituted in the α- or 2-position by an amino group. Most of the amino acids encountered in proteins may be represented by a general structure (Figure 12.1), and differ only in the organic group, R.

About twenty amino acids occur frequently in proteins, although a number of others are occasionally found. The common ones are usually referred to by non-systematic names, often written in abbreviated form (Table 12.1). Note that in two of the common ones the amine group is part of a heterocyclic ring.

Because they contain a basic amino group (Chapter 7) and an acidic carboxyl group (Chapter 8) most amino acids are approximately neutral. However, a few contain a second amine function and are *basic amino acids,* or a second carboxyl group and are *acidic amino acids* (Table 12.1).

In human beings, certain amino acids must be supplied in the daily diet because they cannot be synthesized in sufficient quantity by the human

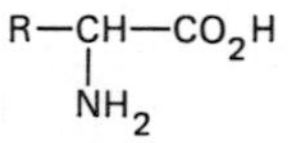

$$R-CH-CO_2H$$
$$|$$
$$NH_2$$

Figure 12.1 An α-amino acid

221

Table 12.1 Common amino acids

Amino acid	Abbreviation	Structure	
Neutral amino acids			
Glycine	Gly	$H-CH-CO_2H$, NH_2	
Alanine	Ala	$CH_3-CH-CO_2H$, NH_2	
Serine	Ser	$HOCH_2-CH-CO_2H$, NH_2	
Cysteine	CySH	$HSCH_2-CH-CO_2H$, NH_2	
Cystine	CyS, CyS	$SCH_2-CH-CO_2H$, NH_2; $SCH_2-CH-CO_2H$, NH_2	
Threonine	Thr	OH; $CH_3CH-CH-CO_2H$, NH_2	Essential
Methionine	Met	$CH_3SCH_2-CH-CO_2H$, NH_2	Essential
Valine	Val	$(CH_3)_2CH-CH-CO_2H$, NH_2	Essential
Leucine	Leu	$(CH_3)_2CHCH_2-CH-CO_2H$, NH_2	Essential
Isoleucine	Ileu	$CH_3CH_2CH-CH-CO_2H$, CH_3 NH_2	Essential
Phenylalanine	Phe	$C_6H_5CH_2CH-CO_2H$, NH_2	Essential
Tyrosine	Tyr	$HO-C_6H_4-CH_2-CH-CO_2H$, NH_2	
Tryptophan	Trp	indole$-CH_2-CH-CO_2H$, NH_2	Essential

Table 12.1

Amino acid	Abbreviation	Structure

Proline   Pro

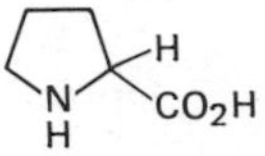

Hydroxyproline   HPro

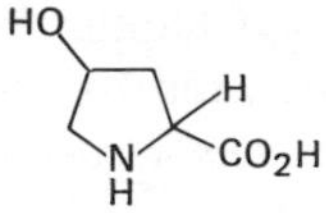

Acidic amino acids

Aspartic acid   Asp

$$HO_2CCH_2\underset{\underset{NH_2}{|}}{C}HCO_2H$$

(Occurs also as the amide asparagine)   Asn

$$H_2NCOCH_2\underset{\underset{NH_2}{|}}{C}HCO_2H$$

Glutamic acid   Glu

$$HO_2CCH_2CH_2\underset{\underset{NH_2}{|}}{C}HCO_2H$$

(Occurs also as the amide glutamine)   Gln

$$H_2NCOCH_2CH_2\underset{\underset{NH_2}{|}}{C}HCO_2H$$

Basic amino acids

Lysine   Lys

$$\underset{\underset{NH_2}{|}}{C}H_2CH_2CH_2CH_2\underset{\underset{NH_2}{|}}{C}HCO_2H \qquad \text{Essential}$$

Arginine   Arg

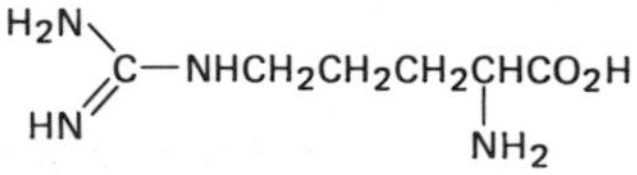

$$\underset{HN}{\overset{H_2N}{\diagup}}C-NHCH_2CH_2CH_2\underset{\underset{NH_2}{|}}{C}HCO_2H$$

Histidine   His

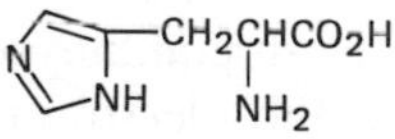

body. These are termed *essential* amino acids, and are indicated in the table. In countries where the population is largely dependent upon a single staple food the lack of some essential amino acids can cause widespread deficiency disease. An example is kwashiorkor, which afflicts millions of children in Asia and Africa when they begin on a diet of rice after being displaced from the mother's breast by a newborn baby.

Question 12.1   Derive the systematic names for (a) serine, (b) valine, (c) lysine (Table 12.1).

With the exception of glycine the amino acids contain an asymmetric carbon atom in the α-position. There are therefore two mirror image

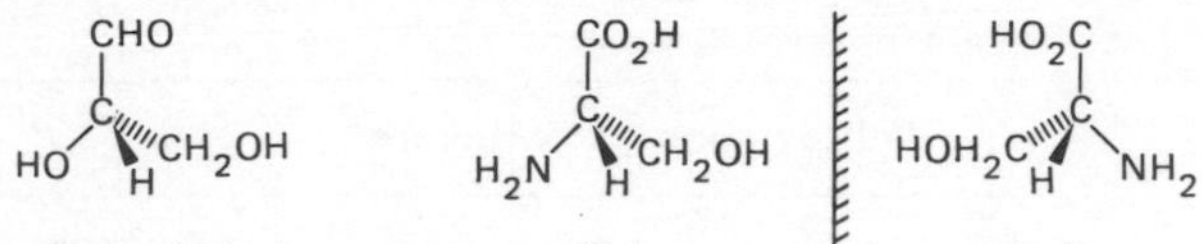

Figure 12.2 Optical isomerism of amino acids

structures (enantiomers—Chapter 2). It is convenient to relate the absolute configuration of amino acids to those of D- or L-glyceraldehyde (Figure 12.2). Except in a very few unusual proteins from micro-organisms, all of the amino acids in natural proteins are of the L-series.

Question 12.2 Use the R/S convention to define the absolute configuration of (a) L-serine (b) L-alanine, and draw the Fischer projection formulas for these molecules (Chapter 2).

12.2 Properties of amino acids

Although we have drawn amino acids as though the two functionalities were independent, in fact they undergo an acid–base interaction to produce a salt-like structure (I, Figure 12.3) called a *zwitterion* (German, *zwitter*=hybrid) since it contains both positive and negative charges. Because of this, amino acids are high melting solids which are often fairly soluble in water, but not in ether or benzene.

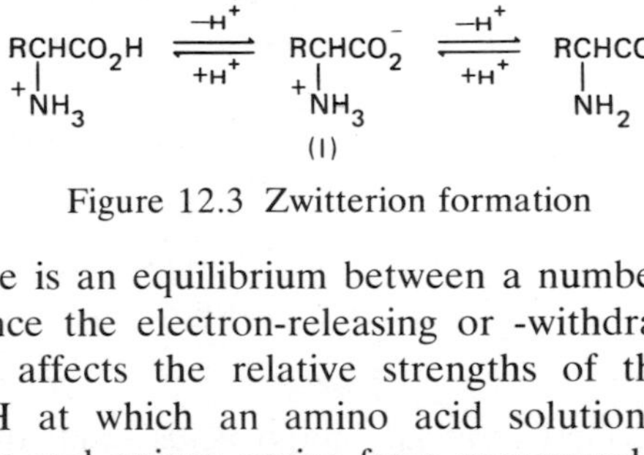

Figure 12.3 Zwitterion formation

In solution there is an equilibrium between a number of different forms (Figure 12.3). Since the electron-releasing or -withdrawing nature of the organic group R affects the relative strengths of the acidic and basic functions, the pH at which an amino acid solution contains an equal number of cations and anions varies from compound to compound. This pH is characteristic of each amino acid and is known as its *isoelectric point*. The isoelectric point of glycine is 6.0. Aspartic acid contains a second carboxyl group and at pH 6.0 a large proportion of its molecules are anionic (Figure 12.4). To reach the isoelectric point a number of these anions must be protonated, and this is only achieved by lowering the pH to 2.8.

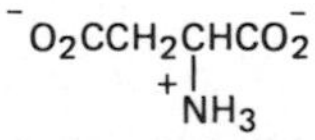

$$^-O_2CCH_2\overset{\overset{\displaystyle +}{|}}{\underset{NH_3}{C}}HCO_2^-$$

Figure 12.4 The aspartate anion

At any given pH different amino acids have slightly different net charges and will therefore move at different rates towards one of the electrodes if an electric field is applied. This provides the basis for their separation by the technique of gel electrophoresis (Chapter 3).

Because of the dual acid–base character of amino acids they are able to neutralize small quantities of other acids or bases without a significant change in the pH of the solution; they are therefore said to act as *buffers*, which is the general term applied to solutions which maintain a constant pH.

Question 12.3 Suggest a value for the isoelectric point of lysine. Towards which electrode will the following amino acids move upon electrophoresis at pH 5.0: (a) glycine; (b) lysine; (c) aspartic acid?

12.3 Reactions of amino acids

At the appropriate pH amino acids undergo the typical reactions of carboxylic acids (Chapter 8) and of amines (Chapter 7). Thus, the carboxyl group may be esterifield with hot, acidified methanol, or the amino group may be converted into an amide with acetyl (ethanoyl) chloride under basic conditions. Amide formation between the carboxyl group of one amino acid and the amino group of a second is a most important process known as *peptide* bond formation. Peptide bonds form the basis of the structure of proteins and are dealt with in detail later.

Reaction with ninhydrin (Figure 12.5) provides a convenient colour test for amino acids, because the products are extensively conjugated and therefore highly coloured. Ninhydrin reagent is used to indicate the positions of amino acid spots on chromatographic plates or papers, or to provide a means of quantitative estimation of amino acids by colorimetry.

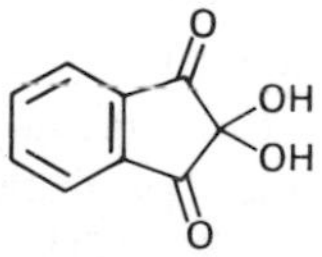

Figure 12.5 The structure of ninhydrin

12.4 Synthesis of amino acids

There are many ways in which amino acids may be synthesized, some very general, others applicable only to one or two specific amino acids. Two of the more general methods are outlined here. In the first, called the Strecker synthesis, ammonia and hydrogen cyanide are added to an aldehyde and the resultant α-aminonitrile is then hydrolysed with aqueous acid (Figure 12.6(a)). The second involves bromination of a carboxylic

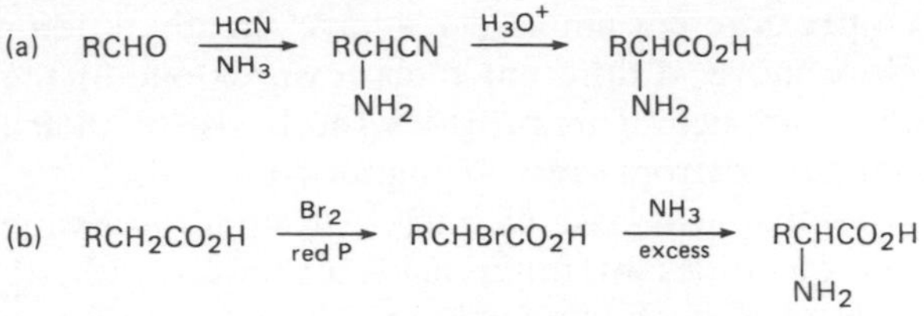

(a) $\quad$ RCHO $\xrightarrow[\text{NH}_3]{\text{HCN}}$ RCHCN $\xrightarrow{\text{H}_3\text{O}^+}$ RCHCO$_2$H
$\qquad\qquad\qquad\qquad\quad$ | $\qquad\qquad\qquad\qquad$ |
$\qquad\qquad\qquad\qquad$ NH$_2$ $\qquad\qquad\qquad\quad$ NH$_2$

(b) $\quad$ RCH$_2$CO$_2$H $\xrightarrow[\text{red P}]{\text{Br}_2}$ RCHBrCO$_2$H $\xrightarrow[\text{excess}]{\text{NH}_3}$ RCHCO$_2$H
$\qquad\qquad\qquad\qquad\qquad\qquad\qquad\qquad\qquad\qquad\qquad\qquad$ |
$\qquad\qquad\qquad\qquad\qquad\qquad\qquad\qquad\qquad\qquad\qquad$ NH$_2$

Figure 12.6 Two methods of synthesizing amino acids

acid, followed by treatment of the α-bromo derivative with a large excess of ammonia (Figure 12.6(b)).

Question 12.4 What is the reason for using a large excess of ammonia in reaction (b) of Figure 12.6?

12.5 Peptides

When two amino acid residues are linked by a peptide bond the product is a *dipeptide*. A *tripeptide* contains three residues, a *tetrapeptide* four (Figure 12.7) and so on. Peptides containing more than about forty residues, with a molecular mass of more than ten thousand, are called proteins. Some enzymes contain as many as three hundred residues.

Although there are only a limited number of naturally occurring amino acids, these can give rise to a large number of peptides and proteins. For example, if we consider only tripeptides constructed from the amino acids listed in Table 12.1, there are twenty choices for each residue, or $20^3 = 8000$ possible tripeptides. For n residues there are 20^n possibilities, and n may be as large as three hundred. Clearly there is scope for an enormous variety of protein structures, as is indeed observed.

A shorthand is necessary for the representation of peptide structures, and the abbreviated names of the constituent amino acids are used for this purpose. Figure 12.8 shows representations of the structure of the naturally occurring tripeptide glutathione (γ-glutamylcysteinylglycine). By convention the amino acid with a free NH$_2$ group (*N-terminal amino acid*) is written on the left and that with a free CO$_2$H group (*C-terminal*) on the right. When this convention alone is not entirely clear (e.g. cyclic peptides), or when emphasis is required, an arrow may be used to indicate the direction of the peptide linkage, thus: —CO$\rightarrow$NH—. (Notice that the

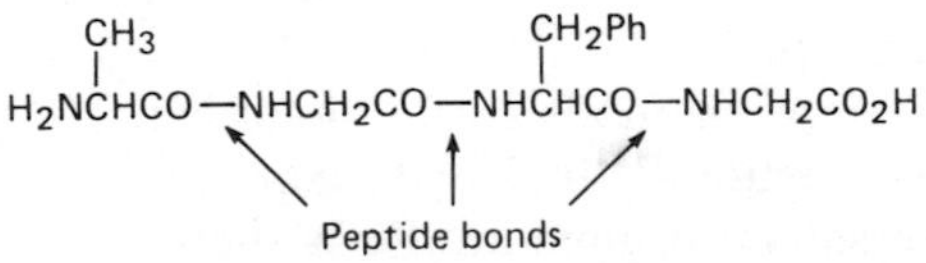

$\qquad\qquad\qquad$ CH$_3$ $\qquad\qquad\qquad\qquad$ CH$_2$Ph
$\qquad\qquad\qquad\qquad$ | $\qquad\qquad\qquad\qquad\qquad$ |
$\quad$ H$_2$NCHCO—NHCH$_2$CO—NHCHCO—NHCH$_2$CO$_2$H

Peptide bonds

Figure 12.7 A tetrapeptide

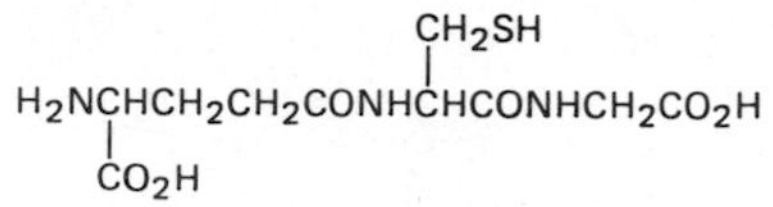

$$\underset{\underset{CO_2H}{|}}{H_2NCHCH_2CH_2CONH\overset{\overset{CH_2SH}{|}}{C}HCONHCH_2CO_2H}$$

H—γ—Glu—CySH—Gly—OH

or H—γ—Glu→CySH→Gly—OH

Figure 12.8 Representations of the
structure of glutathione

first peptide link of glutathione utilizes the γ-carboxyl group of glutamic acid and not the α-carboxyl group; this must be specified in the shorthand.) Since the symbol for the amino acid is generally used only for the part —NHCHRCO— the N-terminal amino acid is given the prefix H— and the C-terminal amino acid is followed by the suffix —OH (Figure 12.8). In the same way derivatives such as a methyl ester or amide can be indicated by adding —OMe or —NH$_2$.

Question 12.5 Draw out the structure of H—Ala—Ser—Met—OH.

12.6 Amino acid composition of peptides and proteins

The first stage in determination of the structure of a protein is the identification and estimation of its constituent amino acids. A pure sample of the protein is required and this in itself presents a considerable problem. In order to obtain a pure sample the researcher usually employs a combination of techniques (Chapter 3) such as gel electrophoresis, ion-exchange chromatography, and density gradient centrifugation (heavy proteins sink faster than light ones). When a protein is homogeneous by these techniques, or when a maximum of biological activity is reached, the protein is considered to be pure. It is then hydrolysed to its constituent amino acids with hot hydrochloric acid, and the hydrolysate is analysed.

The *amino acid analyser* contains a column of ion-exchange resin through which amino acids pass at different rates. The eluate is automatically mixed with ninhydrin solution, which becomes coloured as an amino acid elutes. The colour is monitored by a spectrophotometer and absorbance is automatically plotted against time on a recorder. Since the time spent on the column is characteristic of the amino acid and absorbance is directly related to the quantity present, the recorder trace (e.g. Figure 12.9) immediately indicates proportions of all amino acids in the protein.

Once the approximate molecular mass of the protein has been determined (e.g. by density gradient centrifugation), the relative proportions of amino acids can be converted into absolute numbers of residues in the protein. This is the amino acid *composition* of the protein.

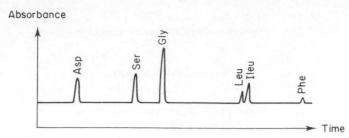

Figure 12.9 Typical recorder trace from amino acid analyser

Question 12.6 A peptide hydrolysate shows four amino acids in the proportions Gly:Ala:Phe:Ser=2:1:1:3; the molecular mass of the peptide is around 1500. What is the amino acid composition of the peptide?

12.7 Primary structure of proteins

The second stage in determination of the structure of a protein is to find out the order in which the amino acids are joined together. This *sequence* of amino acids is called the *primary structure* of the protein. Sanger received the Nobel Prize in 1958 for his determination of the total sequence of insulin, a protein with 51 amino acid residues.

One approach to 'sequencing' employs *end-group analysis,* e.g. the specific identification of the N-terminal and/or C-terminal amino acid residues.

C-terminal analysis may be performed with the aid of the enzyme carboxypeptidase, which specifically cleaves the C-terminal residue from a protein. This amino acid can then be identified. Furthermore, the process can be repeated on the residual protein to give the adjacent residue, and so on to give the sequence.

One method of N-terminal analysis, developed by Sanger, involves reaction of the nucleophilic NH_2 group with 2,4-dinitrofluorobenzene (Figure 12.10). Subsequent hydrolysis of the protein gives an amino acid mixture in which one of the original amino acids has been replaced by a dinitrophenyl derivative. This is the N-terminal amino acid.

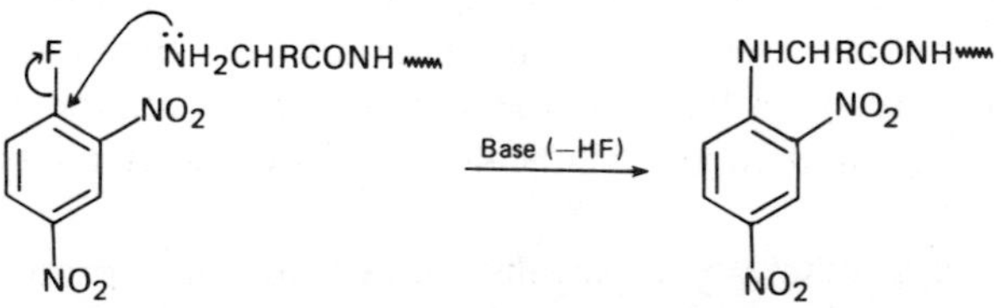

Figure 12.10 Reaction of a peptide terminus
with fluorodinitrobenzene

Question 12.7 Aromatic halides are usually unreactive towards nucleophiles; why is 2,4-dinitrofluorobenzene an exception? Why do the —NHCO— groups not react with this reagent?

Another approach to N-terminal analysis, developed by Edman, involves reaction with phenylisothiocyanate, PhNCS. The product of this reaction is a thiourea (II), which is readily hydrolysed to give an identifiable phenylthiohydantoin (III) under conditions which do not hydrolyse the rest of the protein (Figure 12.11). The whole process can be repeated on the residual protein to obtain the sequence. This process can be automated, the thiohydantoin being identified by gas chromatography (Chapter 3), and the residual protein, after purification by precipitation, being recycled.

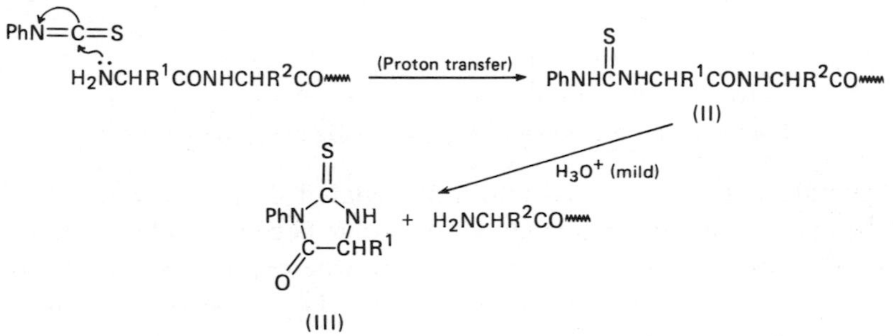

Figure 12.11 Edman's method for labelling N-terminal amino acids

Although in principle the carboxypeptidase and phenylisothiocyanate methods can be used repeatedly to sequence a whole protein, in practice it is practicable only to sequence chains of ten to twenty residues in this way. Consequently, the protein is first subjected to *partial hydrolysis,* which may be performed using mineral acid under conditions less severe than for total hydrolysis, or more conveniently using specific enzymes to cleave peptide bonds adjacent to particular amino acids. The peptides thus produced are separated, purified, and individually sequenced, and then the total sequence is pieced together like a jig-saw puzzle using overlaps in the shorter sequences (Figure 12.12).

Short peptides can also be sequenced by mass spectrometry (Chapter 3).

Peptide A	H – Gly – Phe – Ser – Gly – Ala – OH	
Peptide B	H – Gly – Ala – Val – Thr – Ser – OH	
Peptide C	H – Thr – Ser – Gly – Tyr – OH	
Section of protein	H – Gly – Phe – Ser – Gly – Ala – Val – Thr – Ser – Gly – Tyr – OH	

Figure 12.12 Elucidation of complete protein sequences by overlapping of shorter products of partial hydrolysis

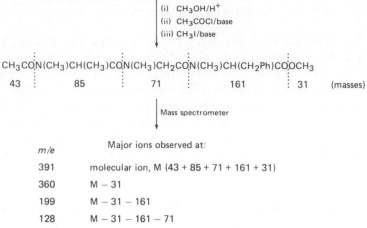

$H_2NCH(CH_3)CONHCH_2CONHCH(CH_2Ph)COOH$

(i) CH_3OH/H^+
(ii) $CH_3COCl/base$
(iii) $CH_3I/base$

$CH_3CON(CH_3)CH(CH_3)CON(CH_3)CH_2CON(CH_3)CH(CH_2Ph)COOCH_3$

| 43 | 85 | 71 | 161 | 31 | (masses) |

Mass spectrometer

m/e	Major ions observed at:
391	molecular ion, M (43 + 85 + 71 + 161 + 31)
360	M − 31
199	M − 31 − 161
128	M − 31 − 161 − 71

Figure 12.13 Sequencing of peptides by mass spectroscopy

The peptide is first converted into a methylated derivative with no NH or OH bonds (no hydrogen bonding), so that it becomes volatile. It is then put into the mass spectrometer and ionized, whereupon it successively loses fragments from the C-terminal end (Figure 12.13). Identification of the fragments lost gives the sequence.

Question 12.9   What sequence of Gly, Ala, and Phe is indicated when the mass spectrum of the derivative shows major ions at m/e values 43, 204, 289, 360, 391?

12.8 Gross structure of proteins

Some proteins and many peptides may be obtained in a crystalline state and have been analysed by X-ray crystallography (Chapter 3). Double Nobel Prize winner Linus Pauling showed that in many compounds the peptide chain is coiled into an α-helix (Figure 12.14(a)), a structure like a clockwise spiral staircase. This structure allows bond angles to have their normal values while avoiding large steric interactions between bulky side chains, which project outwards from the centre of the helix. The structure is maintained by hydrogen bonding between C=O and NH groups on different residues, and there are about 3.6 residues for every turn of the helix. This coiling is known as the *secondary structure* of the protein.

Some proteins (e.g. silk), composed mainly of small amino acids such as glycine and alanine, adopt a different secondary structure called a pleated sheet (Figure 12.14(b)). In these cases a greater degree of hydrogen bonding can be achieved without excessive steric compression.

Further bending of the coiled or pleated chains determines the overall

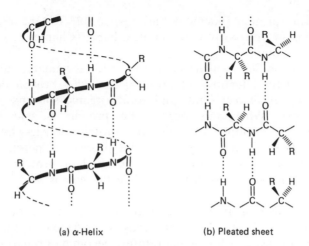

(a) α-Helix   (b) Pleated sheet

Figure 12.14 Secondary structures of peptides and proteins

shape of the protein molecule, which is called its *tertiary structure*. The shape is maintained by weak attractions such as hydrogen bonding, S—S bonding in cystine residues, and *ionic (electrostatic) bridges* between acidic and basic amino acid residues (Figure 12.15). Bends in the chain also occur at the sites of proline and hydroxyproline residues, because of the rigidity imposed by the inclusion of the nitrogen atom in a heterocyclic ring.

The way in which several proteins associate into a large unit is called the *quaternary structure*.

In general, proteins which have roughly linear shapes are insoluble in water and suitable for construction material in skin, muscle, hair, etc. They are classified as *fibrous* proteins. Those with roughly spherical shapes, classified as *globular* proteins, are generally soluble in water and suitable for enzymes, oxygen-carrying proteins, etc. Many of these examples occur not as *simple* proteins, but contain in addition a non-protein portion (a *prosthetic group*) and are known as *conjugated* proteins. For example, the

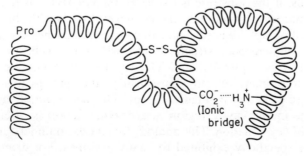

Figure 12.15 Schematic tertiary structure of a protein

oxygen-carrying protein (haemoglobin) responsible for the red colour of blood contains an iron–porphyrin complex, haem (Chapter 13), in addition to the protein, globin.

The weak bonding which maintains the secondary, tertiary, and quaternary structural features of proteins is readily broken down by the action of heat, ultraviolet light, and some organic solvents, or by large changes in pH. When this occurs the protein is said to undergo *denaturation*. For example, albumin, the protein of egg-white, becomes insoluble on heating and coagulates. The process of permanent waving involves chemical reduction of S—S bonds in hair protein, followed by re-oxidation to form new S—S bonds while the hair is held in the desired shape.

Denaturation has a particularly profound effect on the activity of enzymes (Section 12.9). For this reason enzyme-containing washing powders are best used at low temperatures, becoming primarily dependent on their accompanying detergents at high temperature.

12.9 Structure and reactivity of enzymes

Biochemical reactions are catalysed by large protein molecules called *enzymes*. As a result, transformations which in the laboratory require high temperatures, long reaction periods, or extremes of pH, occur rapidly under the conditions prevailing in living systems.

Enzymes are classified according to their function. Thus, an enzyme which catalyses hydrolysis is called a *hydrolase,* whereas a more specific description for one which brings about removal of the C-terminal amino acid residue from a protein is *carboxypeptidase*.

Although enzymes are primarily large proteins, they may contain a non-protein part. In such cases the protein is called the *apoenzyme* and the other part the *coenzyme*. A coenzyme may be a chemically bound prosthetic group or a separate reagent brought into play as reaction proceeds. It may be as simple as the metal ion, Zn^{2+}, or as complex as coenzyme B_{12}, a close relative of vitamin B_{12} (Chapter 13). The coenzyme usually undergoes a reaction complementary to that undergone by the substrate. Thus, if the substrate is oxidized, the coenzyme is reduced; if the substrate eliminates a by-product, the coenzyme reacts with this, and so on.

X-ray crystallography has revealed that several enzymes contain a cavity in their three-dimensional structure into which the substrate can settle, giving a comfortable physical and electrostatic fit. Some flexing of the protein may accompany the process, causing other groups to interact with the substrate (e.g. A and B, Figure 12.16). The *active site* of the enzyme is believed to be in the cavity region and enzyme action is often compared with a *lock and key* situation. This analogy facilitates an understanding of the high substrate specificity exhibited by many enzymes. For example, urease catalyses the hydrolysis of urea, H_2NCONH_2 but has no effect on biuret, $H_2NCONHCONH_2$.

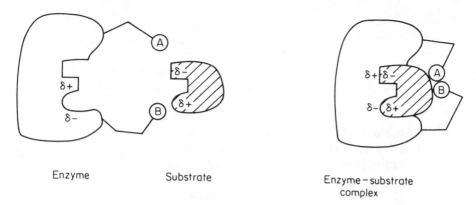

| Enzyme | Substrate | Enzyme–substrate complex |

Figure 12.16 Enzyme action according to the lock and key model

12.10 Synthesis of peptides and proteins

Condensation of any two amino acids, A and B, would give four simple dipeptides (AA, AB, BA, BB) in addition to more complex products. Clearly this is not a satisfactory approach to peptide synthesis. To overcome this problem the peptide is synthesized in a stepwise manner, and in each step the two reactants have one of their functionalities protected. Protecting groups must be chosen such that they can be removed under conditions which do not destroy peptide bonds, and such that one of them can be removed preferentially in the presence of the other.

Carboxyl groups are protected as esters, particularly *t*-butyl esters (removed by mild acid) or methyl esters (removed under alkaline conditions). Amino groups are usually protected as benzyloxycarbonyl derivatives (removed by catalytic hydrogenation) or *t*-butoxycarbonyl derivatives (removed by mild acid). The latter derivatives are prepared by addition of benzyl- or *t*-butyl-chlorocarbonate to the amino acid (Figure 12.17), and are often referred to as Z and *t*-BOC derivatives respectively.

Simply mixing appropriately protected amino acids results in salt formation. The reactants must therefore be activated for condensation to occur. Prior conversion of the carboxyl group to its acid chloride provides effective activation, but this requires an extra step and sometimes results in an unacceptable degree of racemization. A more convenient approach makes use of a condensing agent which is directly added to the mixture of protected amino acids. One such reagent is dicyclohexylcarbodiimide

$$R^1OCOCl + H_2NCHR^2CO_2H \xrightarrow[\text{(—HCl)}]{\text{base}} R^1OCONHCHR^2CO_2H$$

$$R^1 = PhCH_2 \quad \text{(Z derivatives)}$$
$$R^1 = Bu^t \quad \text{(}t\text{-BOC derivatives)}$$

Figure 12.17 Formation of N-protected amino acids

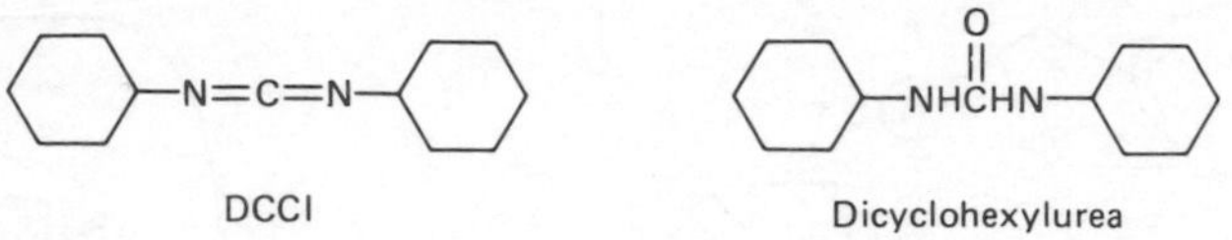

DCCI   Dicyclohexylurea

Figure 12.18 The structure of DCCI and dicyclohexyl urea

(DCCI, Figure 12.18). This first reacts with the free carboxyl group to form an activated derivative, which then reacts with the free amino group to give the protected dipeptide and dicyclohexylurea.

Question 12.9 Suggest a mechanistic pathway for the formation of an amide by reaction of a carboxylic acid, a primary amine, and DCCI.

The next step in building up a large peptide is to remove one of the protecting groups so that the chain can be extended. For example, removal of the N-protecting group, followed by reaction with another N-protected amino acid and DCCI builds up the chain at the N-terminus, leaving the C-terminal residue unaffected. The process is repeated until all necessary residues have been added. The protecting groups are finally removed to give the product (Figure 12.19).

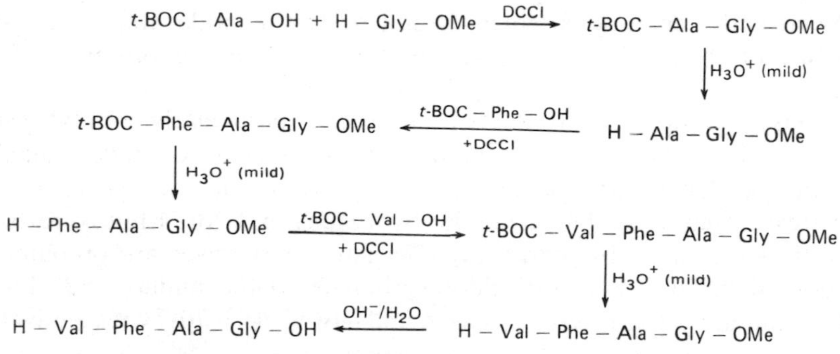

Figure 12.19 A synthetic pathway to a tetrapeptide

Du Vigneaud received the Nobel Prize in 1955 for his syntheses of the hormones oxytocin and vasopressin (Figure 12.20). Oxytocin causes contraction of the uterus; vasopressin regulates excretion of water from the kidneys (Chapter 14).

There are technical difficulties associated with the stepwise building up of very large peptides, and in such cases it is usual to construct smaller peptide 'blocks' of ten to twenty residues which are subsequently joined together.

Clearly this classical approach to the synthesis of large peptides is very tedious and time-consuming. A much quicker, *solid-phase synthesis* has

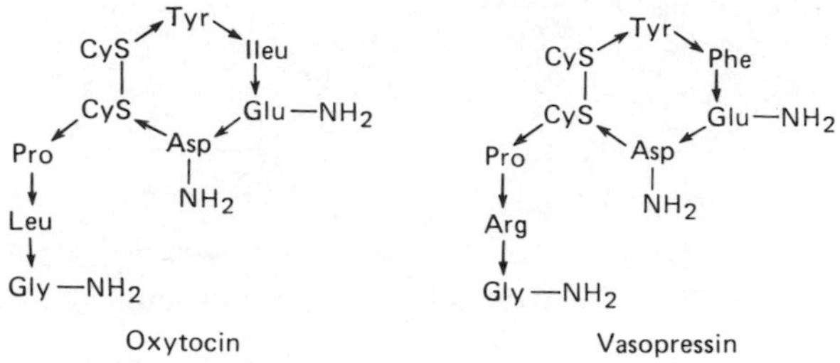

Figure 12.20 The structures of two peptide hormones

been developed by Merrifield. Using this approach the amino acid which will eventually become the C-terminal residue of the peptide is attached to a polymeric resin. The polymer possesses chlorobenzyl groups which react with the free carboxyl group of the N-protected amino acid (Figure 12.21). The classical sequence of deprotection and condensation reactions (Figure 12.19) is then carried out by successively running appropriate solutions through a column of the resin. In this way the peptide chain can be built up very quickly. Finally, the peptide is removed from the resin by acidic cleavage (for example with hydrogen bromide in trifluoroethanoic acid) under conditions which do not cleave peptide bonds. This method for peptide synthesis can be automated.

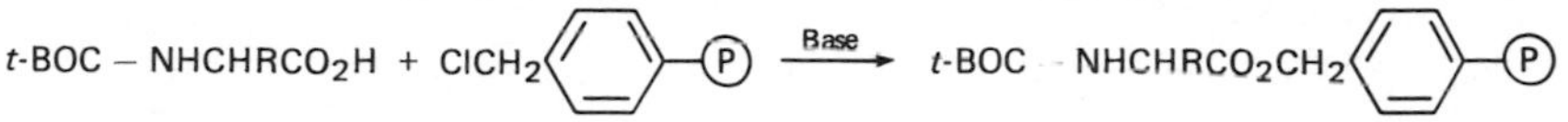

Figure 12.21 Attachment of an N-protected amino acid to a polymeric resin. Solid-phase peptide synthesis technique

Although the solid-phase method is quick and simple there are some disadvantages. Since there is no possibility of intermediate purification steps the final product may be a complex mixture of many very similar peptides which must be separated. For example, even if every condensation step were to proceed in 99% yield, after a hundred residues had been added the percentage of the desired peptide in the mixture would be $0.99^{100} \times 100\% \simeq 35\%$; the remaining 65% would be a mixture of about a hundred other peptides differing in most cases by only a single residue from the desired one. This problem is exacerbated if any hydrolysis of peptide linkages occurs during the deprotection steps. Current research is attempting to overcome these problems.

12.11 Chemical components of nucleic acids

Like proteins, nucleic acids are high molecular mass compounds. Whereas hydrolysis of proteins produces a mixture of amino acids, hydrolysis of nucleic acids yields a mixture containing phosphoric acid, a sugar, and several organic bases. From *ribonucleic acid* (RNA) the sugar is

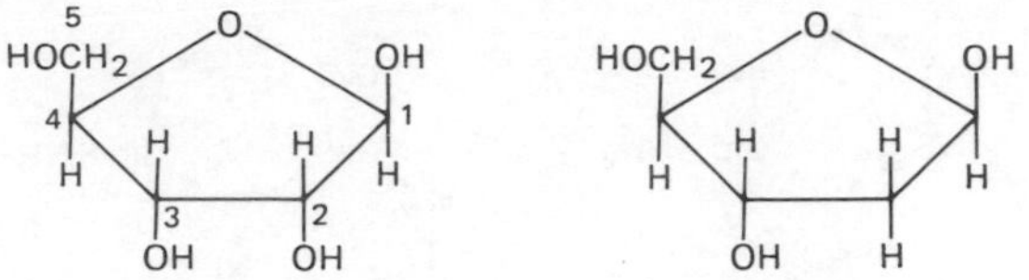

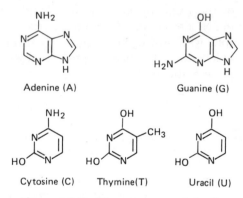

Figure 12.22 The structures of ribose and deoxyribose

ribose whereas from *deoxyribonucleic acid* (DNA) it is *deoxyribose,* which lacks the hydroxy group at carbon atom 2 (Figure 12.22).

The most common bases (Figure 12.23) are adenine, guanine, and cytosine (from both DNA and RNA), thymine (from DNA) and uracil (from RNA). For convenience these are often abbreviated to their initial letters, A, G, C, T, U. Adenine and guanine belong to the general class of heterocyclic bases known as purines, whereas cytosine, thymine, and uracil are pyrimidine bases (Chapter 7).

Adenine (A)   Guanine (G)

Cytosine (C)   Thymine(T)   Uracil (U)

Figure 12.23 The structures of the five
common nucleic acid bases

12.12 Primary structure of Nucleic acids

In the nucleic acids each heterocyclic base is attached via nitrogen to carbon atom 1 of the appropriate sugar molecule. This sugar–base unit is called a *nucleoside*. For example, the nucleoside formed from deoxyribose and adenine is called deoxyadenosine (Figure 12.24(a)).

If each nucleoside unit is attached at the 5'- position (the 5- position of the sugar) to a molecule of phosphoric acid, it becomes a *nucleotide*. For example, the nucleotide formed from cytosine, ribose and phosphoric acid is called cytidylic acid (Figure 12.24(b)).

The names of common nucleosides and nucleotides are listed in Table 12.2.

Nucleotides are the repeating units of nucleic acids in the same way that amino acids are the repeating units of proteins. Adjacent nucleotides are

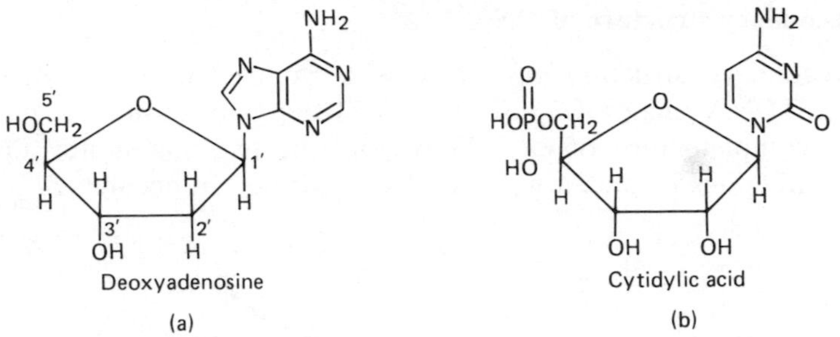

Figure 12.24 The structures of a nucleoside and a nucleotide

Table 12.2

Base	Sugar	Name of nucleoside	Name of nucleotide
Adenine	Ribose	Adenosine	Adenylic acid
Adenine	Deoxyribose	Deoxyadenosine	Deoxyadenylic acid
Guanine	Ribose	Guanosine	Guanylic acid
Guanine	Deoxyribose	Deoxyguanosine	Deoxyguanylic acid
Cytosine	Ribose	Cytidine	Cytidylic acid
Cytosine	Deoxyribose	Deoxycytidine	Deoxycytidylic acid
Thymine	Deoxyribose	Thymidine	Thymidylic acid
Uracil	Ribose	Uridine	Uridylic acid

linked by a bond between the phosphate group of one and the 3'- position of the other. Thus, nucleic acids consist of a sugar–phosphate backbone with projecting base units. For example, a section of a molecule of RNA is illustrated in Figure 12.25.

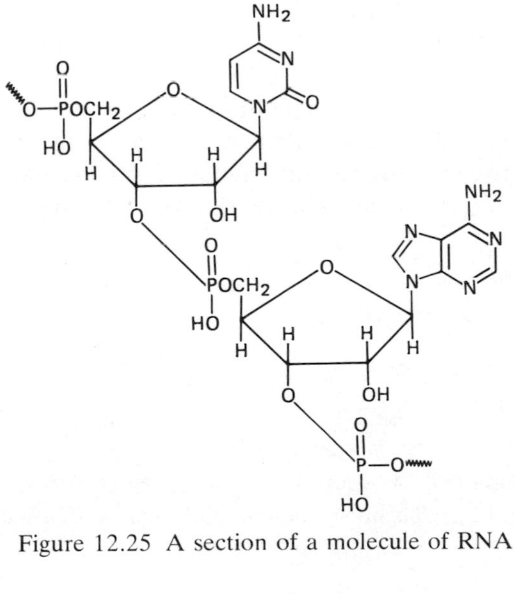

Figure 12.25 A section of a molecule of RNA

12.13 Secondary structure of DNA

The secondary structure of DNA is a double helix comprising two interwoven DNA chains (Figure 12.26). There are about ten nucleotide units per complete turn of each helix, and the two chains are aligned in opposite directions in such a way that base pairing can occur.

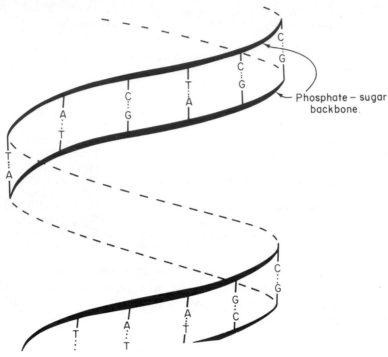

Figure 12.26 A schematic representation of DNA double helix

The two chains form a complementary pair. A thymine residue in one chain is always paired with an adenine residue in the other, and similar pairing exists between cytosine and guanine residues. This is supported by hydrolysis experiments which indicate 1:1 ratios for thymine: adenine and

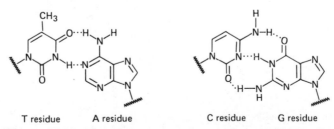

Figure 12.27 Thymine–adenine and guanine–cytosine hydrogen bonding

cytosine:guanine in DNA. Favourable hydrogen bonding is responsible for the specific pairing of appropriate bases (Figure 12.27). Any other pairing produces a less favourable interaction. Crick, Watson, and Wilkins shared a Nobel Prize in 1962 for their work in this area.

Question 12.10 What sequence of bases is complementary to the sequence ACGTAG?

12.14 Replication of DNA

The cells of a living organism are continually undergoing division. When cell division takes place it is necessary for the DNA molecule to be exactly reproduced. This process is called *replication* and is believed to occur in the following way. The two chains of the double helix begin to separate and a new chain is formed alongside each original one. The parent chain acts as a *template,* lining up the nucleotides in such a way that complementary bases are always paired. Thus, when the new chain is produced it is complementary to the parent chain, and therefore the new double helix is identical with the original.

12.15 Biosynthesis of proteins—the role of RNA

Uracil residues form specific hydrogen bonds with adenine residues in the same way as thymine, and uracil in RNA replaces thymine in DNA. With this modification, RNA molecules are constructed on the DNA template in a manner similar to the replication of DNA. However, a particular RNA molecule corresponds to only a section, not the whole, of the DNA chain. Consequently, a single DNA molecule directs the synthesis of many different RNA molecules.

RNA molecules play an important role in the biosynthesis of proteins. There are two main types which are responsible for aligning the amino acids in the appropriate way. These are called *messenger RNA* (m-RNA) and *transfer* (or *soluble*) *RNA* (t-RNA).

Transfer RNA molecules are relatively small (about seventy nucleotide units) and are thus mobile in the cell medium. They contain two essential active sites: (i) a binding site which is specific for a particular amino acid, and (ii) a sequence of three base residues (an *anticodon*) capable of hydrogen bonding with three complementary residues (a *codon*) on the messenger RNA.

Messenger RNA molecules are usually large and can be looked upon as a series of codons. The transfer RNA molecules, together with their associated amino acids, line up alongside the messenger RNA in a way which allows specific hydrogen bonding between the base residues on the codons and anticodons (Figure 12.28). While they are thus held, the amino acids are linked together to form the protein.

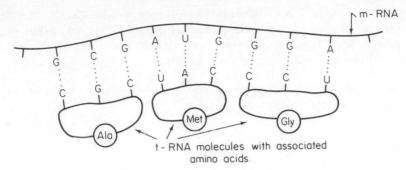

Figure 12.28 The roles of m-RNA and t-RNA in protein biosynthesis

Each messenger RNA molecule is responsible for the synthesis of a particular protein. The section of the DNA chain which gives rise to the messenger RNA molecule is called a *gene*. A *chromosome* consists of two double helices of DNA in a protein sheath.

12.16 Viruses

Viruses are aggregates of protein and nucleic acid material (*nucleoproteins*). Once inside a living cell they interfere with normal protein biosynthesis by directing the synthesis of new proteins corresponding to the viral nucleic acid. Since proteins, especially enzymes, are involved in the entire biochemistry of a living organism, a virus causes tremendous disruption and disease.

Answers to Questions

12.1(a) 2-Amino-3-hydroxypropanoic acid;
(b) 2-amino-3-methylbutanoic acid;
(c) 2,6-diaminohexanoic acid.

12.2(a)

$$H_2N—\overset{CO_2H}{\underset{CH_2OH}{|}}—H$$

L-Serine

According to the priority rules $NH_2 > CO_2H > CH_2OH > H$. Thus, L-serine has the S configuration.

(b)

$$H_2N—\overset{CO_2H}{\underset{CH_3}{|}}—H$$

L-Alanine

$NH_2 > CO_2H > CH_3 > H$. L-Alanine has the S configuration.

12.3 Lysine has a second NH_2 group which is largely protonated at pH 6.0; to bring this back to neutrality base must be added and the isoelectric point is higher, actually 9.6. At pH 5.0 (a) glycine and (b) lysine are protonated and move towards the negative electrode; (c) aspartic acid is anionic at pH 5.0 and moves towards the positive electrode.

12.4   The excess of ammonia is used to minimize polyalkylation (Chapter 7).

12.5
$$\underset{H_2NCHCONHCHCONHCHCO_2H}{\overset{\underset{|}{CH_3}\quad\overset{|}{CH_2OH}\ \overset{|}{CH_2SCH_3}}{}}$$

12.6   Mass of 2 Gly + 1 Ala + 1 Phe + 3 Ser = 150 + 89 + 165 + 315 = 719; measured molecular mass $\sim$ 1500, so all quantities must be doubled. The composition is Gly (4 molecules), Ala (2), Phe (2), Ser (6).

12.7   The presence of strongly electron-withdrawing nitro groups in the *ortho-* and *para-* positions decreases electron density at the C—F bond, thus making the molecule more susceptible to attack by nucleophiles (Chapter 5). Because of overlap of the nitrogen lone pair with the carbonyl group, amides are much less nucleophilic than amines (Chapter 7).

12.8   Phe—Ala—Gly.

12.9

12.10 TGCATC.

CHAPTER 13

Tetrapyrrolic Compounds: Photosynthesis and Respiration

Many vitally important natural compounds contain a linked series of four pyrrole units, although some of these may be in a reduced form. Such *tetrapyrrolic* units occur, for example, in oxygen-carrying proteins such as haemoglobin; in *cytochromes*, the proteins responsible for electron transport in the respiratory chain; in *chlorophylls* and *bacteriochlorophylls*, molecules which are directly implicated in the photosythetic processes in plants and photosynthetic bacteria; in vitamin B_{12}, the anti-pernicious anaemia vitamin; in bile pigments; and in some marine toxins. Figure 13.1 shows a selection of important natural tetrapyrrolic compounds along with an example of a phthalocyanin. Phthalocyanins are non-natural tetrapyrrolic compounds which are synthesized on a large scale for use as dyestuffs.

Two diseases associated with tetrapyrrolic compounds are jaundice and acute porphyria. Jaundice is a yellowing of the skin caused by a build-up of bile pigments in the blood resulting from a malfunction in haem metabolism. Acute porphyria shows itself by a change in the colour of urine to deep red on standing in air and light due to the conversion of colourless bile compounds into coloured ones.

13.1 General Characteristics of Tetrapyrroles

The most important tetrapyrrolic compounds have the four pyrrole units bound in a macrocyclic ring, and except in the vitamin B_{12} series the rings are joined at all points by a single carbon bridge. The basic ring system of haem and the cytochrome prosthetic group is called *porphin* (Figure 13.2(a)) and substituted compounds are known as *porphyrins*. They have a fully conjugated system of π electrons, of which eighteen form a delocalized ring and impart aromaticity to the molecules. The basic ring system of the chlorophylls and bonellin is *chlorin* (Figure 13.2(b)) in which one of the pyrrole rings (ring D) is reduced to the dihydropyrrole level. In the bacteriochlorophylls both ring D and ring B are reduced to the dihydropyrrole level. The chlorophylls and bacteriochlorophylls also have an additional (fifth) ring (see Figure 13.1). All of these compounds retain the aromatic system of eighteen π electrons.

Vitamin B_{12} belongs to a general class of compounds known as *corrins*, of which the simplest is corrin itself (Figure 13.2(c)). In this series two of the rings are joined directly, without a carbon bridge. The corrin molecule contains only six double bonds, and corrins are not aromatic.

Question 13.1 (a) On what basis is it justifiable to refer to porphyrins, chlorins, and bacteriochlorins as aromatic? (b) Why is it that corrins are not aromatic?

All of the extensively conjugated tetrapyrroles are highly coloured. Porphyrins are generally red to purple, chlorins brown or green to blue, and corrins pink. Considerable variation in colour may arise from changes in substituents, or on complexation to a metal.

Complexation of metals, for which all macrocyclic tetrapyrroles show a great tendency, is a very important property of these compounds. Thus, haem is an iron complex, chlorophylls are magnesium complexes, and vitamin B_{12} is a cobalt complex. In the laboratory it is relatively simple to incorporate a variety of other metals.

Since tetrapyrroles contain extensive π systems they are susceptible to attack by electrophiles and may be fairly readily oxidized.

13.2 Photosynthesis

All living organisms require fuel to provide energy to enable them to do work. The fuel required is in the form of organic compounds, which in aerobic organisms are reacted with oxygen from the air in a process known as *respiration* (Section 13.3). These organic compounds are synthesized from carbon dioxide using energy from the sun by a process known as *photosynthesis*. The only organisms capable of carrying out this process are green plants and a few bacteria. Thus animals and non-photosynthesizing plants must take in these organic fuels as food.

Because of its fundamental role as the primary producer of organic fuel, photosynthesis is arguably the most important of all biochemical processes. It is an extremely complex process, involving many different chemical reactions which vary between organisms. It is not well understood, and it is possible only to give a brief outline here. Green plants contain a variety of organic pigments in the chloroplasts. Not only chlorophylls, but also other pigments such as carotenoid compounds are present, and together they provide the facility for absorption of light of almost any wavelength in the visible spectrum. Thus, even on a dull day, what little light is available can be efficiently collected.

Once the light has been absorbed, the pigment molecules are said to be in excited states and readily transfer energy to acceptor molecules whose excited states are lower in energy. Ultimately the molecules with the lowest energy excited states (i.e. longest wavelength absorptions) receive the energy. These are the chlorophylls. In fact, a large proportion of the

I

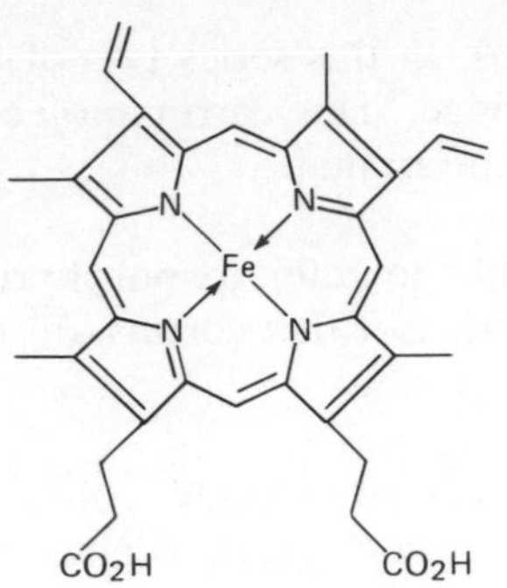

Haem, the prosthetic group of haemoglobin

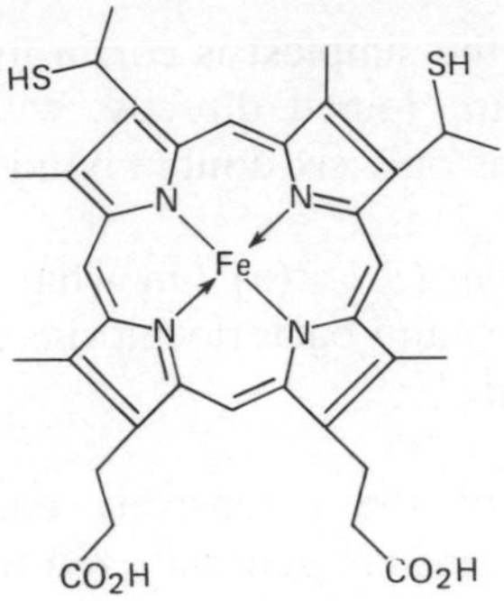

The porphoryrin prosthetic group of cytochrome c

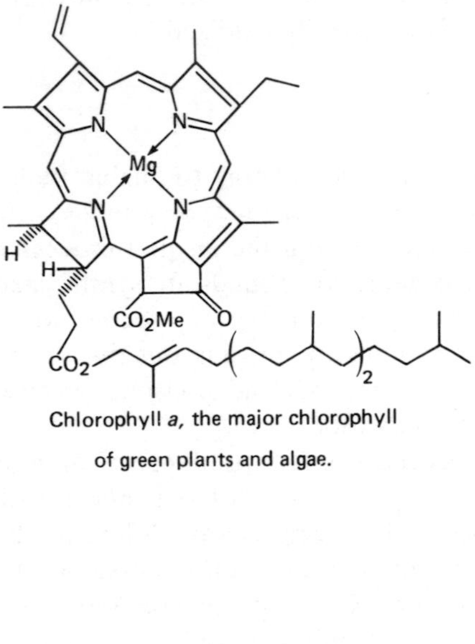

Chlorophyll a, the major chlorophyll

of green plants and algae.

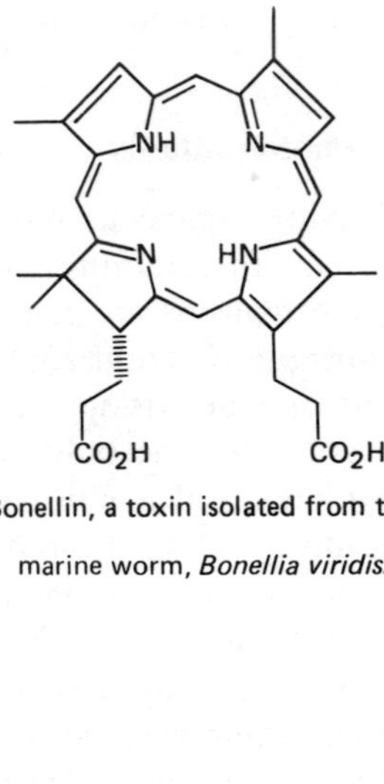

Bonellin, a toxin isolated from the

marine worm, *Bonellia viridis.*

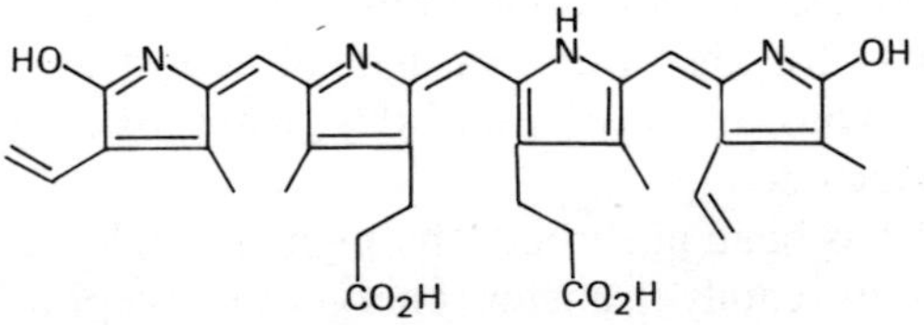

Biliverdin, a green bile pigment.

Figure 13.1

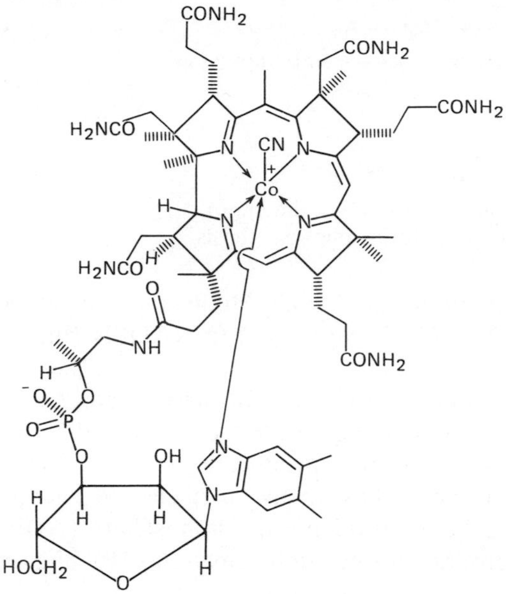

CONH₂ ... (chemical structure labels)

Vitamin B₁₂, the anti-pernicious

anaemia factor

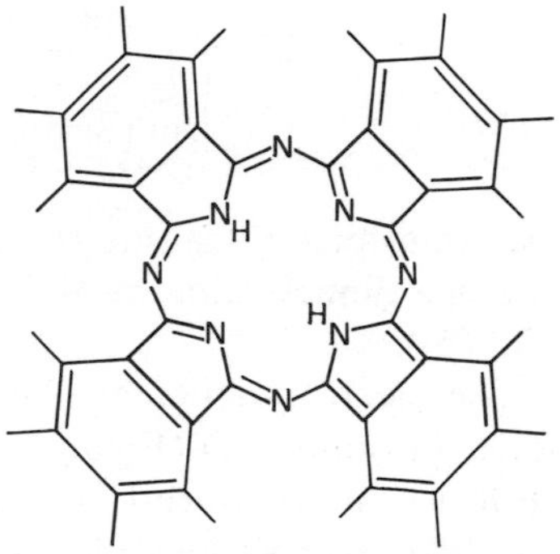

A blue-green phthalocyanine dyestuff

Figure 13.1 The structures of some important tetrapyrrolic compounds

chlorophyll molecules present in the chloroplast is used for light-gathering purposes, but a small proportion, referred to as *reaction centre chlorophyll*, is actually responsible for the primary photochemical reaction. Because of the wavelength of its light absorption ($c.\,700$ nm), reaction centre chlorophyll is referred to as P700. The structure of P700 is not known, though it is closely related to chlorophyll *a* (Figure 13.1). Current speculation centres upon the possibility that P700 is a dimer of chlorophyll *a* monohydrate. For our present purposes the symbol Chl is used to refer to reaction centre chlorophyll, and Chl* to the molecule in its excited state.

In the primary photochemical step the Chl* molecule readily gives up an electron to an acceptor molecule, A (Figure 13.3(a)). The extra electrons

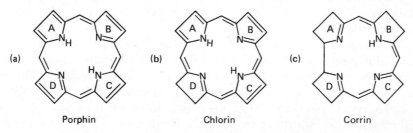

(a) Porphin

(b) Chlorin

(c) Corrin

Figure 13.2 Porphin, chlorin, and corrin ring systems

(a) $Chl^* + A \longrightarrow Chl^+ + A^-$

(b) $2A^- + H^+ + NADP^+ \longrightarrow NADPH + 2A$

(c) $Chl^+ + D \longrightarrow Chl + D^+$

(d) $2D^+ + H_2O \longrightarrow 2D + 2H^+ + \frac{1}{2}O_2$

(e) $ADP + P_i + Energy \longrightarrow ATP$

Figure 13.3 An outline of some of the important reactions occurring during photosynthesis

are transferred via intermediates until they finally reduce nicotinamide adenine dinucleotide phosphate, $NADP^+$ (Chapter 14) to NADPH (Figure 13.3(b)).

The electron-deficient Chl^+ is reduced to Chl by the first member of a series of donors, D (Figure 13.3(c)). Eventually, a positively charged donor molecule is produced which oxidizes water, producing oxygen (Figure 13.3(d)). The various electron transport steps involve a release of energy, which is used to effect a reaction between adenosine diphosphate (ADP) and inorganic phosphate (P_i) to give adenosine triphosphate (ATP) (Figure 13.3(e)).

Using energy from the sun, three vitally important species are produced in the above sequence of reactions, i.e. NADPH, oxygen, and ATP, NADPH (Section 14.3) is a reducing agent, a biological equivalent of $NaBH_4$, and is needed to effect the reduction of carbon dioxide to the oxidation level of carbohydrates. Oxygen is required for the oxidation of carbohydrates, which provides the energy for many biochemical processes, and ATP is the basic energy store for all living things (Section 15.1).

As yet we have said nothing about the actual reactions which occur in the 'fixing' of carbon dioxide. Figure 13.4 outlines the basic reactions. All are catalysed by enzymes, and occur perfectly well in the dark.

The cycle starts when the ATP produced in the photochemical reactions converts ribulose-5-phosphate into its diphosphate. This combines with carbon dioxide forming an unstable six-carbon unit which gives two moles of glyceric acid-3-phosphate. More ATP and NADPH are required to reduce this to glyceraldehyde-3-phosphate and its isomer dihydroxyacetone phosphate. These two three-carbon units are converted into fructose-6-phosphate, which can then follow the normal pathways of carbohydrate metabolism (Chapter 15). Alternatively carbohydrate may be stored as sucrose or starch.

However, only one mole of glyceraldehyde-3-phosphate out of every six follows this pathway. The other five moles must undergo a complex series of reactions to regenerate three moles of ribulose-5-phosphate. This process is represented in schematic form in Figure 13.5. Two moles of glyceraldehyde-3-phosphate (C_3) form one mole of fructose-6-phosphate (C_6) (reaction a) which reacts with a third mole of glyceraldehyde-3-phosphate (C_3) giving xylulose-5-phosphate (C_5) and erythrose-4-phosphate

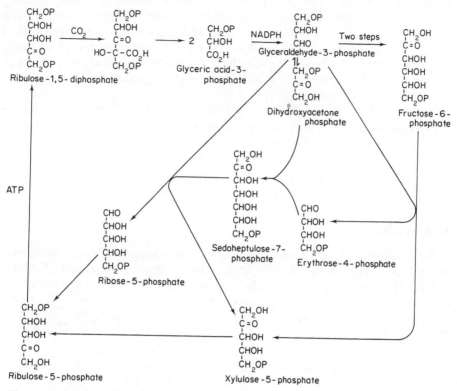

Figure 13.4 The pathways by which carbon dioxide is converted to carbohydrate

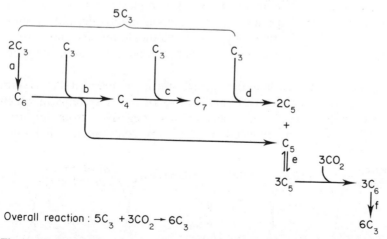

Overall reaction : $5C_3 + 3CO_2 \rightarrow 6C_3$

Figure 13.5 Fixation of carbon dioxide leading to formation of six moles of glyceraldehyde from five moles of the latter

248

(C_4) (reaction b). The latter reacts with a fourth C_3 unit forming sedoheptulose-7-phosphate (C_7) (reaction c). After reaction with a fifth mole of glyceraldehyde-3-phosphate (C_3) this gives ribose-5-phosphate (C_5) and xylulose-5-phosphate (C_5) (reaction d). The three moles of C_5 sugar are isomerized to ribulose-5-phosphate (reaction e), which absorb three moles of carbon dioxide, eventually producing six moles of glyceraldehyde-3-phosphate (reaction f).

The overall result of the photosynthetic sequence is the conversion of carbon dioxide, water, and energy into glucose and oxygen (Figure 13.6).

$$6CO_2 + 6H_2O \longrightarrow C_6H_{12}O_6 + 6O_2$$

Figure 13.6 The overall result of the photosynthetic reaction sequence

13.3 Respiration

Respiration is the process by which aerobic organisms reduce molecular oxygen to water using NADH and reduced flavin coenzymes ($FADH_2$ + $FMNH_2$) produced by the citric acid cycle (Section 15.4). Respiration operates in such a way that the energy which is gradually released is used to synthesize ATP from ADP and inorganic phosphate (Section 15.1).

Tetrapyrrolic compounds, particularly porphyrins, are intimately involved in this process. In higher animals the oxygen must first of all be transported by the circulatory system to the site where it is required. This occurs through the agency of haemoglobin. Oxygen forms a weak complex with the iron atom of the haem (Figure 13.1). When the oxyhaemoglobin reaches an oxygen deficient site the weak complex breaks down, liberating oxygen. Several other species, notably carbon monoxide and cyanide ion, complex more strongly than oxygen to haemoglobin. By thus preventing the formation of oxygen complexes such compounds exhibit considerable toxicity.

Once oxygen has reached the mitochondria of the cell the respiratory chain can operate. This consists of a series of redox systems in which the redox potentials increase (i.e. become less negative) from left to right (Figure 13.7). Thus the reduced form of any component will reduce the

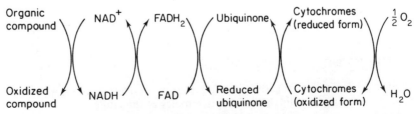

Figure 13.7 The mitochondrial respiratory chain

oxidized form of any system which occurs to the right of it in the diagram, with the liberation of energy which is used to sythesize ATP.

The electron transfer steps involving cytochromes depend upon changes in the oxidation state of the central metal cations (Fe^{3+}/Fe^{2+}; Cu^{2+}/Cu^{+}), whereas those involving ubiquinone depend upon the interconversion of the quinone and dihydroxybenzene forms (Section 6.9). For every mole of NADH oxidized, three of ATP are formed; on the other hand, oxidation of one mole of $FADH_2$ generates two moles of ATP.

Answers to questions

13.1 (a) Aromatic compounds contain a planar ring of $(4n + 2)$ fully conjugated π electrons (Chapter 5): the eighteen π electron tetrapyrrolic systems satisfy this rule (b) The π system of the corrin molecule does not extend completely around the ring.

CHAPTER 14

Other Physiologically Important Compounds

In addition to the major compound classes covered in Chapters 10–13, there are many other compounds with important physiological effects. Although they do not fall naturally into distinct compound classes, they can be categorized as vitamins, hormones, or medicinal compounds according to their physiological role. Vitamins are essential dietary components necessary for the correct functioning of biochemical processes, whereas hormones are synthesized within the body to control these processes. These two classes of substances have in common the fact that they are usually needed in only small quantities. Many natural products have medicinal properties, but in addition to these there are many synthetic drugs which are widely used for the control of disease.

14.1 The vitamins

Like essential amino acids, vitamins cannot be synthesized by animals and must be included in their diets. However, unlike amino acids, the daily requirements are of the order of milligrams rather than grams. Disease arising from vitamin deficiency can be caused by poverty, pregnancy, dieting, or excessive use of processed foods. On the other hand excessive intake of the vitamins A, D, K, and E can result in their accumulation in body fat where they occasionally reach toxic concentrations. Large doses of the water soluble vitamins B and C do not cause problems because they are readily excreted. Vitamins are important because in many cases they are components of coenzymes.

The more important vitamins are listed in Table 14.1 and discussed in more detail below.

14.2 Vitamin A

The structure of vitamin A is given in Figure 14.1. Its role in the function of the retina is discussed in Chapter 8.

14.3 Vitamin B complex

The B vitamins are grouped together for purely historical reasons, since they occur together in a water soluble fraction derived from milk.

250

Table 14.1 The more important vitamins

Code letter		Name	Dietary source	Result of deficiency
A			Dairy products, liver, fish oils, carrots	Gradual loss of vision especially at night
B complex	B_1	Thiamine	Wheat germ, eggs, liver, peas, beans	Beriberi
	B_2	Riboflavin	Milk, liver, yeast, green vegetables	Mouth ulcers, skin rashes
	B_6	Pyridoxine	Liver, yeast, cereals	Anaemia
	B_{12}	Cyanocobalamin	Meat, milk, eggs	Pernicious anaemia
		Nicotinamide	Liver, yeast, milk, vegetables, unpolished rice	Pellagra
		Folic acid	Liver, yeast, green vegetables	Anaemia
C		Ascorbic acid	Citrus fruits, blackcurrants, rosehips, many vegetables	Scurvy
D		Calciferols	Dairy products, fish oils	Rickets in children
E		Tocopherols	Soya bean, wheat germ, green vegetables	Reduced fertility
H		Biotin	Yeast, milk, egg yolk, liver	Dermatitis, weight loss, excessive excretion of NH_3
K			Green leaves	Delayed blood clotting

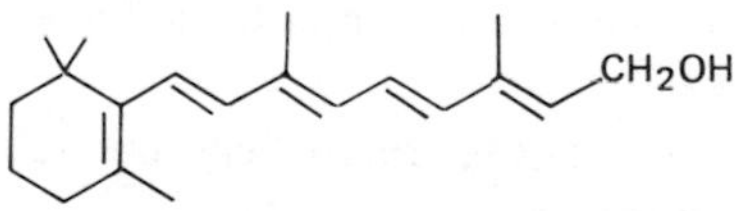

Vitamin A

Figure 14.1 The structure of vitamin A

Thiamine (vitamin B_1), as its pyrophosphate (Figure 14.2(a)), is involved, along with lipoic acid (Figure 14.2(b)), in the decarboxylation of α-keto acids such as pyruvic acid (CH_3COCO_2H). This is important in the formation of acetyl coenzyme A, which is of fundamental importance in the biochemical oxidation of fats and carbohydrates and the biosynthesis of many natural products (Chapter 15).

Riboflavin (vitamin B_2) is a component of flavin mononucleotide (FMN) and flavin adenine dinucleotide (FAD). The latter is an ester of riboflavin

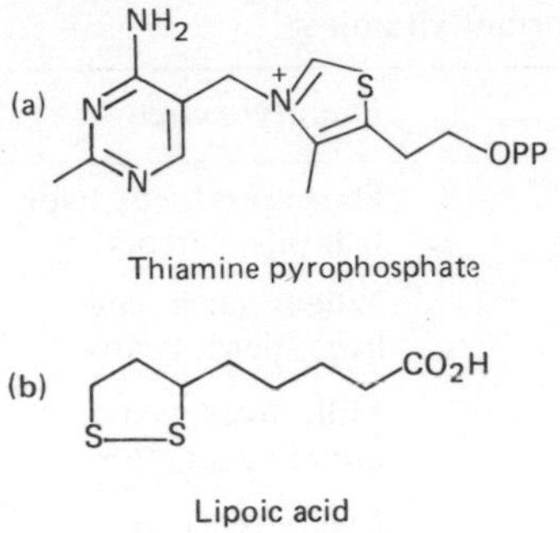

(a)

Thiamine pyrophosphate

(b)

Lipoic acid

Figure 14.2 The structures of thiamine pyrophosphate and lipoic acid

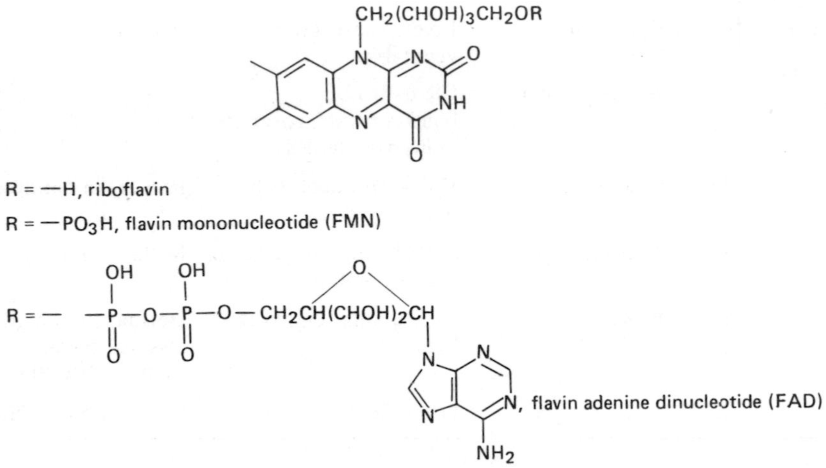

$CH_2(CHOH)_3CH_2OR$

R = —H, riboflavin

R = —PO₃H, flavin mononucleotide (FMN)

R = — $-P-O-P-O-CH_2CH(CHOH)_2CH$ ..., flavin adenine dinucleotide (FAD)

Figure 14.3 The structures of vitamin B_2 and associated compounds

with ADP (Figure 14.3). These flavin coenzymes usually occur as the prosthetic groups of proteins.

They are involved in oxidation–reduction reactions in which two hydrogen atoms are reversibly added to the riboflavin part of the molecule (Figure 14.4). This is important, for example, in the respiratory chain (Section 13.3).

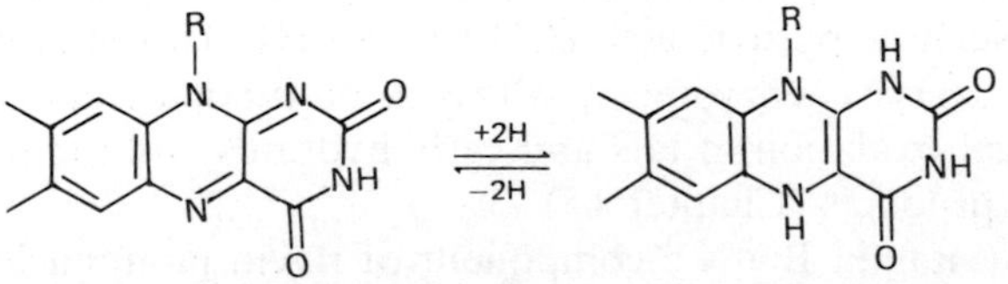

Figure 14.4 Oxidized and reduced forms of riboflavin

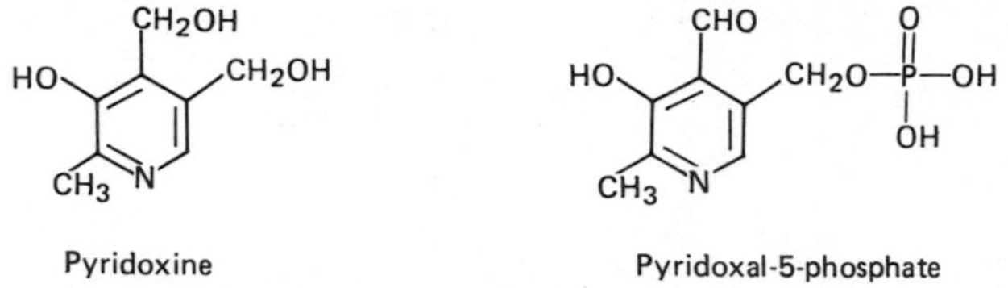

Pyridoxine   Pyridoxal-5-phosphate

Figure 14.5 The structures of pyridoxine and pyridoxal-
5-phosphate

Pyridoxine (vitamin B_6) is converted *in vivo* to pyridoxal-5-phosphate (Figure 14.5), which is involved in the deamination and transamination of amino acids.

The structure of cyanocobalamin (vitamin B_{12}) is given in Figure 13.1. In the body it is converted into coenzyme B_{12}. The Nobel Prize for Chemistry in 1964 went to Dorothy Crowfoot Hodgkin for determining the structure of vitamin B_{12} by X-ray diffraction techniques. The laboratory synthesis of vitamin B_{12} by R. B. Woodward and A. Eschenmoser represents one of the major achievements of organic synthesis.

Nicotinamide forms part of nicotinamide adenine dinucleotide (NAD^+) and its phosphate ($NADP^+$), (Figure 14.6), both of which are involved in oxidation–reduction reactions (Chapters 13 and 15).

The function of NAD^+ and $NADP^+$ as oxidizing agents formally involves abstraction of H^- from an organic compound by the pyridinium unit (Figure 14.7).

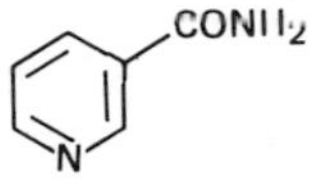

Nicotinimide

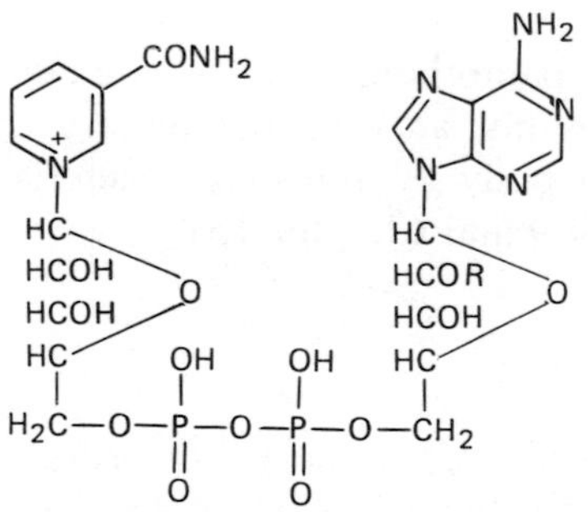

R =   H, NAD^+

R = PO_3H, $NADP^+$

Figure 14.6 The structures of the
coenzymes derived from nicotinamide

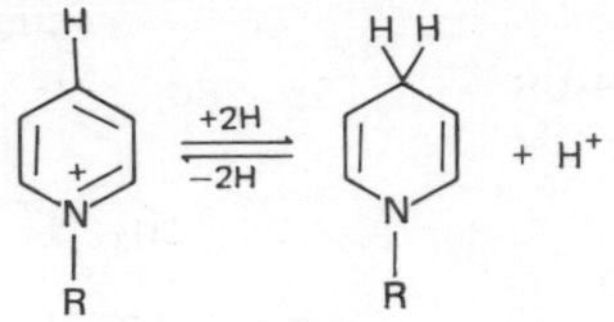

Figure 14.7 Oxidized and reduced forms of the nicotinamide coenzymes

The tetrahydro derivative of folic acid (Figure 14.8) is important in the transfer of single carbon units (—CH$_3$, —CH$_2$OH or —CHO) to other molecules.

Such processes are involved, for example, in the biosynthesis of purine and pyrimidine bases which are constituents of nucleic acids (Chapter 12).

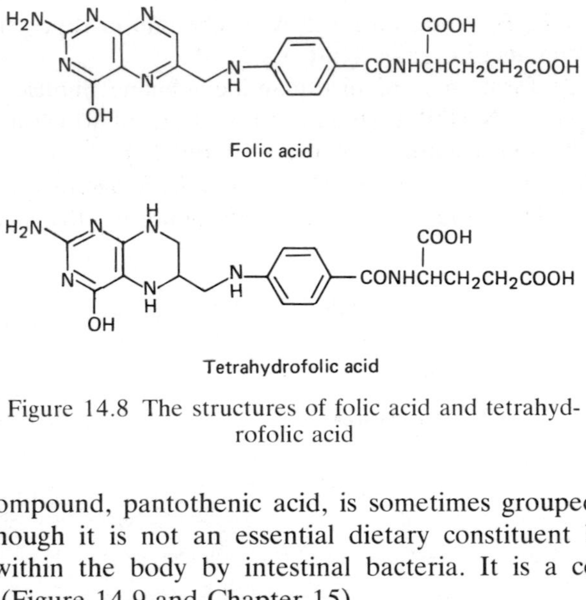

Folic acid

Tetrahydrofolic acid

Figure 14.8 The structures of folic acid and tetrahydrofolic acid

Another compound, pantothenic acid, is sometimes grouped with the B vitamins, although it is not an essential dietary constituent because it is synthesized within the body by intestinal bacteria. It is a constituent of coenzyme A (Figure 14.9 and Chapter 15).

14.4 Vitamin C

Ascorbic acid (vitamin C) is a good reducing agent and is easily converted to dehydroascorbic acid (Figure 14.10). The reaction is reversed by mild reducing agents. The biochemical role of vitamin C is not well understood, though it is undoubtedly associated with its reducing properties. The γ-lactone ring of ascorbic acid is readily hydrolysed on heating in neutral or alkaline solution, resulting in a reduced vitamin C content in cooked vegetables.

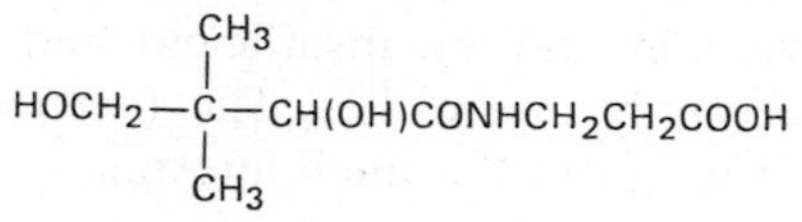

Pantothenic acid

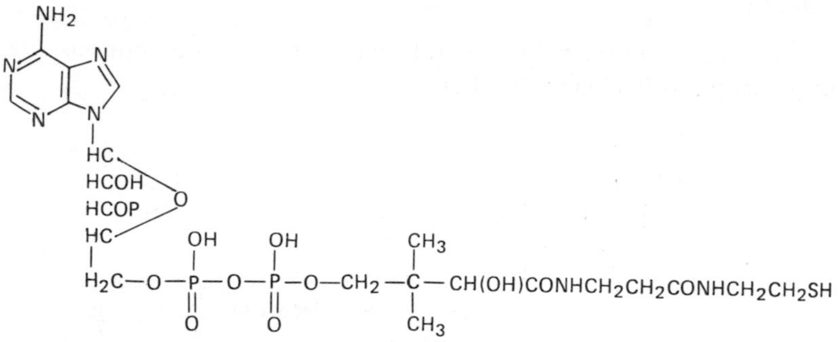

Coenzyme A

Figure 14.9 The structures of pantothenic acid and coenzyme A

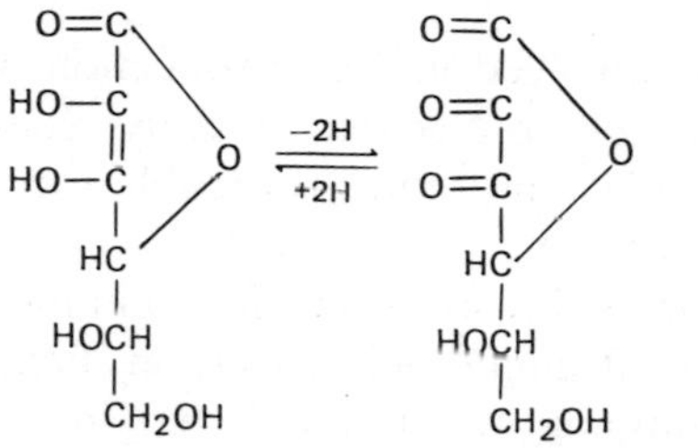

Ascorbic acid   Dehydroascorbic acid

Figure 14.10 The behaviour of ascorbic acid as a reducing agent

14.5 Vitamin D

Several D vitamins are known, the most common being calciferol (D_2) and cholecalciferol (D_3) (Figure 14.11). They are formed from sterols in mammalian skin by exposure to sunlight. However, in temperate climates

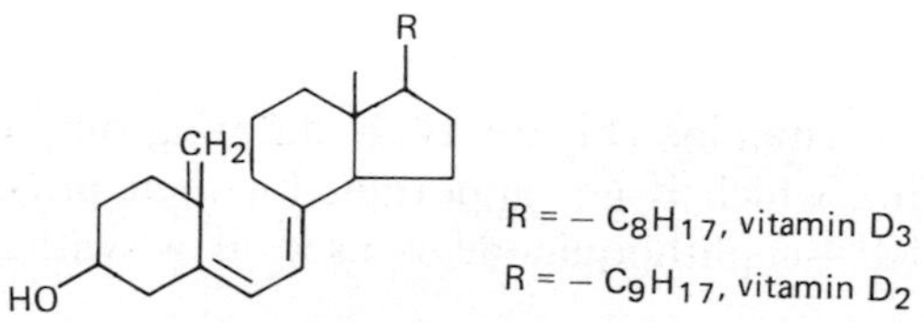

$R = - C_8H_{17}$, vitamin D_3
$R = - C_9H_{17}$, vitamin D_2

Figure 14.11 The structures of vitamin D_2 and vitamin D_3

during the winter formation of the vitamin may be insufficient and it is thus necessary to ensure an adequate supply in the diet. The function of vitamin D is to promote absorption of Ca^{2+} from the small intestine.

14.6 Vitamin E

Vitamin E again consists of several closely related compounds, the α-, β-, and γ-tocopherols (Figure 14.12).

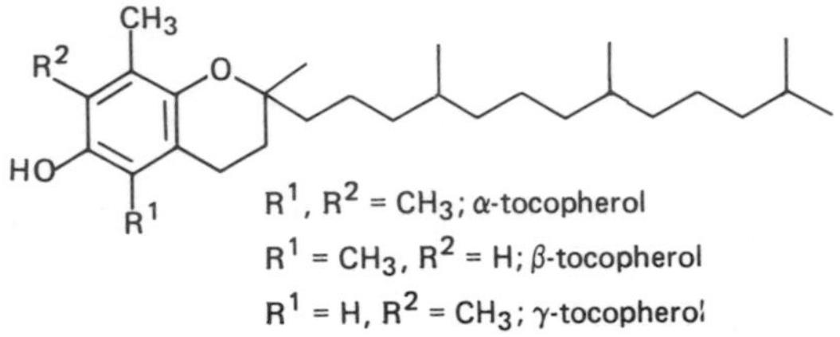

$R^1, R^2 = CH_3$; α-tocopherol
$R^1 = CH_3, R^2 = H$; β-tocopherol
$R^1 = H, R^2 = CH_3$; γ-tocopherol

Figure 14.12 The structures of the tocopherols

14.7 Vitamin H

Biotin (vitamin H) is involved in the introduction of carbon dioxide into organic molecules, e.g. the conversion of acetyl coenzyme A to malonyl coenzyme A (Chapter 15). The biotin reversibly forms an adduct with the carbon dioxide (Figure 14.13).

In higher animals it is involved in the elimination of the ammonia formed by deamination of amino acids. Carbon dioxide, carried by biotin, combines with the ammonia and is converted in several steps into urea which is excreted.

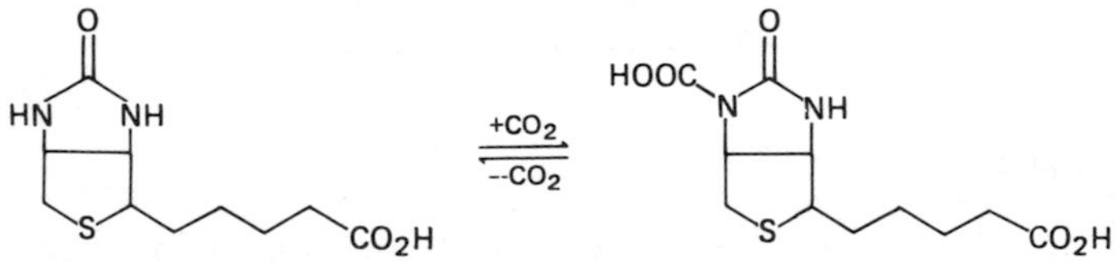

Biotin

Figure 14.13 The reversible addition of carbon dioxide to biotin

14.8 Vitamin K

There are two K vitamins (Figure 14.4) differing only in the size of the isoprene side chain, which is not important for their physiological activity. Indeed, 2-methyl-1,4-naphthoquinone is used as a synthetic substitute for the vitamin.

Another series of quinones, the ubiquinones (Figure 14.15), often referred to collectively as coenzyme Q, are involved in electron transport reactions in the respiratory chain (Section 13.3).

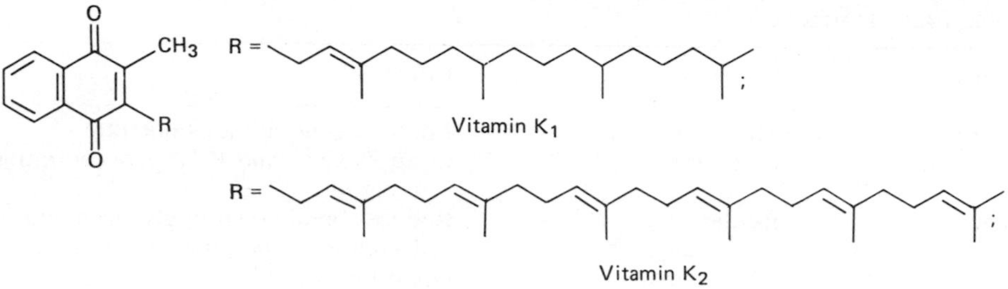

R = H; 2-methyl-1,4-naphthoquinone

Figure 14.14 The structures of vitamin K_1, vitamin K_2 and
2-methyl-1,4-naphthoquinone

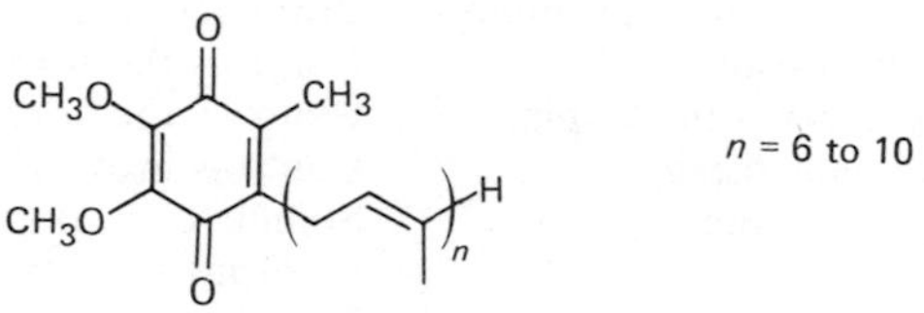

Coenzyme Q (ubiquinones)

Figure 14.15 The structures of the
ubiquinones

14.9 Hormones

Hormones are chemical messengers, produced within the endocrine glands of animals, that stimulate or inhibit metabolic activity in other tissues or organs. Regulatory compounds in plants are sometimes also referred to as hormones.

In mammals, hormones are distributed by the blood stream or lymphatic system. Some hormones affect cell membrane permeability and control the movement of substances, whereas others alter enzyme kinetics or control protein synthesis.

Structurally most hormones are derived either from amino acids or steroids. The more important hormones are listed in Table 14.2.

14.10 The thyroid and parathyroid hormones

Thyroxine (Figure 14.16) is synthesized in the thyroid from tyrosine residues in the protein thyroglobulin, iodine being introduced by electrophilic attack on the aromatic rings.

Overactivity of the thyroid gland results in hyperthyroidism, characterized by nervous restlessness and loss of weight caused by stimulation of the general metabolism. Underproduction of thyroxine (hypothyroidism) reduces metabolic rate and retards mental, skeletal, and sexual development in children.

258

Table 14.2 Hormones

Source	Hormone	Effect
Thyroid	Thyroxine	Controls general metabolic rate
Parathyroid	Parathyrin	Controls Ca^{2+} and PO_4^{3-} concentration in blood
Pancreas	Insulin	Reduces breakdown of glycogen and decreases blood glucose levels
	Glucagon	Opposite to insulin
Adrenal medulla	Adrenaline	Increases breakdown of glycogen to glucose
	Noradenaline	Causes constriction of skeletal muscle and cardiac arterioles
Adrenal cortex	Cortisone	Controls glycogen and protein metabolism
	Aldosterone and others	Promotes retention of Na^+
Pituitary	Thyrotropin	Controls the thyroid
	Adrenocorticotropin	Controls the adrenal cortex
	Somatotropin	Controls growth
	Lactotropin	Stimulates hormone production by the corpus luteum and lactation
	Interstitial cell stimulating hormone	Stimulates testosterone and progesterone production
	Follicle stimulating hormone	Promotes growth of gonads
	Oxytocin	Causes contraction of uterus
	Vasopressin	Causes contraction of capillaries and arterioles
Ovary	Oestrogens	Cause proliferation of uterine lining prior to ovulation
Corpus luteum	Progesterone	Prepares uterus for implantation of fertilized ovum and maintains pregnancy
	Relaxin	Causes relaxation of pelvic ligament during childbirth
Testes	Testosterone	Causes development of sperm and secondary sexual characteristics

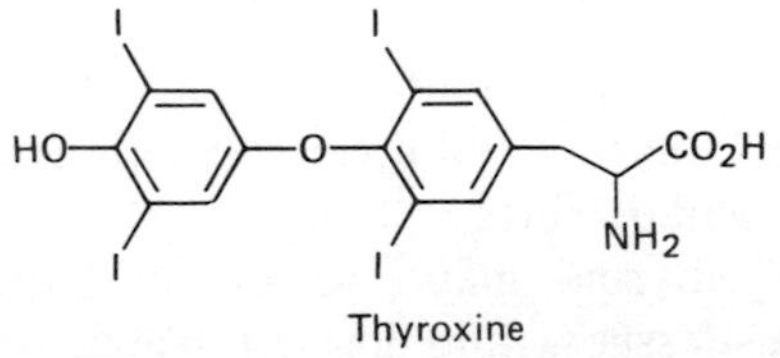

Thyroxine

Figure 14.16 The structure of thyroxine

The parathyroids are four very small glands at the back of the thyroid. The hormone they produce is a polypeptide with a molecular mass of about 100,000. Excess hormone (hyperparathyroidism) results in the loss of calcium from bones and its increased concentration in the blood, and also in increased excretion of phosphate. Hypoparathyroidism has the opposite effect and can be treated with vitamin D (Section 14.5).

14.11 The pancreatic hormones

The pancreas produces two polypeptide hormones, insulin and glucogen, which control glucose levels in the bloodstream.

A deficiency of insulin causes diabetes mellitus. This results in increased breakdown of glycogen in the liver and impaired transport of glucose into the cells, causing a building up of glucose in the bloodstream and its excretion in the urine. Because insufficient blood glucose reaches the cells, fat is metabolized to provide energy, and the breakdown products accumulate eventually causing coma and death if untreated. Treatment is by injection of insulin, but this must be carefully controlled since excess insulin lowers the blood sugar level too far, resulting in sweating, palpitation, unconsciousness, and finally death.

Glucagon has the opposite effect to insulin, i.e. it raises the blood sugar level.

14.12 The adrenal hormones

The adrenal glands, situated just above the kidneys, consist of an inner cortex and an outer medulla. The two most important hormones of the medulla are adrenaline (epinephrine) and noradrenaline (norepinephrine) (Figure 14.17).

Adrenaline has the effect of increasing blood glucose levels by promoting breakdown of glycogen. It increases heart rate and dilates skeletal arterioles, while constricting subcutaneous arterioles. In general, noradrenaline has the opposite effect, causing constriction of both skeletal and cardiac arterioles.

Unlike those of the medulla the hormones of the cortex are essential to life. Surgical removal of the cortex, or loss of its function caused by

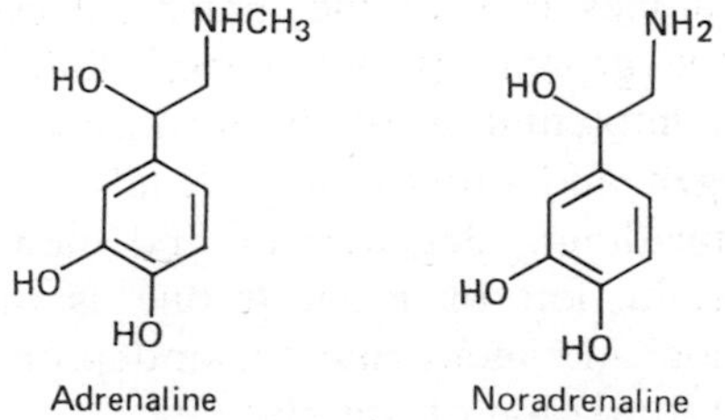

Adrenaline   Noradrenaline

Figure 14.7 The structures of adrenaline and noradrenaline

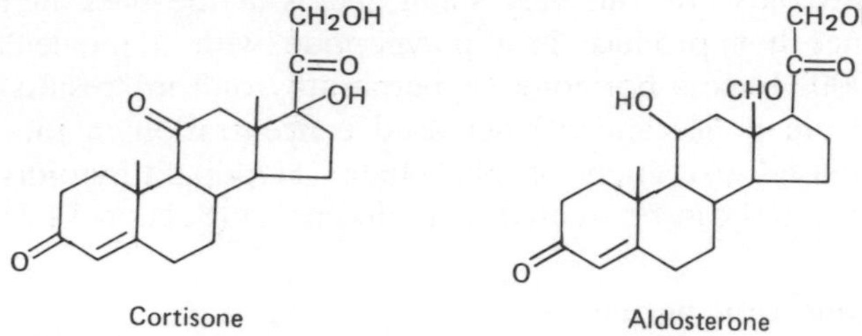

Figure 14.18 The structures of cortisone and aldosterone

Addison's disease, results in decreased excretion of K^+ and water, increased excretion of Na^+, Cl^-, and HCO_3^- and decreased levels of liver and muscle glycogen. Death usually occurs within a week if the condition is untreated. The adrenal cortex produces many hormones, including cortisone, which is concerned with glycogen metabolism, and aldosterone, which controls the ionic balance (Figure 14.18). Both are members of a class of compounds known as *steroids*, a class to which cholesterol (Section 10.3) and the sex hormones (Section 14.13) also belong.

14.13 The pituitary hormones

The pituitary gland or hypophysis, is a small gland situated at the base of the brain. It consists of anterior, middle, and posterior lobes. Hormones are produced in the posterior and anterior lobes and are all polypeptides. Oxytocin and vasopressin (see Sction 12.10) are produced in the posterior lobe, and the other six are produced in the anterior lobe. The pituitary gland is often considered to be the master gland since the hormones produced regulate the action of the other endocrine glands. This is particularly important in relation to the production of female sex hormones during the menstrual cycle and pregnancy, since the follicle stimulating hormone and interstitial cell stimulating hormone are intimately involved in controlling the levels of oestrogens and progesterone.

14.14 The sex hormones

The female sex hormones control the menstrual cycle and pregnancy. Oestrogens facilitate the growth of the uterine lining and progesterone prepares the uterus for implantation of the fertilized ovum. If fertilization does not occur oestrogen and progesterone levels decrease, and without progesterone the uterine lining degenerates and menstruation ensues. If fertilization occurs, production of progesterone is maintained and this inhibits further ovulation and menstruation, and promotes embedding of the fertilized ovum and formation of the placenta.

At the onset of labour the corpus luteum secrets relaxin, which enables the pelvic ligaments to relax and allows passage of the baby through the

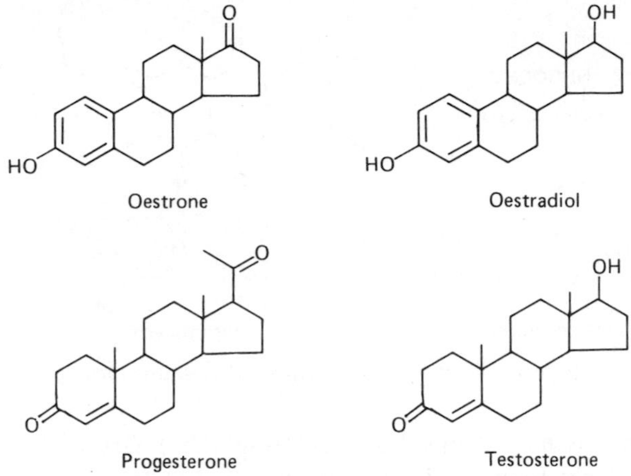

Figure 14.19 The structures of four sex hormones

birth canal. After parturition a sharp fall in the level of ovarian hormones allows increased formation of lactotropin, thus stimulating milk production.

In males, the mature testes produce testosterone, which is responsible for development of secondary male sexual characteristics and the maturation of sperm.

The sex hormones are all members of a class of compounds known as *steroids*, which have a fused system of three six-membered rings and a five-membered ring. Two examples of oestrogens, together with progesterone and testosterone, are given in Figure 14.19.

Oestrogenic hormones and their synthetic analogues can be used to regulate the level of female sex hormones artificially, and are therefore widely used in contraceptive pills.

14.15 Medicinal compounds

The range of compounds used for medicinal purposes is vast and only a few representative examples can be given here. Some others are included where appropriate in other chapters.

One of the most important classes of medicines is the antibiotics, i.e. compounds which kill microorganisms or stop their growth. Examples include the penicillins and cephalosporins (Chapter 7 and 15), the tetracyclines (Chapter 5), and chloramphenicol and erythromycin (Figure 14.20). Sulphonamides (Chapter 7), though not strictly speaking antibiotics, stop bacteria multiplying by competing with them for the vitamin, folic acid.

The production of most commonly used antibiotics involves microbial fermentation, although the initial products may subsequently be modified chemically. Chloramphenicol is now manufactured by chemical means, though it is a natural product.

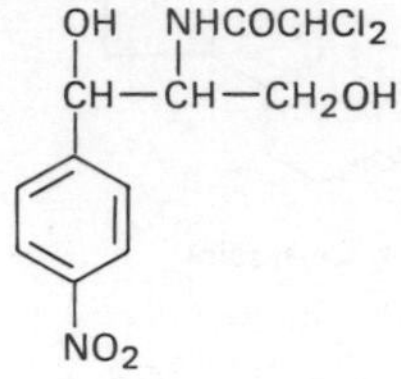

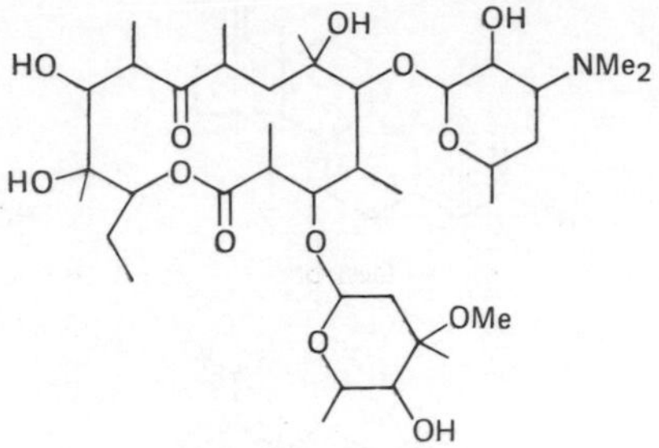

262

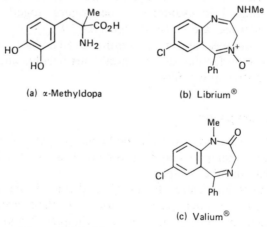

Chloramphenicol   Erythromycin

Figure 14.20 The structures of two antibiotics

Many cardiovascular drugs (i.e. affecting the heart and blood vessels) are chemical derivatives of the so-called catecholamines, of which adrenaline and noradrenaline (Section 14.12) are examples. For example, α-methyldopa (Figure 14.21(a)) is a useful antihypertensive agent. Some tranquillizers act by competing with or affecting the action of catecholamines on the central nervous system. Two of the milder tranquillizers which find widespread application are Librium® and Valium® (Figures 14.21(b) and 14.21(c) respectively). Hallucinogenic, hypnotic, and sedative drugs also act on the central nervous system.

(a) α-Methyldopa   (b) Librium®

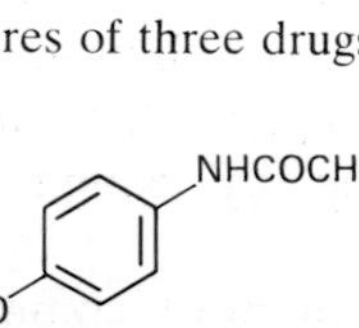

(c) Valium®

Figure 14.21 The structures of three drugs

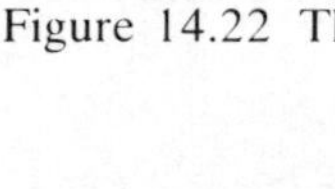

(a) Aspirin 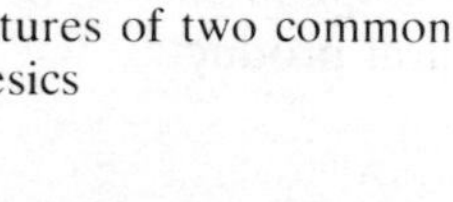   (b) Paracetamol

Figure 14.22 The structures of two common
analgesics

Analgesics are compounds which alleviate pain without significantly impairing consciousness. The most widely used is aspirin (Figure 14.22(a)); others include paracetamol (acetominophen, Figure 14.22(b)) and morphine and its analogues (Section 6.9).

One of the major areas in which new drugs are required is for the treatment of cancer. Several naturally occurring compounds have significant anti-tumour activity. but are difficult to obtain in sufficiently large amounts. Some synthetic anti-cancer drugs can help in controlling certain tumours, but most have severe side effects and must be used with caution.

CHAPTER 15

Metabolism and Biosynthesis

In this book it is possible to touch only briefly upon some of the more important aspects of the chemical reactions occurring in living organisms. These processes by which chemical compounds are synthesized and degraded are known collectively as *metabolism*, comprising *catabolism* (degradation) and *anabolism* (synthesis). *Biosynthesis* is the general word used to describe the pathways by which molecules are synthesized in nature.

Metabolism may be conveniently subdivided into primary and secondary metabolism. The primary processes occur almost universally throughout the plant and animal kingdoms, whereas secondary processes have a much more restricted distribution, often producing metabolites of no obvious importance to the organism.

Although many of the reactions involved in metabolism appear to be rather complex in many respects, they resemble reactions encountered in previous chapters. For example, phosphorylation of hydroxyl groups converts them into good leaving groups, so that the adjacent carbon atom becomes susceptible to nucleophilic attack (Chapter 6), whereas oxidation and reduction reactions are of widespread importance.

The relationships between the various metabolic and biosynthetic pathways are summarized in Figure 15.1.

15.1 Energy in living organisms

All living organisms, whether plant or animal, require energy. This energy is obtained by the controlled oxidation of fats and carbohydrates (or proteins if excess is present in the diet). For example, the complete oxidation of one mole of glucose results in the evolution of 2814 kJ of energy (Figure 15.2). The same total amount of energy is liberated whether glucose is oxidized by direct combustion or in a stepwise fashion by glycolysis, the citric acid cycle and the respiratory chain (Sections 15.2, 15.4, and 13.3).

Since the conversion of adenosine diphosphate (ADP) into adenosine triphosphate (ATP) (Figure 15.3) is endothermic, this process can be used to absorb the energy liberated during the stepwise oxidation of glucose and other substrates.

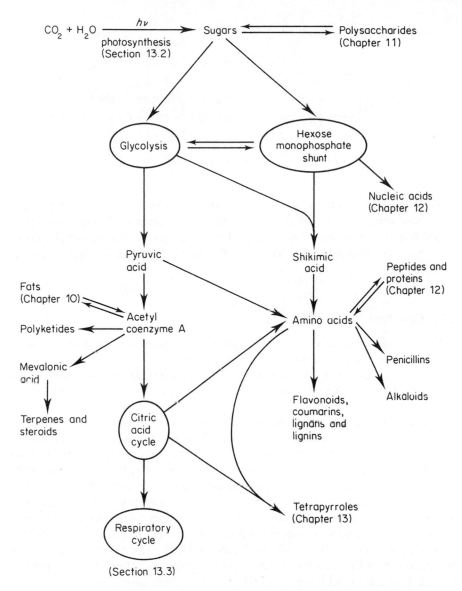

Figure 15.1 The relationships between the metabolic and biosynthetic pathways

$$C_6H_{12}O_6 + 6O_2 \longrightarrow 6CO_2 + 6H_2O$$

$$\Delta H = -2814 \text{ kJ mol}^{-1}$$

Figure 15.2 The complete oxidation of
glucose

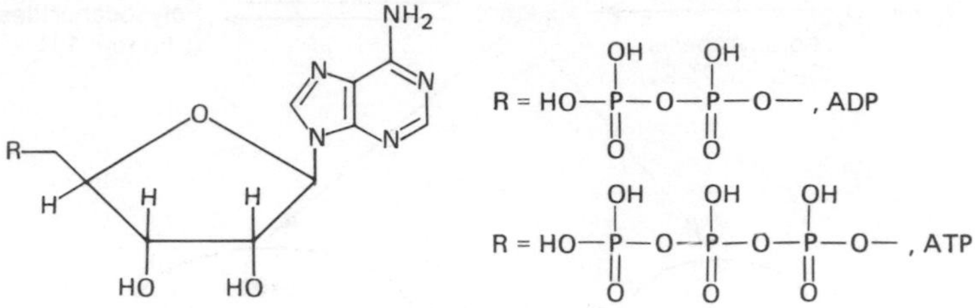

Figure 15.3 The structures of ADP and ATP

ATP is highly reactive and effects phosphorylation of many compounds. Once phosphorylated, these compounds are destabilized and themselves become more reactive. They may then undergo reactions which would otherwise be endothermic. Thus ATP can be considered as an 'energy store'.

15.2 Glycolysis

Glucose and fructose, or such compounds which can be hydrolysed to these, are the major nutritional carbohydrates.

The first stage in the oxidation of these hexoses to carbon dioxide and water is their breakdown into three-carbon units by a process known as *glycolysis* (Figure 15.4).

Glucose or fructose is converted into fructose-1,6-diphosphate which is cleaved into two interconvertible three-carbon units, dihydroxyacetone phosphate and glyceraldehyde-3-phosphate. This latter step is a retro-aldol reaction (i.e. a reverse of the aldol condensation, Chapter 8) and the steps from glucose to the two three-carbon units are an exact reversal of some steps in the dark reaction of photosynthesis (Section 13.2). The two three-carbon units are converted into two moles of pyruvic acid by a sequence of several steps during which four moles of ADP are converted into ATP and two moles of NAD^+ into NADH. Taking into account the two moles of ATP used to phosphorylate the hexose, there is a net gain of two moles each of ATP and NADH over the whole glycolysis process.

When adequate supplies of oxygen are present to reoxidize the NADH in the respiratory cycle (Section 13.3), the pyruvic acid enters the citric acid cycle (Section 15.4) and is completely broken down to carbon dioxide. If oxygen is absent the NADH reduces pyruvic acid to lactic acid (Figure 15.5). For example, lactic acid is produced in muscle tissue during excessive exertion and this gives rise to characteristic muscle pains. On resting the oxygen supply is restored and the lactic acid is reoxidized to pyruvic acid which enters the citric acid cycle.

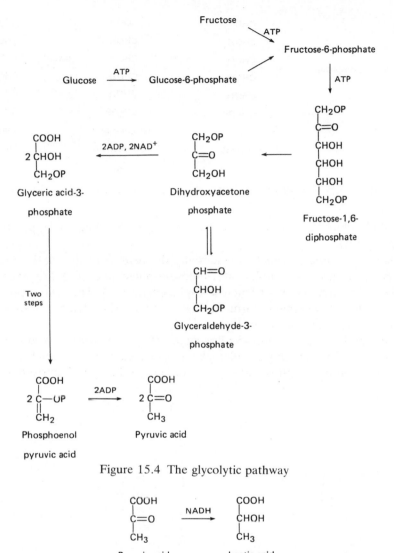

Figure 15.4 The glycolytic pathway

Figure 15.5 The reduction of
pyruvic acid to lactic acid

15.3 The hexose monophosphate shunt

The main alternative metabolic pathway of glucose and fructose is
known as the *pentose phosphate cycle* or *hexose monophosphate shunt*.
Fructose-6-phosphate is isomerized to glucose-6-phosphate which is
oxidized to gluconic acid-6-phosphate (C_6). This loses carbon dioxide,
leaving ribulose-5-phosphate (C_5). During these oxidations two moles of

Fructose-6-phosphate ⟶

$$
\begin{array}{c}
\text{CHO} \\
| \\
\text{CHOH} \\
| \\
\text{CHOH} \\
| \\
\text{CHOH} \\
| \\
\text{CHOH} \\
| \\
\text{CH}_2\text{OP}
\end{array}
$$

Glucose -6- phosphate

$\xrightarrow{\text{NADP}^+}$

$$
\begin{array}{c}
\text{COOH} \\
| \\
\text{CHOH} \\
| \\
\text{CHOH} \\
| \\
\text{CHOH} \\
| \\
\text{CHOH} \\
| \\
\text{CH}_2\text{OP}
\end{array}
$$

Gluconic acid -6-phosphate (C_6)

$\xrightarrow{\text{NADP}^+}$

$$
\begin{array}{c}
\text{CO}_2 \\
+ \\
\text{CH}_2\text{OH} \\
| \\
\text{C}=\text{O} \\
| \\
\text{CHOH} \\
| \\
\text{CHOH} \\
| \\
\text{CH}_2\text{OP}
\end{array}
$$

Ribulose-5- phosphate (C_5)

Figure 15.6 The conversion of fructose-6-phosphate to ribulose-5-phosphate

NADPH, which can enter the respiratory cycle (Section 13.3), are formed (Figure 15.6).

Then, by a process which is essentially the reverse of the 'dark reactions' of photosynthesis, six moles of ribulose-5-phosphate (C_5) are eventually converted to five moles of fructose-6-phosphate (C_6). The whole process is represented in schematic form in Figure 15.7. The structures of all the sugars involved are shown in Figure 13.4.

Two moles of ribulose-5-phosphate (C_5) isomerize to xylulose-5-phosphate (C_5) and ribose-5-phosphate (C_5) (reactions a) which combine, via sedoheptulose-7-phosphate (C_7) and glyceraldehyde-3-phosphate (C_3)

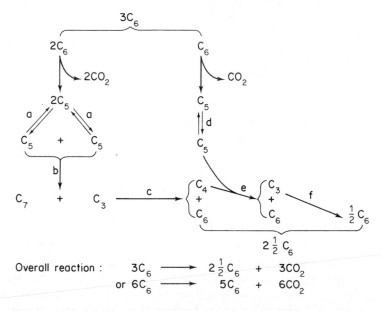

Overall reaction : $3C_6 \longrightarrow 2\tfrac{1}{2}C_6 + 3CO_2$

or $6C_6 \longrightarrow 5C_6 + 6CO_2$

Figure 15.7 Schematic representation of the hexose monophosphate shunt

(reaction b), to give fructose-6-phosphate (C_6) and erythrose-4-phosphate (C_4) (reaction c). Another mole of ribulose-5-phosphate (C_5) is isomerized to xylulose-5-phosphate (C_5) (reaction d) which reacts with the erythrose-4-phosphate (C_4) to give fructose-6-phosphate (C_6) and glyceraldehyde-3-phosphate (C_3) (reaction e). Two moles of the latter can combine giving a further mole of fructose-6-phosphate (C_6) (reaction f). Overall the net result is that out of six moles of hexose which pass through the cycle five are reformed and one is completely oxidized to carbon dioxide.

The importance of the pathway is that it is an alternative to glycolysis and the citric acid cycle for the complete oxidation of hexoses to carbon dioxide, and that it provides precursors for other metabolic pathways. Ribose-5-phosphate is a source of ribose and deoxyribose, which are components of nucleic acids (Chapter 12), ATP (Section 15.1), NAD^+, and FAD (Section 14.2). Erythrose-4-phosphate and phosphoenol pyruvic acid (produced by glycolysis, Section 15.2) combine together to give shikimic acid, which is the precursor of aromatic amino acids (Section 15.8).

15.4 The citric acid cycle

The citric acid cycle, also known as the tricarboxylic acid cycle or Kreb's cycle, is the means by which pyruvic acid is completely broken down into three moles of carbon dioxide (Figure 15.8). It is also important as a source of intermediates used for the biosynthesis of amino acids (Section 15.8).

Pyruvic acid is converted into acetyl coenzyme A with loss of one mole of carbon dioxide. Acetyl coenzyme A reacts with oxaloacetic acid in an aldol-like reaction (chapter 8) to give citric acid, which is then converted via a series of other tricarboxylic acids into oxalosuccinic acid. This sequentially loses two moles of carbon dioxide to give succinic acid as its coenzyme A derivative, which is subsequently converted in a series of steps back into oxaloacetic acid. The oxaloacetic acid can then react with a further mole of acetyl coenzyme A to repeat the cycle.

As a consequence of these reactions four moles of NADPH or NADH and one mole of $FADH_2$ are formed. It is when these are oxidized in the respiratory cycle (Section 13.3) that the energy obtained by oxidation of pyruvic acid is finally available in a biochemically useful form as ATP.

Question 15.1 (a) Draw curly arrows to indicate the mechanism of the aldol condensation of oxaloacetic acid with acetyl coenzyme A. (b) Explain why oxalosuccinic acid readily undergoes decarboxylation.

15.5 Fat Metabolism

Triglycerides (Chapter 10) are an important part of the diet of animals since they can be broken down with liberation of considerable energy.

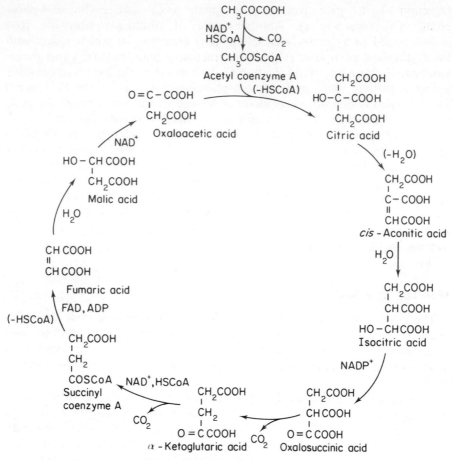

Figure 15.8 The citric acid cycle

They also have an important storage function in that any food eaten in excess of immediate requirements can be deposited as fat.

The first stage in the metabolism of ingested fats is their enzyme catalysed hydrolysis to glycerol and fatty acids. The glycerol is phosphorylated and oxidized to glyceraldehyde-3-phosphate which can enter the glycolysis pathway (Section 15.2). Degradation of the fatty acids proceeds by the sequence shown in Figure 15.9, culminating in a reverse Claisen-type reaction in which coenzyme A displaces acetyl coenzyme A, leaving a fatty acid with two fewer carbon atoms.

Question 15.2 Draw a mechanism for the retro-Claisen reaction shown in the last step of Figure 15.9.

Since most naturally occurring fatty acids have even numbers of carbon atoms, the process shown in Figure 15.9 is repeated until they are

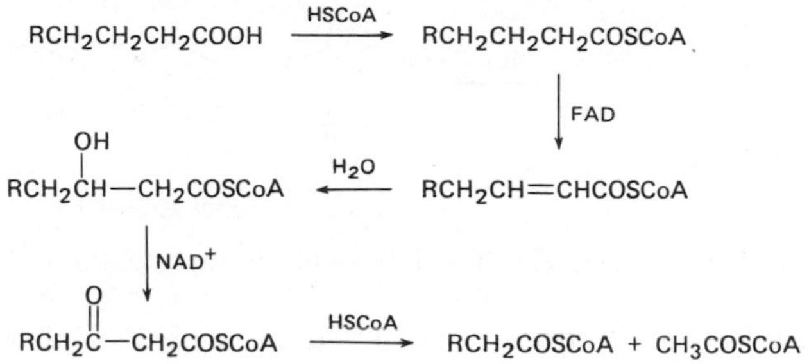

HSCoA
RCH2CH2CH2COOH
RCH2CH2CH2COSCoA
FAD
OH
H2O
RCH2CH—CH2COSCoA
RCH2CH=CHCOSCoA
NAD+
O
HSCoA
RCH2C—CH2COSCoA
RCH2COSCoA + CH3COSCoA

Figure 15.9 Degradation of a fatty acid

converted entirely into acetyl coenzyme A. If, however, the acid contains an odd number of carbon atoms the last unit is propionyl coenzyme A, which is then converted into succinyl coenzyme A. Both acetyl coenzyme A and succinyl coenzyme A can enter the citric acid cycle (Section 15.4).

Fatty acid biosynthesis can proceed by either of two pathways. One is essentially a reversal of the breakdown process whereas the other, more important, pathway involves prior carboxylation of acetyl coenzyme A (Section 14.7) to give the more reactive malonyl coenzyme A (Figure 15.10).

In this pathway the malonyl group is first transferred to the thiol group of an acyl carrier protein (AcP-SH). An acetyl coenzyme A molecule undergoes a similar transfer and the two units, while attached to the proteins, react together to give the acetoacetyl derivative. This is then reduced in a series of steps to butanoyl-SAcP, and successive additions of two-carbon fragments in a similar way give rise to fatty acids containing even numbers of carbon atoms.

Question 15.3 Suggest a mechanism for the reaction of malonyl-SAcP with acetyl-SAcP.

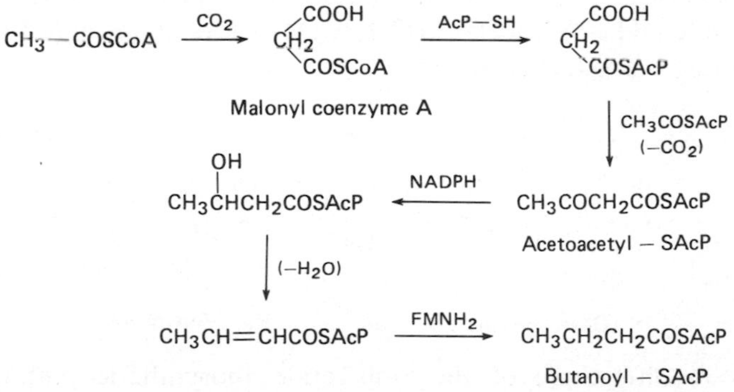

CH3—COSCoA
CO2
COOH
CH2
COSCoA
Malonyl coenzyme A
AcP—SH
COOH
CH2
COSAcP
CH3COSAcP
(−CO2)
OH
CH3CHCH2COSAcP
NADPH
CH3COCH2COSAcP
Acetoacetyl – SAcP
(−H2O)
CH3CH=CHCOSAcP
FMNH2
CH3CH2CH2COSAcP
Butanoyl – SAcP

Figure 15.10 Fatty acid biosynthesis by the malonate pathway

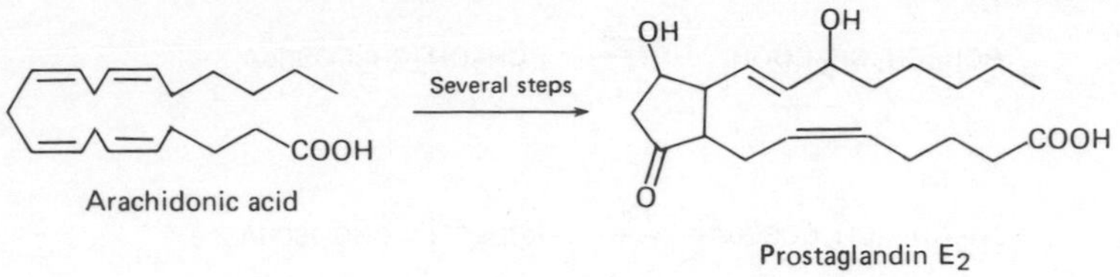

Figure 15.11 The biosynthesis of prostaglandin E_2

Unsaturated fatty acids arise by one of two mechanisms. Dehydrogenation of saturated acids, in the presence of air and NAD^+, is possible in several organisms. In bacteria and in animal cells unsaturated fatty acids may also arise by elongation of the $\alpha\beta$-unsaturated fatty acids formed as intermediates in the biosynthesis of saturated fatty acids (Figure 15.10).

An important group of physiologically active compounds known as prostaglandins are biosynthesized by cyclization of polyunsaturated fatty acids (e.g. Figure 15.11).

15.6 Polyketides

Besides fatty acids, there are several other classes of compounds formed from acetate and malonate. These arise by successive condensations of acetyl coenzyme A with several molecules of malonyl coenzyme A to give polyketones (Figure 15.12), which subsequently cyclize to give the products, which are known as *polyketides*.

Cyclization may proceed via Claisen or aldol condensations, giving six-membered rings which enolize to aromatic products (e.g. Figure 15.13). The aromatic products may be further modified subsequently in a number of ways, such as methylation of the OH groups or substitution on the aromatic ring.

Question 15.4 Griseophenone and alternariol are biosynthesized from the same polyketone intermediate. Draw the polyketone and indicate those carbon atoms of griseophenone and alternariol which would become radioactively labelled if $[1\text{-}^{14}C]$ acetate $(CH_3{}^*CO_2^-)$ were administered to cultures of the organisms producing them.

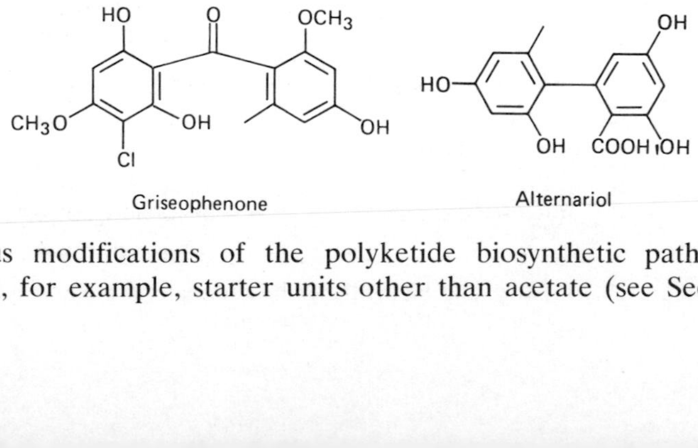

Griseophenone   Alternariol

Various modifications of the polyketide biosynthetic pathway occur, involving, for example, starter units other than acetate (see Section 15.9),

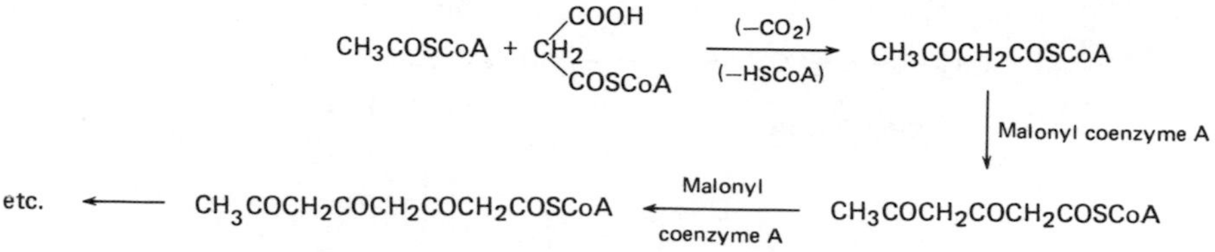

Figure 15.12 The formation of polyketones from malonyl coenzyme A

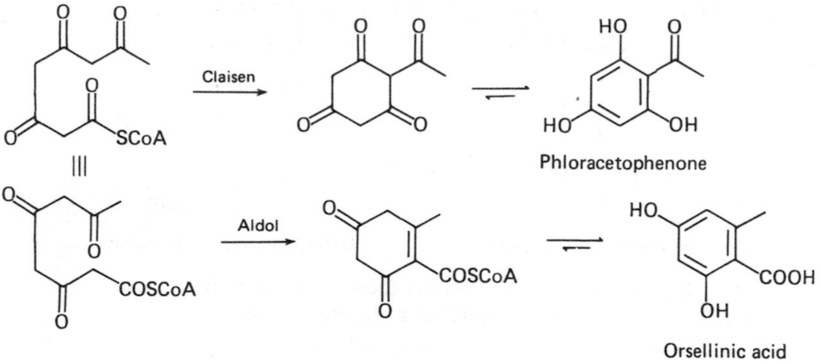

Figure 15.13 Examples of the cyclization of a polyketone to form polyketides

or subsequent modification of aromatic rings. Vitamins E and K (Sections 14.6 and 14.8) arise via alkylation of aromatic precursors by geranyl pyrophosphate (Section 15.7).

15.7 Terpenes and steroids

Many natural products appear to be built up from C_5 units related to isoprene (2-methylbuta-1,3-diene). The biosynthesis of these compounds involves an aldol-type condensation of acetyl coenzyme A with acetoacetyl coenzyme A, followed by a series of steps resulting ultimately in isopentenyl pyrophosphate and its isomer, dimethylallyl pyrophosphate (Figure 15.14).

Dimethylallyl pyrophosphate, like other allyl compounds possessing good leaving groups, is susceptible to attack by nucleophiles (Chapter 9). Thus, it reacts with the double bonds of isopentenyl pyrophosphate giving geranyl pyrophosphate (Figure 15.15). This is also an allylic pyrophosphate and can add on other isopentenyl pyrophosphate units forming farsenyl and geranyl pyrophosphates. Further polymerization eventually produced rubber (Chapter 9).

Compounds containing 10, 15, or 20 carbon atoms and derived from isopentenyl pyrophosphate are called mono-, sesqui-, and diterpenes respectively. Some monoterpenes have structures which are simply related to geranyl pyrophosphate (e.g. citronellol), whereas others have considerably more complex structures resulting from subsequent cyclization

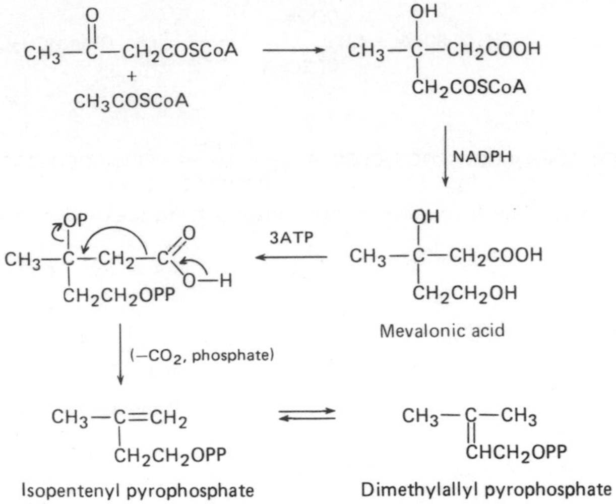

CH$_3$—C—CH$_2$COSCoA → CH$_3$—C—CH$_2$COOH
+
CH$_3$COSCoA

Figure 15.14 The biosynthesis of isopentenyl and dimethylallyl pyrophosphates

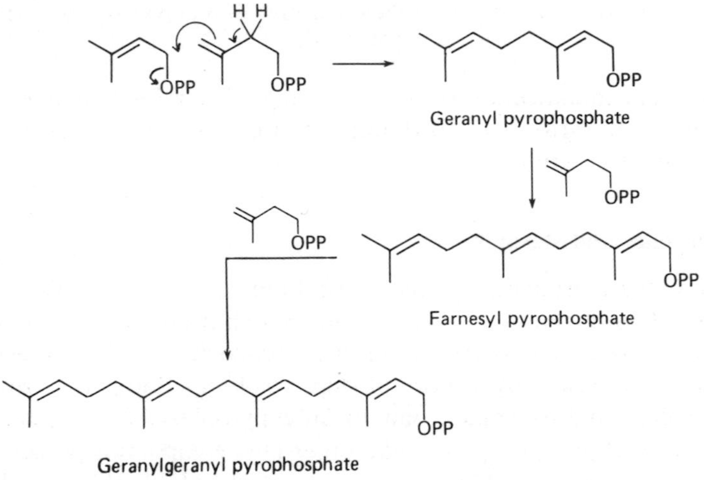

Isopentenyl pyrophosphate   Dimethylallyl pyrophosphate

Mevalonic acid

Geranyl pyrophosphate

Farnesyl pyrophosphate

Geranylgeranyl pyrophosphate

Figure 15.15 Formation of geranyl pyrophosphate

and modification (e.g. α-pinene, camphor, and γ-terpinene, Figure 15.16). Many sesqui- and diterpenes also have cyclic structures. Terpenes often have powerful odours, such as those associated with certain plants and plant extracts.

In all of the terpenes mentioned so far the isoprene units are linked 'head to tail'. 'Tail to tail' linkages occur when two C_{15} or C_{20} units are linked to give triterpenes (e.g. squalene) and carotenoids (e.g. phytoene) respectively. Squalene, first identified in shark liver, is the acyclic

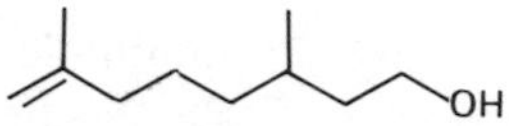

Citronellol

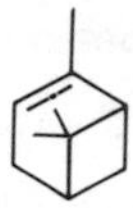

α-Pinene

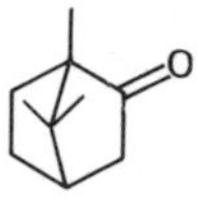

Camphor

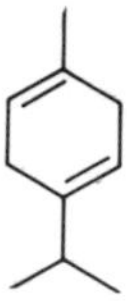

γ-Terpinene

Figure 15.16 The structures of some important terpenes

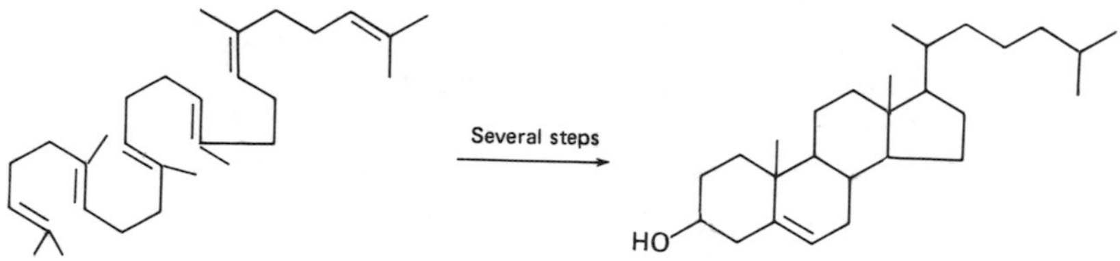

Squalene

Cholesterol

Figure 15.17 The structures of squalene and cholesterol

intermediate involved in the formation of steroids (Section 14.14), such as cholesterol (Figure 15.17).

Phytoene is transformed in plants into other carotenoids such as β-carotene (Figure 15.18). Carotenoids are red and yellow pigments occurring in carrots, tomatoes, and many other plants.

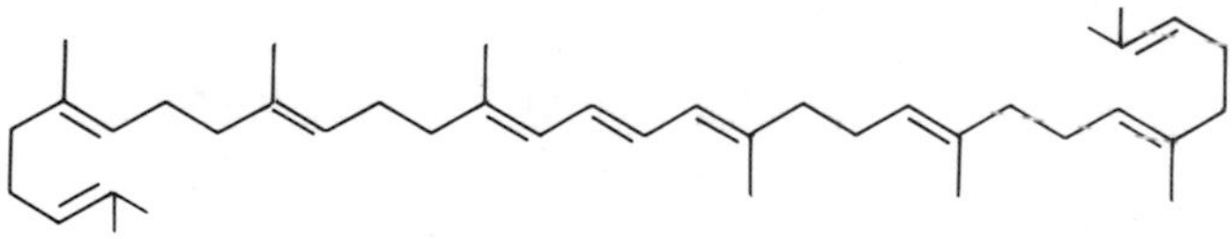

Phytoene

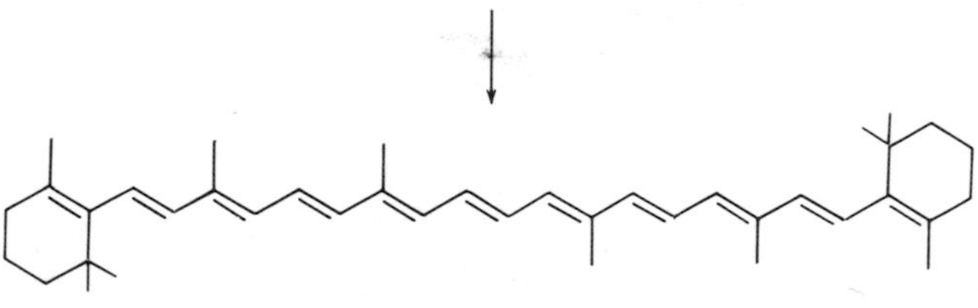

β-Carotene

Figure 15.18 The structures of phytoene and β-carotene

15.8 Amino acid metabolism

Several amino acids arise directly by amination of the α-keto acid intermediates involved in the citric acid cycle (Section 15.4). For example, pyruvic acid is interconvertible with alanine, oxaloacetic acid with aspartic acid, and α-ketoglutaric acid with glutamic acid (Figure 15.19).

$$CH_3COCOOH \longrightarrow CH_3CH\begin{smallmatrix}NH_2\\COOH\end{smallmatrix}$$

Pyruvic acid   Alanine

$$HOOCCH_2COCOOH \longrightarrow HOOCCH_2CH\begin{smallmatrix}NH_2\\COOH\end{smallmatrix}$$

Oxaloacetic acid   Aspartic acid

$$HOOCCH_2CH_2COCOOH \longrightarrow HOOCCH_2CH_2CH\begin{smallmatrix}NH_2\\COOH\end{smallmatrix}$$

α-Ketoglutaric acid   Glutamic acid

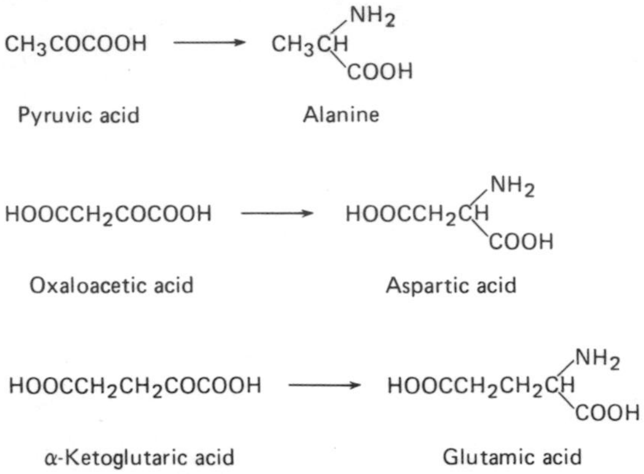

Figure 15.19 Biosynthesis of three amino acids from citric acid cycle metabolites

Serine, glycine, and cysteine arise only slightly less directly from glyceric acid-3-phosphate (Figure 15.20).

The amination step in all of these reactions involves the formation of a Schiff's base (chapter 8) between the keto acid and pyridoxamine-5-phosphate (Figure 15.21), itself formed from pyridoxal-5-phosphate (Section 14.3) and an amino acid.

The aromatic amino acids, phenylalanine and tyrosine, arise by reaction of two moles of phosphoenol pyruvate and erythrose-4-phosphate (Section 15.3) as shown in Figure 15.22.

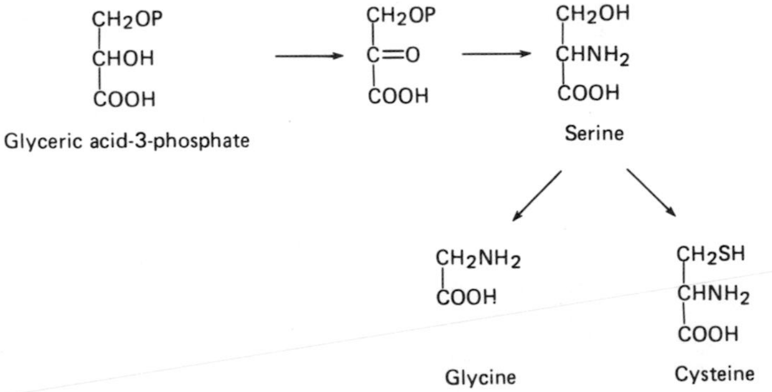

Figure 15.20 Biosynthesis of three amino acids from glyceric acid-3-phosphate

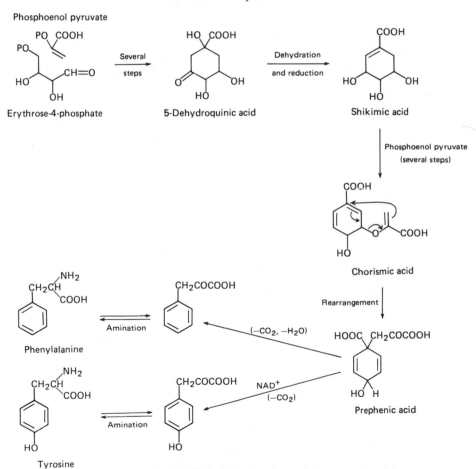

Figure 15.21 The involvement of pyridoxamine-5-phosphate in amino acid biosynthesis

Figure 15.22 Biosynthesis of phenylalanine and tyrosine

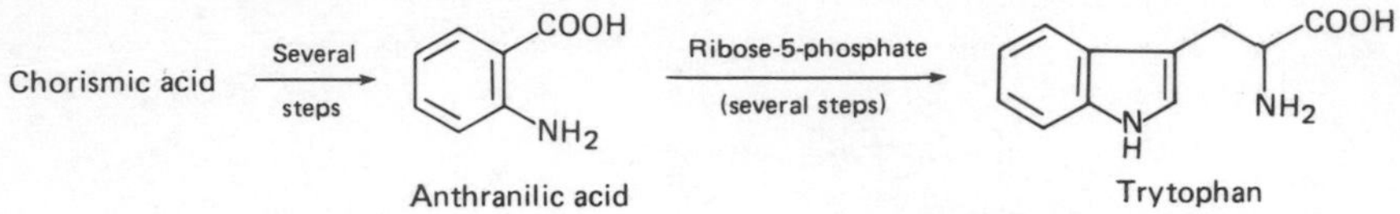

Figure 15.23 Biosynthesis of tryptophan from chorismic acid

Various other metabolites are synthesized from intermediates in this pathway. Shikimic acid, for example, is the precursor of gallic acid, an important component of tannins, while chorismic acid gives rise to anthranilic acid, a precursor of tryptophan (Figure 15.23).

15.9 Flavonoids, coumarins, lignans, and lignin

Deamination of phenylalanine or tyrosine gives cinnamic acid or 4-hydroxycinnamic acid respectively. Hydroxylation of the aromatic ring, isomerization of the double bond, and cyclization gives coumarin or umbelliferone (fgure 15.24).

The coumarin nucleus is also present in the synthetic anticoagulant, Warfarin® (Figure 15.25), which has widespread application as a rodenticide.

Cinnamic and 4-hydroxycinnamic acids are also precursors of lignin and the lignans. Lignin, along with cellulose, forms the main structural material of wood, whereas lignans are small molecules with structures related to that of lignin. These products are formed by oxidative coupling reactions of phenols such as coniferyl alcohol (Figure 15.26).

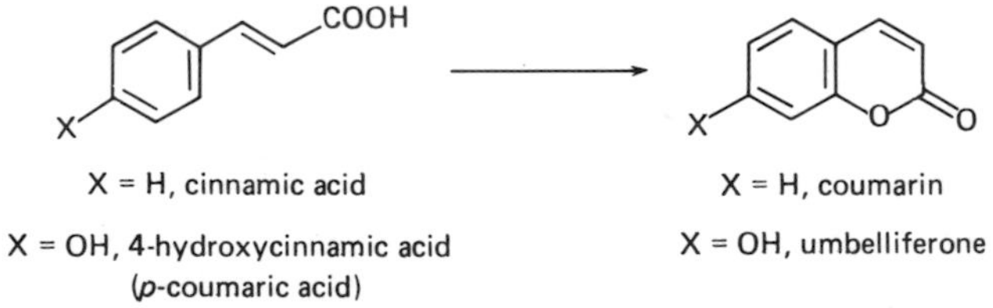

X = H, cinnamic acid   X = H, coumarin

X = OH, 4-hydroxycinnamic acid   X = OH, umbelliferone
(p-coumaric acid)

Figure 15.24 The biosynthesis of coumarin and umbelliferone

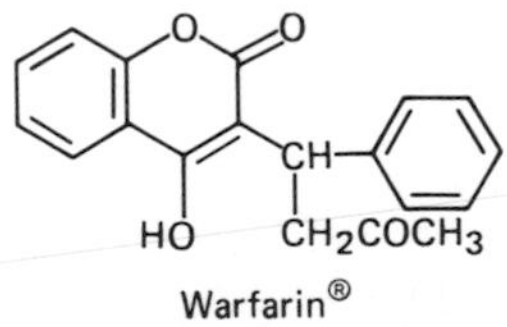

Warfarin®

Figure 15.25 The structure of Warfarin*

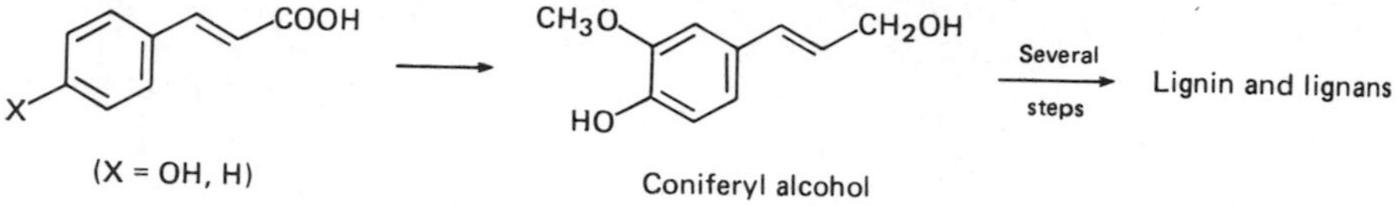

Figure 15.26 The biosynthesis of lignin and lignans

Another widespread series of plant metabolites, the flavonoids, are formed from a cinnamic acid derivative by successive reactions with malonyl coenzyme A to give a second aromatic ring (cf. Section 15.6). The biosynthesis proceeds via intermediate chalcones (e.g. Figure 15.27).

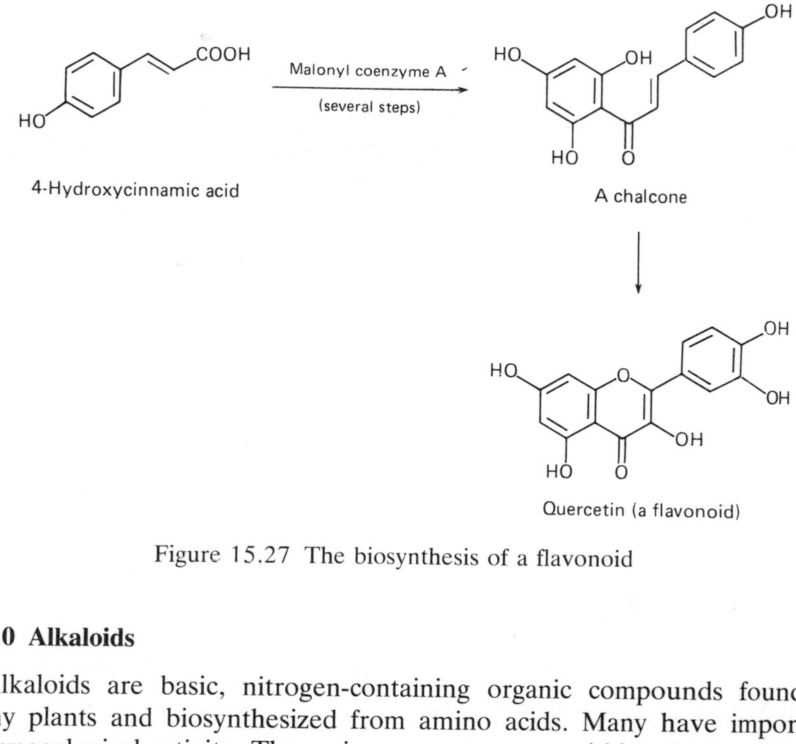

Figure 15.27 The biosynthesis of a flavonoid

15.10 Alkaloids

Alkaloids are basic, nitrogen-containing organic compounds found in many plants and biosynthesized from amino acids. Many have important pharmacological activity. The main precursors are ornithine, lysine, aspartic acid, phenylalanine, tyrosine, and tryptophan. For example, ornithine and nicotinic acid give the important tobacco component nicotine (Figure 15.28(a)). Phenylalanine and tyrosine give rise to relatively simple derivatives such as ephedrine (Figure 15.28(b)) or by a more complex series of reactions to isoquinoline alkaloids such as reticuline and morphine (Figure 15.28(c); see Section 6.9).

Tryptophan gives rise to another group of alkaloids known as indole alkaloids, of which ergotamine and other lysergic acid derivatives are examples.

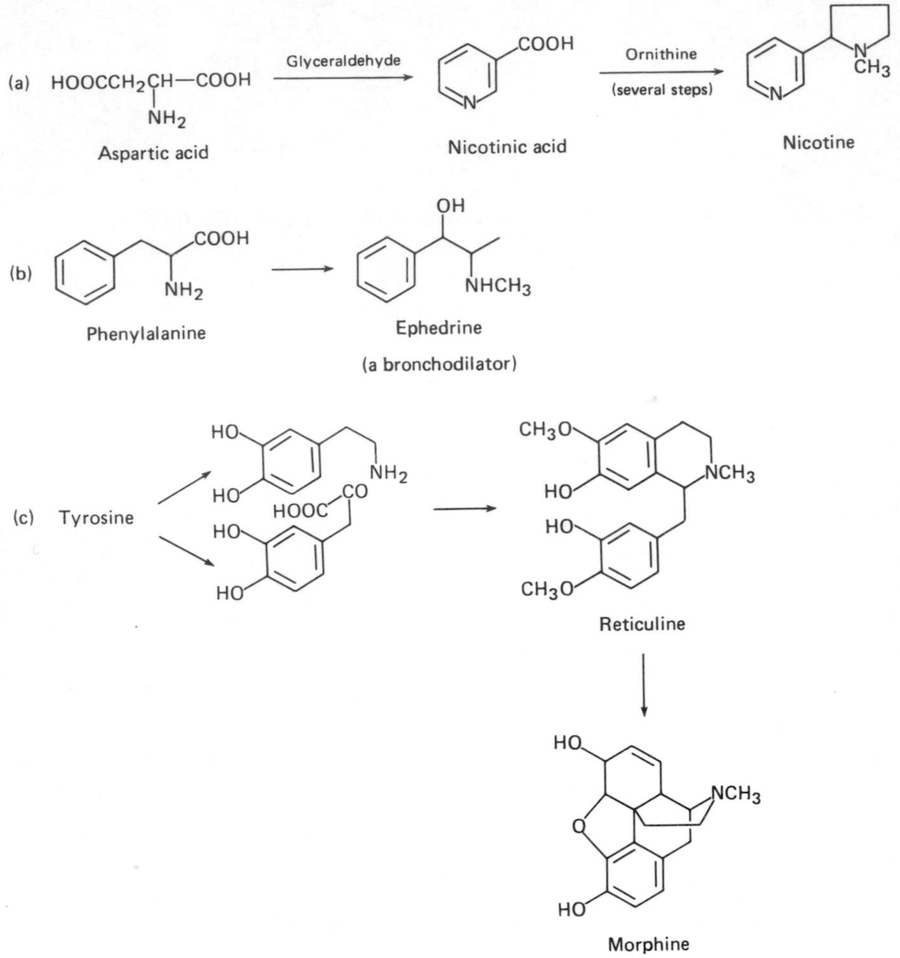

Figure 15.28 The biosynthesis of some alkaloids

15.11 Penicillins and cephalosporins

Penicillins and cephalosporins, which are mould metabolites, arise by condensation of L-valine, L-cysteine, and α-aminoadipic acid (Figure 15.29).

Introduction of phenylacetic, phenoxyacetic, or other acids to the fungal cultures results in replacement of the α-aminoadipoyl side chain by other groups, giving alternative products such as benzyl or phenoxymethyl penicillins (penicillins G and V respectively). In recent years semisynthetic penicillins have been developed in which the side chain of benzyl penicillin is replaced chemically by other groups. These are less susceptible to hydrolysis by stomach acid and so can be given by mouth rather than by

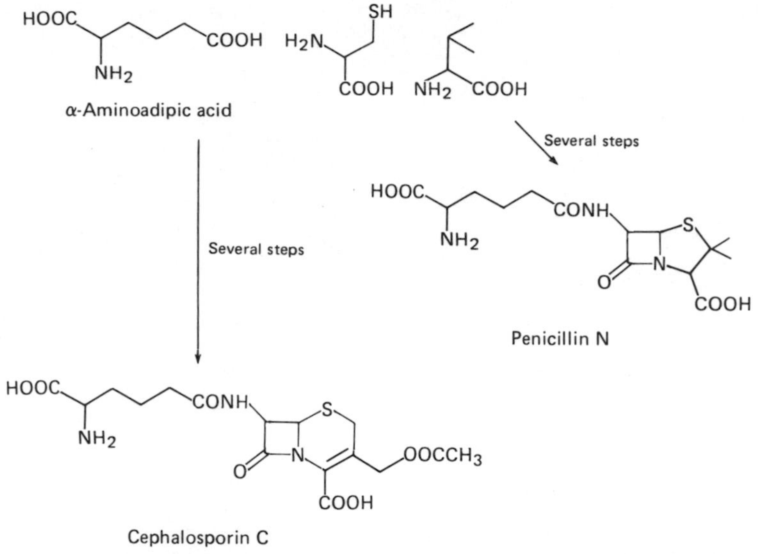

Figure 15.29 The biosynthesis of cephalosporin C and penicillin N

injection. They are also active against many more types of bacteria and are referred to as broad spectrum antibiotics.

15.12 Tetrapyrrolic compounds

Tetrapyrroles are biosynthesized from succinyl coenzyme A and glycine which undergo a Claisen-type condensation to give a β-keto acid which readily decarboxylates yielding δ-aminolevulinic acid. Two molecules of δ-aminolevulinic acid condense to give porphobilinogen (Figure 15.30), which is found in the urine of people suffering from acute porphyria.

Four porphobilinogen molecules combine together with elimination of the benzylic amino groups, giving a linear tetrapyrrole which cyclizes with

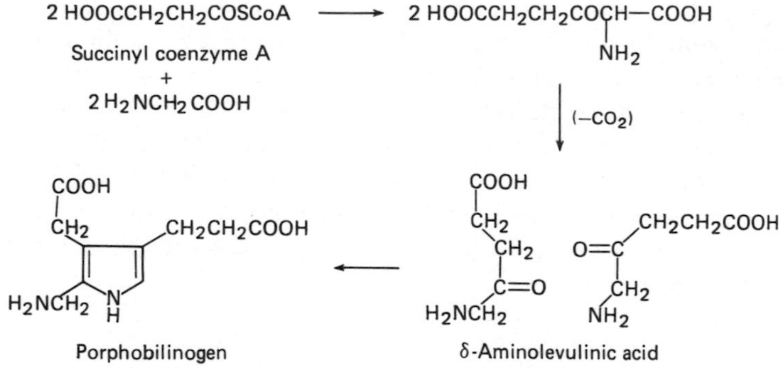

Figure 15.30 The biosynthesis of porphobilinogen

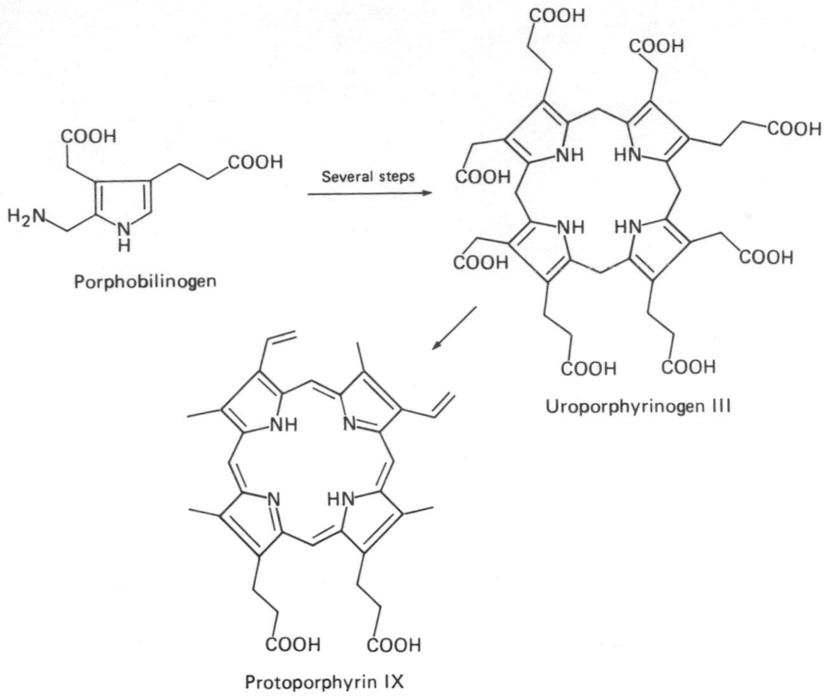

Figure 15.31 The biosynthesis of protoporphyrin IX

isomerization to form uroporphyrinogen III. Decarboxylation and dehydrogenation then yield protoporphyrin IX (Figure 15.31).

Conversion of protoporphyrin IX to haem merely involves incorporation of Fe^{2+}, and cytochrome prosthetic groups require only trivial additional modification. Conversion to chlorophyll a, however, requires incorporation of Mg^{2+} and some further modifications including esterification of the ring D carboxylic acid unit with phytol (a terpene; see Section 15.7). Vitamin B_{12} involves considerable modification of the basic tetrapyrrolic macrocycle, but inspection of its structure (Figure 13.1) shows the same basic arrangement of side chains. The major changes are the extrusion of a bridging carbon atom and the addition of several extra methyl groups, which are known to be derived from the amino acid methionine.

Bile pigments result from *degradation* of haem. The basic route involves loss of iron followed by opening of the macrocycle with loss of the bridging carbon atom between rings A and B. Successive reduction steps and a final reoxidation step give bilin, which is responsible for the characteristic brown colour of faeces.

Answers to questions

15.1 (a)

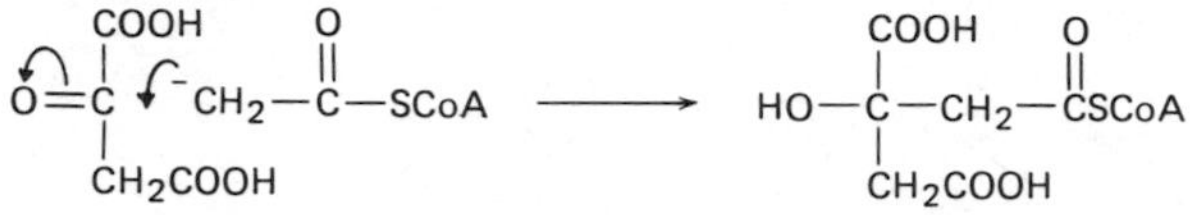

Citric acid coenzyme A derivative

15.1 (b) One of the carboxyl groups in oxalosuccinic acid is positioned β to a carbonyl group, and such acids readily decarboxylate (see Chapter 9).

15.2

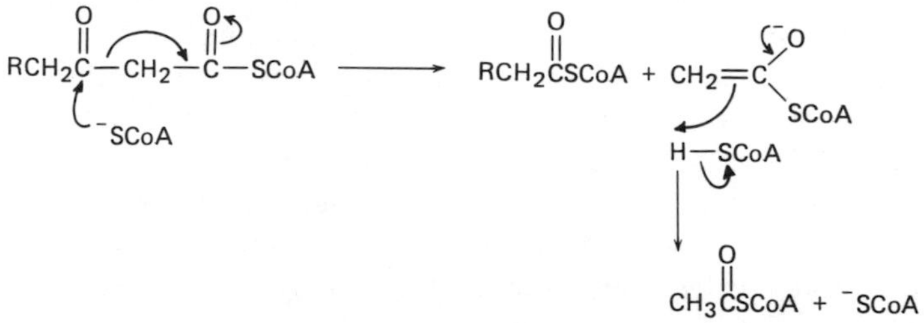

15.3

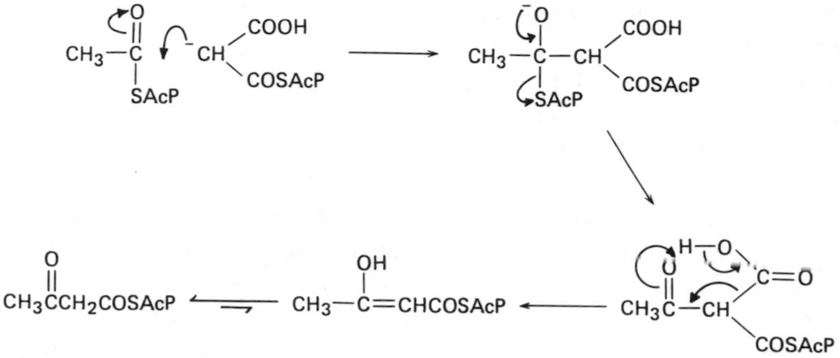

15.4

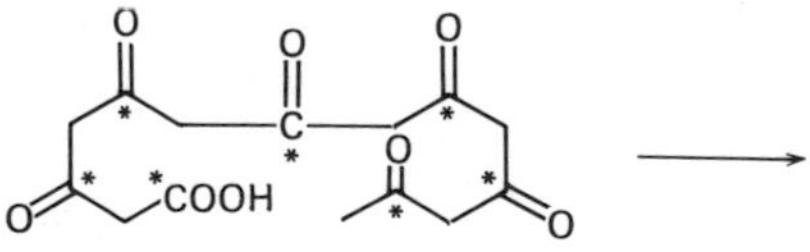

Griseophenone

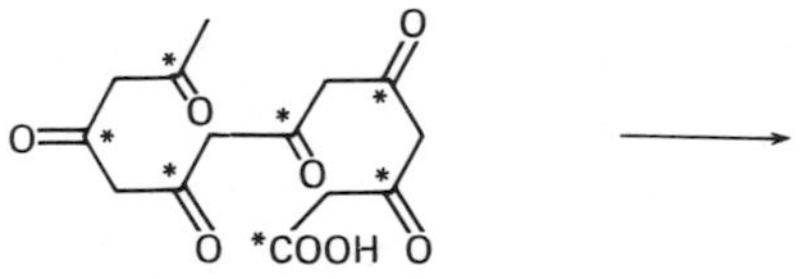

Alternariol

APPENDIX

Nomenclature

Due to the complexity and variety of organic compounds it is essential to be able to assign an unambiguous name to each compound. A set of rules have been defined by the International Union of Pure and Applied Chemistry (IUPAC) to enable unambiguous systematic names to be assigned. However, many organic compounds also have trivial (non-systematic) names which are widely used in the literature. For example, trivial names are assigned to many compounds isolated from natural sources before their structures are determined, and the trivial names are often retained. Trivial names are also used for many compounds whose structures are so complex that their systematic names are unwieldy (e.g. quinine, morphine, glucose). Finally many simple compounds (e.g. acetone, acetylene) have trivial names which are so well established that their systematic names are rarely used.

In this book we have adopted the policy of using the names by which compounds are *usually* known. However, in the case of compounds which are usually referred to by their *trivial* names we have also given their systematic names in brackets in order that the reader should become familiar with both names for the compound. In the case of compounds which have trivial names by which they are *occasionally* known the trivial names are given in brackets, after the systematic names. In this appendix the general rules used to derive systematic names are summarized.

A.1 alkanes

Straight-chain alkanes are named according to the number of carbon atoms which they contain. The names of some straight-chain alkanes are given in Table 1 along with the names of the corresponding alkyl groups which are derived by removal of a hydrogen atom from one of the terminal carbon atoms.

Branched-chain alkanes are named as derivatives of the straight-chain alkane corresponding to the longest carbon chain in the molecule. All other groups are then named as substituents of the main chain. The position of a substituent is specified by numbering the carbon atoms of the main chain in such a way that the carbon atoms carrying substituents have

Table A.1 Straight-chain alkanes and alkyl groups

C_nH_{2n+2}	Name of alkane	C_nH_{2n+2}	Name of alkyl group	Abbreviation
CH_4	Methane	CH_3—	Methyl	Me
C_2H_6	Ethane	C_2H_5—	Ethyl	Et
C_3H_8	Propane	C_3H_7—	Propyl	Pr
C_4H_{10}	Butane	C_4H_9—	Butyl	Bu
C_5H_{12}	Pentane	C_5H_{11}—	Pentyl	
C_6H_{14}	Hexane	C_6H_{13}—	Hexyl	
C_7H_{16}	Heptane	C_7H_{15}—	Heptyl	
C_8H_{18}	Octane	C_8H_{17}—	Octyl	
C_9H_{20}	Nonane	C_9H_{19}—	Nonyl	
$C_{10}H_{22}$	Decane	$C_{10}H_{21}$—	Decyl	
$C_{11}H_{24}$	Undecane	$C_{11}H_{23}$—	Undecyl	
$C_{12}H_{26}$	Dodecane	$C_{12}H_{25}$—	Dodecyl	

the lowest possible numbers (e.g. Figure A.1). Substituent chains are cited in alphabetical order.

Cyclic alkanes are named by adding the prefix 'cyclo' to the name of the straight-chain alkane containing the same number of carbon atoms. In cycloalkanes carbon atom 1 is that which bears the substituent which comes first alphabetically (e.g. Figure A.2).

Question A.1 Draw structural formulas of (a) 2,2-dimethyl-4-propylheptane, and (b) 1,3-diethylcyclopentane.

A semisystematic method of nomenclature is sometimes used in naming alkyl groups, particularly branched ones. Using this system the CH_3CH_2-CH_2— group is called a *normal* or *n*-propyl group (Pr^n), while $(CH_3)_2$-

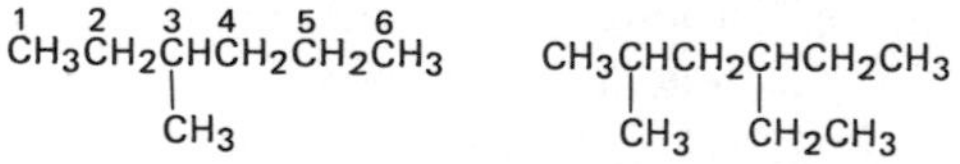

3-Methylhexane (*not* 4-methylhexane or 2-ethylpentane)

4-Ethyl-2-methylhexane

Figure A.1

1,2-Dimethylcyclopropane

1-Ethyl-3-methylcyclohexane

Figure A.2

$$CH_2{=}CH_2$$

Ethene

$$\overset{4}{C}H_3\overset{3}{C}H_2\overset{2}{C}H{=}\overset{1}{C}H_2$$

British name: but-1-ene (*not* but-3-ene)

American name: 1-butene (*not* 3-butene)

$$\overset{1}{C}H_3\overset{2}{C}H{=}\overset{3}{C}H\overset{4}{C}H_3$$

But-2-ene

(or 2-butene)

$$CH_2{=}CH{-}CH{=}CH_2$$

Buta-1,3-diene

(or 1,3-butadiene)

Cyclopentene

Figure A.3

CH— is called an *iso*-propyl group (Pr^i). Similarly the $CH_3CH_2CH_2$-CH_2— group can be called the *n*-butyl group (Bu^n), $CH_3CH_2CHCH_3$ the *secondary* or *s*-butyl group (Bu^s), $(CH_3)_2CHCH_2$— the *iso*-butyl group (Bu^i) and $(CH_3)_3C$— the *tertiary* or *t*-butyl group (Bu^t).

A.2 alkenes

Alkenes are named by replacing the ending '-ane' of the corresponding alkane by '-ene'. Dienes are named by replacing the ending '-ane' by '-adiene'. The position of each double bond is indicated by giving the number of the first carbon atom of the double bond. The numbering is chosen to make the numbers as small as possible (Figure A.3).

Alkenes with alkyl substituents are named as for the corresponding saturated compounds except that the main chain is always chosen to include the double bond(s), even if this is not the longest chain. The numbering is chosen to give the lowest possible numbers to the double bond(s) (Figure A.4).

The groups $CH_2{=}CH$— and $CH_2{=}CHCH_2$— are commonly referred to as the vinyl and allyl groups respectively. Other unsaturated groups can be named by replacing the last letter '-e' of the alkene name by '-yl' and indicating the position of attachment by a number.

A.3 alkynes

Alkynes are named by replacing the ending '-ane' of the corresponding

$$\overset{4}{C}H_3\overset{3}{C}H_2\diagdown\underset{CH_3}{\overset{2}{C}}{=}\overset{1}{C}H_2$$

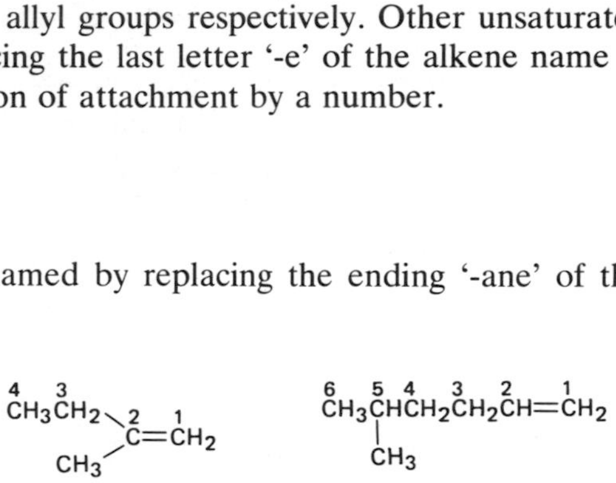

2-Methylbut-1-ene

(or 2-methyl-1-butene)

$$\overset{6}{C}H_3\overset{5}{C}H\overset{4}{C}H_2\overset{3}{C}H_2\overset{2}{C}H{=}\overset{1}{C}H_2$$
$$|$$
$$CH_3$$

5-Methylhex-1-ene

(or 5-methyl-1-hexene)

Figure A.4

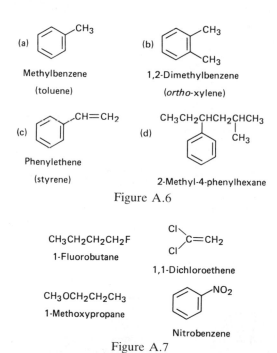

$$HC\equiv CH$$

Ethyne

(acetylene)

$$CH_3CH_2CH_2C\equiv CCH_3$$

Hex-2-yne

(or 2-hexyne)

$$CH_3CHC\equiv CH$$
$$\;\;\;\;\;|$$
$$\;\;\;\;CH_3$$

3-Methylbut-1-yne

(or 3-methyl-1-butyne)

$$CH_3CHC\equiv CCHCH_2CH_3$$
$$\;\;\;\;|\;\;\;\;\;\;\;\;\;\;|$$
$$\;\;\;CH_3\;\;\;\;\;CH_2CH_3$$

5-Ethyl-2-methylhept-3-yne

(or 5-ethyl-2-methyl-3-heptyne)

Figure A.5

alkane by '-yne'. The position of the triple bond is indicated by numbering the carbon atoms in exactly the same way as for alkenes (Figure A.5).

A.4 arenes

Aromatic compounds are usually named as derivatives of the parent aromatic hydrocarbons. Thus, the systematic name for toluene (Figure A.6(a)) is methylbenzene. Disubstituted benzene derivatives are frequently referred to as *ortho* (1,2-), *meta* (1,3-), and *para* (1,4-) derivatives (e.g. Figure A.6(b)). In some cases the benzene ring is itself considered as a substituent (e.g. Figure 6(c) and (d)), in which case the C_6H_5— group is called the phenyl group (Ph). The $C_6H_5CH_2$— group is called the benzyl group.

(a) Methylbenzene (toluene)

(b) 1,2-Dimethylbenzene (*ortho*-xylene)

(c) Phenylethene (styrene)

(d) 2-Methyl-4-phenylhexane

Figure A.6

$$CH_3CH_2CH_2CH_2F$$

1-Fluorobutane

1,1-Dichloroethene

$$CH_3OCH_2CH_2CH_3$$

1-Methoxypropane

Nitrobenzene

Figure A.7

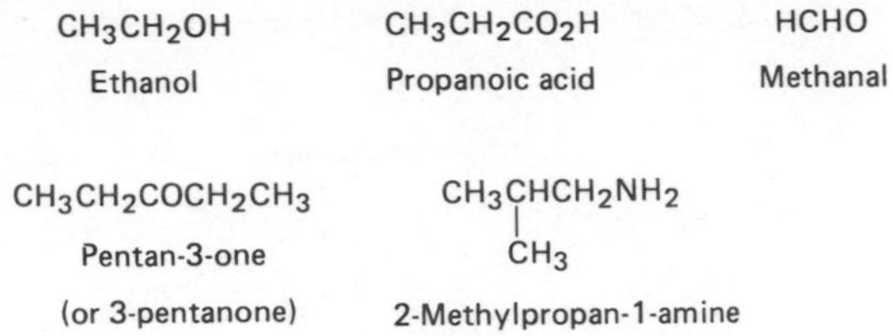

$$CH_3CH_2OH \qquad CH_3CH_2CO_2H \qquad HCHO$$

Ethanol   Propanoic acid   Methanal

$$CH_3CH_2COCH_2CH_3 \qquad CH_3\overset{|}{C}HCH_2NH_2$$

Pentan-3-one $\overset{|}{C}H_3$

(or 3-pentanone)   2-Methylpropan-1-amine

Figure A.8

A.5 Functional groups

Alkyl halides, nitro compounds, and ethers are named by adding the appropriate prefix to the name of the parent hydrocarbon (Figure A.7).

Depending upon the overall structure of a compound, the presence of other functional groups is indicated by using an appropriate prefix or suffix (Table A.2). If only one such group is present then the appropriate suffix is used in place of the final 'e' in the name of the hydrocarbon containing the same number of carbon atoms (e.g. Figure A.8). If two or more such groups are present then the highest ranking group (as listed in Table A.2) is regarded as the principal group and is represented by its suffix. All other functional groups present are then identified by an appropriate prefix and listed in alphabetical order. Thus a compound containing both —NH_2 and —OH groups is regarded as an aminoalcohol rather than as a hydroxy-amine, whereas a compound containing both —CN and —OH groups is named as a hydroxynitrile rather than as a cyanoalcohol.

Question 2 Give systematic names to the following compounds:

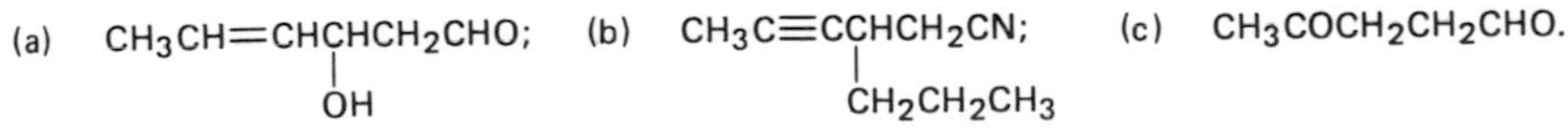

(a) $CH_3CH{=}CHCHCH_2CHO;$ (b) $CH_3C{\equiv}CCHCH_2CN;$ (c) $CH_3COCH_2CH_2CHO.$
        $\overset{|}{O}H$ $\overset{|}{C}H_2CH_2CH_3$

Table A.2 Names of functional groups in order of decreasing priority

Group	Prefix name	Suffix name	Alternative suffix
—CO_2H	Carboxy-	-oic acid	(-carboxylic acid)
—SO_3H	Sulpho-	-esulphonic acid	
—CO_2R	Alkoxycarbonyl-	-oate	(-carboxylate)
—SO_3R	Alkoxysulphonyl-	-esulphonate	
—COCl	Chloroformyl-	-oyl chloride	(-carbonyl chloride)
—$CONH_2$	Carbamoyl-	-amide	(-carboxamide)
—CN	Cyano-	-enitrile	(-carbonitrile)
—CHO	Oxo- (or formyl-)	-al	(carbaldehyde)
$\overset{\displaystyle O}{\underset{\displaystyle \|}{-C-}}$	Oxo-	-one	
—OH	Hydroxy-	-ol	
—SH	Mercapto-	-ethiol	
—NH_2	Amino-	-amine	

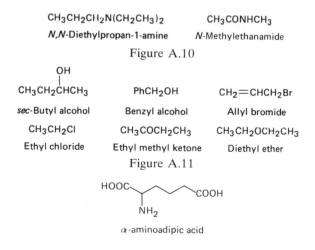

Cyclobutane carbaldehyde	Cyclohexane carboxylic acid	Naphthalene-2-carbonitrile

Figure A.9

The alternative suffixes shown in brackets in Table A.2 are less common but afford a more convenient way of naming derivatives of cycloalkanes and arenes (Figure A.9). In this case the suffix is added to the full name of the hydrocarbon containing one less carbon atom.

Secondary and tertiary amines and amides are named as derivatives of the corresponding primary amines and amides. The additional substituents attached to the nitrogen atom are listed before the name of the parent amine or amide and a capital N is used to indicate that the groups are attached to nitrogen rather than carbon (e.g. Figure A.10).

An older method of naming some compounds, which may also be encountered but which is no longer favoured, uses the name(s) of the appropriate alkyl or aryl group(s) followed by a word to describe the type of functional group present (e.g. Figure A.11).

As well as locating substituents on a chain by numbers, a system of Greek letters is sometimes encountered. Thus the first carbon atom from the main functional group is the α-carbon, the second is the β-carbon etc. (e.g. Figure A.12).

Finally it is worth noting that some trivial names of well-known compound have been adopted as systematic names (e.g. phenol PhOH, benzoic acid PhCO$_2$H, benzaldehyde, benzamide, benzonitrile). Table A.3 lists a number of other compounds whose trivial names though not adopted as systematic names, are nevertheless commonly used.

$CH_3CH_2CH_2N(CH_2CH_3)_2$ $CH_3CONHCH_3$

N,N-Diethylpropan-1-amine   *N*-Methylethanamide

Figure A.10

OH
|
$CH_3CH_2CHCH_3$ $PhCH_2OH$ $CH_2{=}CHCH_2Br$

sec-Butyl alcohol   Benzyl alcohol   Allyl bromide

CH_3CH_2Cl $CH_3COCH_2CH_3$ $CH_3CH_2OCH_2CH_3$

Ethyl chloride   Ethyl methyl ketone   Diethyl ether

Figure A.11

$HOOC$ ⌒⌒ $COOH$
|
NH_2

α-aminoadipic acid

Question 3 Give the systematic names of the compounds listed in Table A.3.

In the same way that alkyl groups are named by replacing the ending '-ane' of an alkane by '-yl' so acyl groups (RCO—) are named by replacing the ending '-ic' of an acid by '-yl' (e.g. propanoyl, CH_3CH_2CO—, benzoyl, PhCO—, acetyl, CH_3CO—).

Table A.3 Some compounds having commonly used trivial names

	Formula	Trivial name
(a)	$CHCl_3$	Chloroform
(b)	CCl_4	Carbon tetrachloride
(c)	$(CH_3)_2CHOH$	*Iso*propanol
(d)	$(CH_3)_3COH$	*Tert*-butanol
(e)		*Ortho*-cresol
(f)	HCO_2H	Formic acid
(g)	HCHO	Formaldehyde
(h)	CH_3CO_2H	Acetic acid
(i)	CH_3CHO	Acetaldehyde
(j)	CH_3CN	Acetonitrile
(k)	$CH_3CH_2CO_2H$	Propionic acid
(l)	CH_3CH_2CHO	Propionaldehyde
(m)	$CH_3CH_2CH_2CO_2H$	Butyric acid
(n)	$CH_3CH_2CH_2CHO$	Butyraldehyde
(o)	CH_3COCH_3	Acetone
(p)	CH_3COCl	Acetyl chloride
(q)	$PhCOCH_3$	Acetophenone
(r)	HO_2C—CO_2H	Oxalic acid
(s)	$HO_2CCH_2CO_2H$	Malonic acid
(t)	$HO_2CCH_2CH_2CO_2H$	Succinic acid
(u)		Maleic acid
(v)		Fumaric acid
(w)	$CH_3CH(OH)CO_2H$	Lactic acid
(x)	$HO_2CCH(OH)CH(OH)CO_2H$	Tartaric acid
(y)		Citric acid
(z)	$PhNH_2$	Aniline

Answers to questions

1. (a)

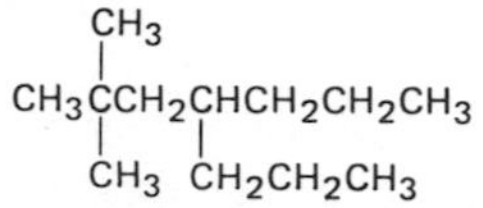

$$CH_3\underset{\underset{\displaystyle CH_3}{|}}{\overset{\overset{\displaystyle CH_3}{|}}{C}}CH_2\underset{\underset{\displaystyle CH_2CH_2CH_3}{|}}{CH}CH_2CH_2CH_3$$

(b)

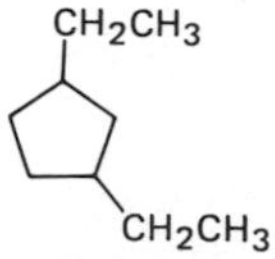

2. (a) 3-Hydroxyhex-4-enal; (b) 3-propylhex-4-ynenitrile;
   (c) 4-oxopentanal.

3. (a) trichloromethane; (b) tetrachloromethane; (c) propan-2-ol; (d) 2-methylpropan-2-ol; (e) 2-methylphenol; (f) methanoic acid; (g) methanal; (h) ethanoic acid; (i) ethanal; (j) ethanenitrile; (k) propanoic acid; (l) propanal; (m) butanoic acid; (n) butanal; (o) propanone; (p) ethanoyl chloride; (q) 1-phenylethanone; (r) ethanedioic acid; (s) propanedioic acid; (t) butanedioic acid; (u) Z-but-2-enedioic acid; (v) E-but-2-enedioic acid; (w) 2-hydroxypropanoic acid; (x) 2,3-dihydroxylbutanedioic acid; (y) 3-carboxy-3-hydroxypentanedioic acid; (z) benzenamine.

Index

acetal 152, 209
acetaldehyde 149, 183
acetanilide 129
acetic acid 70, 144, 159, 160
acetic anhydride 137, 144,
acetoacetyl coenzyme A 27.
acetominophen 263
acetone 51
acetyl chloride 146, 225
acetylcholine 125, 126
acetyl coenzyme A 161, 251, 256,
 265, 269, 270, 271, 272, 273
archiral molecules (definition of) 23
acid anhydride 113
acid dissociation constant 46, 159
cis-aconitic acid 270
acrolein 181
acrylonitrile 182
activation energy 14
active site (on enzyme) 232
acyclic alkanes 59
acyl chloride 113, 127, 175, 186
acylation 89
Addison's disease 260
addition reaction 17, 80
adenine 236, 237, 238
adenosine 237
adenosine diphosphate (ADP) 246,
 248, 252, 264, 266, 267, 270
 triphosphate 247, 248, 249, 264,
 266, 267, 269, 274
S-adenosyl methionine 107, 108
adenylic acid 237
adipic acid 164
adrenal cortex 258, 259
 medulla 258, 259
adrenaline 258, 259, 262
adrenocorticotropin 258
aerosol propellants 100
aglycone 209
alanine 230, 276
alcohols 1, 13, 75, 106, 108, 109,
 110–113, 117, 127, 156, 157, 159,
 166, 175, 176, 205, 213
aldehydes 13, 110, 111, 130, 131,
 139, 185, 213, 225
aldohexose 205
aldol condensation 161, 162, 173,
 174, 185, 189, 269, 272, 273
aldose 205, 211, 213
aldosterone 258, 260
aldrin 178, 179
alkaloids 122, 265
alkenes 13, 75, 84, 88, 108, 109, 111,
 117, 125, 175
alkenyl chlorides 184
alkoxides 106, 112, 118
alkyd resins 191, 201
alkyl diazanum salt 128
alkyl halides 13, 75, 105, 106, 107,
 108, 109, 124, 138, 139, 173, 186
alkyl lithium 73
alkyl polyethylene glycol 202
alkyl trimethylammonium bromide
 202
alkylation 88
alkynes 13, 75, 106, 110, 117, 175
allenes 175
D-allose 206
allyl alcohols 188
 amines 188
 anion 12
 cation 12
 chlorides 188
allylic bromination 66
almonds 209
D-altose 206
amides 13, 51, 126, 131, 139, 153,
 176
amines 13, 130, 159, 175
amino acid analyser 227
amino acids 43, 121, 130, 150, 158,
 174, 188, 250, 257, 265, 269
amino carbonyl compounds 174, 188

α-aminoadipic acid   280, 281
2-amino-2-deoxyglucose   204, 208
β-D-2-amino-2-deoxyglucose   209
δ-amino levulinic acid   281
ammonia   8, 71, 123, 138, 140, 153, 158, 159, 225, 226
ammonium cyanate   2
amygdalin   209, 210
amylase   218
amylopectin   218
amylose   218
anabolism   264
anaemia   251
anaesthetics   121
analysis, elemental   46
aniline   81, 123, 124, 129
anisole   91, 93
anomeric carbon atoms   205
anthracene   81, 85, 95, 96
anthranilic acid   278
antibiotics   221, 261, 281
antibodies   221
antibonding orbitals   6, 7, 8, 49
anticancer drugs   263
anticodon   239
antifreeze   70, 100
antiseptics   102
aphrodisiacs   146
apoenzyme   232
D-arabinose   206
arachidonic acid   272
arginine   223
aromatic diazonium salts   128
aromatic hydrocarbons   59
aryl halides   13
ascorbic acid   251, 254, 255
asparagine   223
aspartic acid   223, 276, 279, 280
asphalt   74
aspirin   262, 263
asymmetric carbon atom   24
atactic polymerization   73
atomic orbital   5, 6
atropine   136
auroglaucin   89
axial position   37
azeotropic distillation   152
azobenzene   42, 133
azoxybenzene   133
azulene   95

bacon fat   197
bacteriochlorophylls   242
bakelite   102, 192, 193

barbiturates   132
base dissociation constant   46, 123
beans,   251
beef tallow   197
beeswax   203
Benedict's reagent   163, 211
benzaldehyde   79, 133, 149, 158, 162, 209
benzamide   144
benzene   60, 79, 84, 85, 89
benzene diazonium chloride   116
benzene-1,4-dicarboxylic acid   96
benzenesulphonyl chloride   126
benzenonium ion   85
benzil   183
benzilic acid   183
benzoic acid   134, 146, 160
1,2-benzoquinone   113
1,4-benzoquinone   113
benzoyl chloride   144
benzyl chlorocarbonate   233
benzyl halides   105
benzyloxycarbonyl derivatives   233
benzyl penicillin   280
beriberi   251
bile pigments   242, 282
bilin   282
biliverdin   244
biosynthesis   187, 197, 251, 269, 271, 273
biotin   251, 256
bitumen   74
biuret   132, 232
blackcurrants   251
blood   221
boiling points   17
bond angles   3, 4
   lengths   4
bonding orbitals   6–7, 8–10, 49
bonellin   242, 244
borane   68
Brady's reagent   155
bromamide   132
bromination   66, 81, 115, 162, 225
bromine   62
bromine water   65, 116, 211
bromo benzene   48, 134
(E)-3-bromobut-2-enoic acid   32, 33
(Z)-3-bromobut-2-enoic acid   32, 33
bromohydrins   65
bromonium ion   64
2-bromopropane   66
N-bromosuccinimide   66
buffers   225

builders (in detergents)   202
butanal   144, 146
butane   35, 36, 61, 74, 102
butan-1-ol   102, 146
butanoic acid   159, 160, 198
butanone   54, 144, 146
but-1-ene   61
cis-but-2-ene   61
trans-but-2-ene   61
cis-butenedioic acid   22,
trans-butenedioic acid   22
t-butoxycarbonyl derivatives   233
butter   146, 197, 198, 200
t-butyl chloride   17
t-butyl chlorocarbonate   233
t-butylbenzene   95
but-1-yne   61
but-2-yne   61

caffeine   138
Cahn–Ingold–Prelog Rules 30–33
calciferol   251, 255
calcium carbide   75
camphor   217, 274, 275
canonical forms   11
caprolactam   192
carbamic acid   132
carbanions   14, 106, 130, 132, 181,
   186
carbenium ions   14
carbodiimides   175, 176
carbohydrates   100, 126, 130, 174,
   188, 246, 251
carbon dioxide   166
   monoxide   168, 248
   tetrachloride   3, 60, 62
carbonium ions   14, 66, 67, 85, 88,
   103–105
carbonyl group   11
carboxylic acid   13, 112, 126, 175, 225
carboxypeptidase   232
carnauba wax   203
carotene   50
β-carotene   180, 275
carotenoids   180, 243, 274
carrots   181, 251, 275
catabolism   264
catalysis, free radical   72, 200
catalyst, Lindlar's   71
   metal oxide   97
   mineral acid   112
   nickel   71, 165
   organometallic   179
   palladium   71

partially poisoned   71, 165
$PdCl_2$, $CuCl_2$ 168
phosphoric acid   117
platinum   71
silica alumina   74
silver   70, 118
Ziegler–Natta   72
catalytic cracking   74, 75
   hydrogenation   71, 131, 165, 166,
      199, 233
catecholamines   262
celluloid   217
cellulose   1, 204, 217, 278
cellulose acetate   217
   trimitrate   217
central nervous system   262
cephalosporin   190, 261
cephalosporin C   281
cereals   251
cerebrosides   209, 210
chalcone   279
chemical shifts   53
chiral molecules (definition of)   23
chloral hydrate   189
chloramine *T*   88
chlorin   242, 243, 245
chlorinase   93
chlorination   93, 97
chlorine   62
chlorine water   65
chloroacetic acid   160
chlorobenzene   102, 118, 134
2-chlorobuta-1,3-diene   179
2-chlorobutanoic acid   160
3-chlorobutanoic acid   160
4-chlorobutanoic acid   160
α-chlorocarbonyl compounds   188
chloroethane   102
chloroform   3, 42, 62, 100, 101
chlorohydrins   65
chloromethane   15, 62
1-chloro-2-methyl propane   64
2-chloro-2-methyl propane   17, 63, 64
1-chloropropane   64
2-chloropropane   63, 64
3-chloroprop-1-ene   188
chlorophyll   1, 50, 122, 242, 243,
chlorophyll a   244, 282
chlorosulphonic acid   129
chlorotetracycline   87
cholecalciferol   255
cholesterol   200, 260, 275
choline   125
chorismic acid   277, 278

chromatography 43–45
  column 44
  gas 45, 229
  gel permeation 44
  high performance liquid 44
  ion exchange 44, 227
  paper 43, 225
  thin layer 44
chromic acid 113
chromosome 221, 240
cinnamaldehyde 143
cinnamic acid 278, 279
cinnamon 143
citrate ion 163
citric acid 2, 161, 162, 269, 270
  cycle 248, 264, 265, 266, 276
citronellol 273, 275
citrus fruits 251
Claisen condensation 161, 162, 173,
    174, 183, 185, 189, 191, 272, 281
Clemmenson reaction 90, 165
coal tar 19
cocaine 136
codon 239
coenzyme A 102, 161, 254, 255
coenzyme B$_{12}$ 111, 232, 253
coenzyme Q 256, 257
coenzymes 232, 250
coke 75
configuration 22, 104
conformations, anti 35
  eclipsed 35
  gauche 35
  of molecules 35–38
  staggered 35
conformers 35
  boat 36–37
  chair 36–37
  crown 37–38
  envelope 37–38
  skewboat 36–37
Congo red 135
coniferin 209, 210
coniferyl alcohol 278, 279
conjugated dienes 177–180
  diynes 73
  systems 11
  molecules 49
  proteins 231
contraceptive pill 261
copper acetylide 73
copper(I) bromide 135
copper(I) chloride 134
copper(II) chloride 168

copper(I) cyanide 135, 139
copper(I) halides 117
copper(I) oxide 163
corn 218
corpus luteum 258, 260
corrin 243, 245
cortisone 260
cotton 217
coumarins 265, 278
Couper 2
covalent bonds 7
cows 217
Crick 239
Crowfoot Hodgkin, D. 253
crude oil 19, 74
crystallization 40
cumere 118
cumulated double bonds 175
curly arrows 15
cyanide ion 248
cyanocobalamin 251, 253
cyanohydrins 157, 158, 214
cyclic hydrocarbons 59
cycloalkanes 60
cyclobutane 61
cyclohexane 168, 207
cyclohexanone 164, 168
cyclohexene 59, 64
cyclopentane 61
cyclopentanone 164
cyclopentene 61
cyclopropane 59, 60, 61, 62, 66
cis-cyclopropane-1,2-dicarboxylic
    acid 23, 24
trans-cyclopropane-1,2-dicarboxylic
    acid 23
cyclopropanone 51
cysteine 102, 222, 276, 280
cystidine 237
cystine 222
cytidylic acid 236, 237
cytochrome 242, 248, 282
cytochrome c 244
cytosine 236, 237, 238, 239

Dacron® 96, 143, 191
dairy products 251
DDT 79, 80, 100, 101
decoupling (n.m.r.) 54
deer 217
dehydration 118, 139
dehydroascorbic acid 254, 255
dehydroquinic acid 277
delocalization 82, 104–105

denaturation 232
density gradient centrifugation 227
deoxyadenosine 236, 237
deoxyadenylic acid 237
deoxycytidine 237
deoxycytidylic acid 237
deoxyguanosine 237
deoxyguanylic acid 237
deoxyribose 204, 236, 237, 269
2-deoxyribose 208
dermatitis 251
deshielding effect 53
detergents 202
deuteriochloroform 52
diabetes mellitus 259
diamines 174, 188
diastase 117
diastereosisomers 28, 33, 34, 213
diazonium ion 127
  salt 116, 117, 133
trans-1,2-dibromocyclohexane 64
dicarbonyl compounds 174, 176,
  182–183, 185, 186, 187, 189,
  190
dichlorodifluoromethane 100, 101
cis-1,2-dichloroethene 22
trans-1,2-dichloroethene 22
dichloromethane 62, 100, 101
dicyclohexylcarbodiimide 154, 233,
  234
dicyclohexyl urea 234
dicldrin 178, 179
Diels–Alder reaction 178, 179
dienes 174–180
1,4-dienes 185
dienophile 178
diethyl ether 60, 101, 102
  malonate 181, 186
  oxalate 183
digestion 150
Digitalis sp. 210
digitoxigenin 211
digitoxin 210
dihalides 174, 188
dihalogenoalkenes 65
dihydroxyacetone phosphate 246, 247,
  266, 267
1,2-dihydroxybenzene 114
1,4-dihydroxybenzene 114
(−)-2,3-dihydroxybutanedioic acid 27
1,2-dimethoxyethane 103
dimethylallyl pyrophosphate 273, 274
dimethylamine 123, 124
N,N-dimethylaniline 123, 124

2,4-dimethyl cyclobutane-1,3-
  dicarboxylic acid 25
cis-1,3-dimethylcyclopentane 23
trans-1,3-dimethylcyclopentane 23
dimethyl ether 2
dimethyl formamide 144
N,N-dimethylmethanamide 144
2,2-dimethylpropane 61
dimethyl terephthalate 191
2,4-dinitrofluorobenzene 228
2,4-dinitrophenylhydrazine 155, 180
2,4-dinitrophenylhydrazones 155
diols 174, 188, 189
1,1-diols 151
1,2-diols 69, 70
dipole 11
dipole moment 18
directing effects (of aromatic
  substituents) 92
distillation 41–42
disulphides 115
diterpenes 273, 274
Domagk, G. 169
double helix 238, 239
doublets, 53
drugs 121
Du Vigneaud 234

Edman 229
eggs 251
E isomer 32–33
electron 3, 6
  density 5
  diffraction 3, 56, 79
electronegativity 10–11
electronic effects 17
  theory 3
electrophiles 15, 64, 135, 137, 173,
  186
electrophilic attack 66, 70
  substitution 84–95
electrophoresis 46, 225, 227
elimination 17, 75, 108–109, 125, 175
eluate 44
E_1 mechanism 108–109
E_2 mechanism 108–109
emulsin 216
enamine 174, 184
enantiomers 24
  resolution of 33–34
end group analysis 228
endocrine glands 257
enediols 211
enolate anion 147, 160

enols   147, 174, 184
enzymes   150, 209, 210, 216, 221,
    232, 240
ephedrine   279, 280
epinephrine   259
epoxides   70, 111, 118
equatorial position   37
equilibrium   151, 152
ergotamine   279
erythromycin   261, 262
D-erythrose   206
erythrose-4-phosphate   246, 247, 268,
    269, 276, 277
Eschenmoser   253
essential amino acids   223
esters   13, 51, 130, 139
ethanal   149, 183
ethan-1,2-diol   152
ethane   2, 18, 35, 61, 74
ethanoic acid   144, 159
ethanoic anhydride   214
ethanol   1, 2, 100, 117, 152, 165, 183
ethanoyl chloride   146, 225
ethene   9, 15, 50, 61, 72, 75, 168, 177
ether   42, 13
ethers   106, 107, 112, 184
ethyl acetate   42, 144
ethyl acetoacetate   161, 186, 187
ethylbenzene   88, 102
S-ethyl ethanethioate   144
ethyl ethanoate   144
ethyl 3-oxobutanoate   161
ethyl thioacetate   144
ethylene glycol   100, 101, 152, 191,
    192
ethylene oxide   101, 102
ethyne   10, 61, 69, 75

farnesyl pyrophosphate   274
fats   100, 126, 143, 219, 251, 259, 265
fatty acids   270, 271, 272
Fehling's reagent   163, 204, 205, 211
fermentation   19, 117
fertilization   260
fertilizer   132
fibrous proteins   231
filtration   40
fingernails   221
Fischer projection formulas   26–28,
    205
fish oils   251
flame ionization detector   45
flavin adenine denucleotide
    (FAD)   251, 252, 269, 270, 271
reduced form (FADH$_2$)   248, 249,
    269
flavin mononucleotide (FMN)   251,
    252
    reduced form (FMNH$_2$)   248,
        271
flavonoids   265
flavourings   147
flax   217
fluorine   63
fluorobenzene   134
folic acid   251, 254, 261
formaldehyde   102, 110, 151, 162,
    192, 212
formic acid   159, 160, 212
formula, empirical   47
    molecular   47
fractional distillation   41–42, 74
fractionating column   41
fragmentation patterns   48
free radical chain process   63, 97
free radicals   14
Friedel Crafts reaction   88, 89, 90, 97,
    166
α-D-fructofuranose   208
β-D-fructofuranose   208
α-D-fructopyranose   208
β-D-fructopyranose   208
fructose   204, 208, 266, 267
D-fructose   205
fructose-1,6-diphosphate   266, 267
fructose-6-phosphate   246, 247, 267,
    268, 269
fructoside   209
fuel oil   74
fumaric acid   22, 270
functional groups   12–13, 51
furan   85
furanose structure   208

D-galactose   206, 215
gallic acid   114, 115, 278
gas chromatography   198
Gattermann–Koch reaction   90
gene   240
geranyl pyrophosphate   273, 274
geranylgeranyl pyrophosphate   273,
    274
gliotoxin   116, 117
globin   232
globular proteins   231
glucagon   258, 259
D-glucaric acid   212
D-gluconic acid   212

gluconic acid-6-phosphate   267, 268
glucose   117, 153, 183, 204, 209,
    219, 258, 259, 264, 265, 266, 267
D-glucose   163, 205, 206, 207, 209,
    212, 213, 214, 215, 216
α-D-glucose   205, 207
β-D-glucose   205, 207
glucose-6-phosphate   267, 268
α-glucoside   209
β-glucoside   209
glutamic acid   223, 276
glutaric acid   164
glutathione   226, 227
D-glyceraldehyde   205, 206, 224
L-glyceraldehyde   224
(+)-glyceraldehyde   30–32
(−)-glyceraldehyde   30–32
glyceraldehyde-3-phosphate   166, 246,
    247, 248, 266, 267, 268, 269
glyceric acid-1,3-diphosphate   166
glyceric acid-3-phosphate   246, 247, 267
glycerol   100, 101, 143, 191, 196, 270
glycine   222, 223, 224, 230, 276, 281
glycogen   218, 219, 258, 259
glycolysis   265, 270
Grignard reagents   73, 75, 110–111,
    118, 156, 157, 159, 166
guanine   236, 237, 238, 239
guanosine   237
guanylic acid   237
D-gulose   206

haem   122, 232, 244, 248, 282
haemoglobin   232, 242, 244
hair   221
halogenation   62–66, 87
halogenoalkanes   75
halogenoalkencs   174
halogenocarbonyl compounds   174,
    188
halogens   62
halohydrins   118
hardening (of vegetable oils)   185, 200
Haworth formulas   207
α-helix   230, 231
hemiacetal   152, 153, 190, 205, 209
hemiketal   152, 153, 190, 209
heptane   97
heredity   221
hexan-1,6-diamine   192
hexan-1,6-dioic acid   164, 192
hexose   205
hexose monophosphate shunt   265
hides   115

Hinsburg test   126
histidine   138, 223
Hofmann elimination reaction   125
    rearrangement reaction   132, 139,
    176
homologous series   59, 60
hormone, interstitial cell
    stimulating   258, 260
    follicle stimulating   258, 260
    sex   143, 196, 260, 261
hormones   221
Hückels rule   83
human fat   198
Hund's rule   6
hybrid orbitals   7
hybridization   7–10
hydration   117
hydrazine   165
hydrazobenzene   133
hydrazones   128, 130, 139
hydriodic acid   107
hydroboration   68, 111, 117
hydrobromic acid   107
hydrochloric acid   150, 154, 159, 165,
    214, 227
hydroformation   97
hydrogen   6, 8, 168
hydrogen bonds   18, 122, 231
hydrogen bromide   235
hydrogen chloride   158, 175
hydrogen cyanide   157, 210, 214, 225
hydrogen halides   66, 117
hydrogen peroxide   68
hydrogenation   75, 80
hydrolase   232
hydrolysis   100, 116, 118, 131, 132,
    140, 150, 166, 176, 204, 209, 232,
    235, 238
hydroxyazobenzene   116
3-hydroxybenzene-1,2-dicarboxylic
    acid   96
3-hydroxybutanal   161
hydroxycarbonyl compounds   174, 188
4-hydroxycinnamic acid   278, 279
β-hydroxyesters   111
3-hydroxyindole   209
hydroxylamine   140, 155, 187
hydroxylation   95
2-hydroxy-3-methylbutanedioic acid   28,
    32
hydroxyproline   136, 223
(−)-2-hydroxypropanoic acid   23
(+)-2-hydroxypropanoic acid   23
hyperconjugation   92

D-idose 206
imidazole 138
imines 128, 130, 139, 155
indican 210
indigo 1
indigotin 209
inductive effect 11
initiation step 63
insulin 258, 259
intramolecular reactions 153
inversion 216
iodine 62
iodine value 199
 number 199
iodobenzene 134
iodoform reaction 163
ionic bonds 7
 bridges 231
ions, fragment 48
 molecular 48
i.r. spectra, of alcohols, phenols &
 chloroalkanes 103
 of alkenes 60
 of alkynes 60
 of amines 123
 of carbonyl compounds 147
iron(III) chloride 187
isocitric acid 270
isocyanates 132, 175, 176
isoelectric point 224
isoleucine 222
isomerism, geometrical 21–23, 60
 optical 21, 23–34
isomers 2
 cis 22
 meso 28
 optical 24
 trans 22
iso-octane 74
isopentenyl pyrophosphate 89, 273,
 274
isoprene 178, 179, 273
isotactic polymerization 72
isothiocyanate 175, 176

jaundice 242
jute 217

Kekulé 2, 79, 82
kerosine 74
ketals 152, 209
ketene 175
β-keto acids 186, 191
keto-enol tautomerism 147

β-keto esters 186
α-ketoglutaric acid 187, 270, 276
ketohexose 205
ketones 13, 51, 110, 111, 130, 131,
 139, 185
ketose 205, 211, 213
Krebs cycle 269
kwashiorkor 223

lactation 258
lactic acid 2, 158, 165, 183, 266,
 267
(+) lactic acid 23, 24, 26, 27, 32, 33,
 34
(−) lactic acid 23, 26, 33, 34
lactones 190
lactotropin 258
lactose 204, 215, 216
lanolin 203
lard 197, 198
Lavoisier, A. 2
Le Bel 2
lead tetraethyl 74
leather 115
leaving groups 103, 264
lecithins 125, 197
leucine 222
Lewis acid 87, 88
Librium® 262
light petroleum 60, 74
lignans 265, 278
lignin 114, 265, 278
linoleic acid 197, 198, 200
linolenic acid 197, 198, 200
linseed oil 185
lipids 196
lipoic acid 251, 252
lithium aluminium hydride 34, 131,
 164, 165
liver 180, 251
lock and key model 232, 233
lone pair electrons 8, 11
LSD 122
lubricating oil 74
lycopene 180
lysergeric acid 279
lysine 223, 279
D-lyxose 206

maleic acid 22
 anhydride 22
malic acid 270
malonic acids 186
malonic esters 186

malonyl coenzyme A   186, 187, 256,
    271, 272, 273, 279
malt   117
maltase   216
maltose   117, 204, 215, 216, 218
mandelonitrile   158, 209
D-mannose   206
margarine   71, 196, 199, 200
mass spectrometry   47–49, 229, 230
mass spectrum   48
meat   251
mechanisms of organic
    reactions   14–16
melanin   114
melting points   17, 19
menstrual cycle   260
menstruation   260
mercury(II)acetate   67, 68
    sulphate   68
Merrifield   235
mesomeric effect   11
messenger RNA   239
metal alkynes   156
methanal   102, 110, 151, 212
methane   3, 59, 61, 62, 75
methanoic acid   212
methanol   2, 18, 100, 101, 122, 152,
    209, 225
methionine   89, 222, 282,
methoxybenzene   91, 94, 102
methoxybenzoic acid   160
methoxyethane   102
methyl acetate   146, 157
methylamine   122, 123, 124
N-methylaniline   123, 124
methylbenzene   95
methylbenzenes   166
S-methyl benzothioate   153
2-methylbuta-1,3-diene   178, 273
2-methylbutane   61
3-methylbut-1-ene   68
α-methyldopa   262
methylene blue   42
methyl cation   16
methyl chloride   62
methyl ethanoate   146
methyl formate   53
methyl-α-D-glucopyranoside   209
methyl-β-D-glucopyranoside   209
6-methyl-2-hydroxybenzoic acid   96, 97
methyl iodide   214
2-methyl-1,4-naphthoquinone   256, 257
4-methylphenol   102
N-methyl propanamide   144

2-methylpropane   61, 63, 64
2-methylpropene   61
methyl salicylate   143
methyl thiolbenzoate   153
mevalonic acid   265
micelles   201, 202
Michael reaction   181
milk   251
mitochondria   248
mixed melting points   19
models, ball & stick   4
    space filling   4
molecular orbital theory   5–7, 81–82
monoterpenes   273
morphine   1, 114, 122, 263, 279, 280
mull   52
multiplet   53
multiplicity   53
muscle   221
muscone   143
musk deer   143
mutarotation   207
myristic acid   197, 198

naphthalene   95
natural gas 74
Newman projection formulas   35
nicotine   136, 279, 280
nicotinic acid   280
nicotinimide   251, 253, 254
nicotinimide adenine dinucleotide
    (NAD+), 121, 122, 166, 253, 266,
    267, 269, 270, 271, 272
nicotinimide adenine dinucleotide,
    reduced form (NADH), 166, 183,
    248, 249, 266, 267, 269
nicotinimide adenine dinucleotide
    phosphate (NADP+) 246, 253,
    268, 270
nicotinimide adenine dinucleotide
    phosphate reduced form (NADPH)
    246, 247, 268, 269, 271, 274
ninhydrin   189, 225
nitration   85, 140
nitrenes   132
nitric acid   85, 86, 115, 137, 164, 212
nitriles   13, 106, 111, 131, 139, 140,
    166
nitrobenzene   85, 91
4-nitrobenzoic acid   160
nitro-compounds   13
nitromethane   133
nitronium ion   86
N-nitrosamine   127, 128

nitrosobenzene 133
N-nitrosoammonium ion 127
nitrosonium ion 127
nitrous acid 127, 140
nomenclature (of stereoisomers) 30–33
non-bonding electrons 8, 49
noradrenaline 258, 259, 262
norepinephrine 259
nucleic acids 121, 122, 138, 265, 269
nucleophile 14, 15, 73, 103–105, 112, 124, 139, 145, 148, 149, 182, 186, 273
nucleophilic addition 148, 161, 173, 182
   attack 131, 164, 180, 181
   substitution 124, 148, 161
nucleoproteins 240
nucleoside 236
nucleotide 236
nylon 143, 164, 168, 192

octane 97
octane rating 74
all-trans-octa-2,4,6-triene 50
oestradiol 261
oestrogens 258, 260, 261
oestrone 261
oil, coconut 197
   cod liver 197, 198
   corn 197, 198
   cottonseed 197
   linseed 197, 198, 200
   of wintergreen 143
   olive 197, 198
   rapeseed 197
   sunflower 197
   whale 197
oils 143, 168, 185
oleic acid 197, 198
oligosaccharide 204
opsin 166, 167
optical activity 23, 26
   isomerism 23–34
   rotatory dispersion (o.r.d.) 26
orbital 5, 6
organoboranes 111
organocadmium reagents 111
organolithium reagents 111
Orlon® 72
ornithine 190, 191, 279, 280
orsellinic acid 273
osazones 213
osmium tetroxide 69, 70

osones 213
ovulation 258, 260
oxaloacetic acid 161, 269, 270, 276
oxalosuccinic acid 187, 269, 270
oxidation 117, 118, 121, 164, 166, 167, 198, 211, 264
oximes 128, 130, 139, 155
OXO process 168
oxygenase 116
oxyhaemoglobin 248
oxytocin 234, 258, 260
ozone 69, 70, 179
ozonolysis 70

paint 196
palladium(II)chloride 168
palmitic acid 197, 198
Paludrin® 79, 80
pantothenic acid 254, 255
papaverine 102
paracetamol 262, 263
parathyrin 258
partition coefficient 42
Pauling 230
peas 251
pellagra 251
penicillin 190, 261, 265
penicillin G 190, 280
penicillin N 281
penicillin V 280
Penicillium glaucum 34
penta-2,3-diene 29
pentane 61, 102, 146
pentanedioic acid 164
pentan-1-ol 103
pentose 205
pentose phosphate cycle 267
peptides 154, 176, 265
peracids 118
perbenzoic acid 70
perfumes 147
pericyclic reactions 178
periodic acid 212
periodic table 10
pernicious anaemia 251
perspiration 146
pesticides 100
petroleum 74, 97
phenanthrene 85, 95
phenol 13, 101, 102, 107, 134
phenol formaldehyde polymers 192
phenoxides 106, 112, 116, 118
phenoxymethylpenicillin 280
phenylacetic acid 280

phenylalanine   95, 222, 276, 277, 278, 279, 280
phenylcyanide   134
N-phenylethanamide   129
phenylhydrazine   128, 187, 213
phenylhydrazones   128, 213
phenylhydroxylamine   133
phenyl isocyanate   176
phenyl isothiocyanate   229
phenyl magnesium bromide   157
phenyloxyacetic acid   280
phenylthiohydantoin   229
phloracetophenane   273
phosphate esters   106
phosphoenol pyruvic acid   267, 269, 276, 277
phospholipids   196, 197
phosphonamides   126
phosphonyl chlorides   126
phosphoric acids   108, 117, 235
phosphorus halides   107, 117
photosynthesis   204, 265, 266, 268
phthalic acid   192
phthalocyanin   242, 245
phytoene   274
phytol   282
pi (π) bonds   9, 61, 81, 144, 145, 180
   electrons   64
   orbitals   9, 81
picric acid   112
α-pinene   274, 275
piperidine   136
pleated sheet   230, 231
polaroid   25, 26
polyacrylonitrile   72
polyamides   143, 192
polyesters   70, 143, 191
polyethene   72
polyisoprene   179
polyketide   161, 186, 187, 265
polymerization   178, 179
poly(methylmethacrylate)   72
polypropene   72
polypropylene   72
polysaccharides   265
polystyrene   72, 79, 80
polytetrafluoromethane   72
polythene   1
polyurethane   176, 192, 201
polyvinylchloride   72
porphin   242, 245
porphobilinogen   281, 282
porphyria   242, 281
porphyrin   242, 243, 244, 248

potassium dichromate   163
   hydroxide   125, 199
   iodide   117, 135
   permanganate   60, 69, 70, 80, 96, 163, 164, 166
potatoes   218
pregnancy   258, 260
prephenic acid   277
prism, Nicol   25
progesterone   258, 260, 261
proline   136, 190, 191, 223
propagation step   63
propanamide   146
propane   3, 5, 59, 61, 63, 64, 74
propane-1,2,3,-triol   191
propanoic acid   23, 146
propanone   68
propene   59, 60, 61, 67, 72
propionyl coenzyme A   271
propyl anion   12
propyl cation   12
1-propyl radical   64
2-propyl radical   64
propyne   59, 61, 62, 67, 68
prostaglandin E_2 272
prostaglandins   143
prosthetic group   231, 252
proteins   1, 121, 122, 126, 131, 143, 150, 154, 176, 265
protoporphyrin IX   282
purine   121, 138
pyranose structure   208
pyridine   83, 84, 85, 121, 136, 137, 138
pyridoxal-5-phosphate   155, 253, 276, 277
pyridoxamine-5-phosphate   276, 277
pyridoxine   251, 253
pyrimidine   138
pyrolysis   19
pyrrole   83, 84, 85, 121, 136, 137
pyrrolidine   136, 137
pyruvic acid   117, 166, 183, 265, 266, 267, 269

quantum mechanics   3, 5
quaternary ammonium hydroxides   125
   salt   125
   structure (of proteins)   231
quercetin   279
quinine   122
quinones   182, 249
quinuclidine   121

racemic mixture 26
Raney nickel 165
rate constant 17
rate determing step 14
rate equation 16
reaction centre chlorophyll 245
reaction coordinate 14
reaction mechanism 14
rearrangement reaction 17
recrystallization 40
reducing sugars 211
reduction 118, 121, 164, 264
Reformatsky reagent 111, 156, 157
refridgerants 100
relaxin 258, 260
resonance 145
resonance hybrid 11, 160
resonance theory 81,
resorcinol 101, 102
respiratory chain 256, 264
  cycle 265, 266, 268
reticuline 114, 279, 280
retina 180
11-*cis*-retinal 166, 167
trans-retinal 166, 167
retinene 180
retro-aldol reaction 266
R_f value 43
rhodopsin 166, 180
riboflavin 251, 252
ribose 236, 237, 269
D-ribose 206
ribose-5-phosphate 247, 248, 268,
ribulose-5-phosphate 246, 247, 248,
  267, 268
ribulose-1,5-diphosphate 247
rice 218, 251
rickets 251
rosehips 251
Rosenmund reduction 165
rubber 179, 273

saccharin 88
Sandmeyer reaction 135
Sanger 228
saponification number 199
  value 199
saturated carbon atom 3
  compounds 59
Scheele, C. W. 2
Schiffs base 128, 130, 155, 276
scurvy 25i
secondary structure (of proteins)
  230

sedoheptulose-7-phosphate 247, 248,
  268
semicarbazide 128, 155
semicarbazones 155
Sephadex® 44
serine 222, 276
D-serine 224
L-serine 224
sesquiterpenes 273, 274
shielding effect 53
shikimic acid 97, 265, 269, 277, 278
sigma (σ) bonds 7, 61, 81, 144, 145
silicones 201
silk 230
silver oxide 214
singlets 53
skin 221
S_N2 attack 188
S_N1 mechanism 103–105, 134
S_N2 mechanism 103–105
soap 1, 196
sodamide 73
sodium 165
  alkyl benzene sulphonate 202
  alkyl sulphonate 202
  amalgam 213
  borohydride 67, 164, 165, 181,
    213, 246
  cyanide 139
  ethoxide 161, 181
  hydrogen carbonate 159
  hydroxide 125, 154, 159, 161
  nitrite 127
solid state synthesis (of proteins) 234
soluble RNA 239
somatotropin 258
sorbitol 212, 213
soya bean 251
sp hybrid atoms 62
sp hybrid orbitals 10
sp^2 hybrid atoms 62
sp^2 hybrid orbitals 9
sp^3 hybrid atoms 62
sp^3 hybrid orbitals 8
spectrophotometer 227
spectroscopy, i.r. 50–52
  n.m.r. 52–55, 83, 84
  u.v. 49–50
spermaceti wax 202, 203
sphingolipids 196, 197
sphingomyelin 197
spin quantum numbers 6
squalene 274
starch 204, 217, 218, 246

stearic acid   197, 198
steroids   196, 257, 260, 261, 265, 275
sterols   255
stomach   150
Strecker synthesis   158, 225
streptidine   211
streptomycin   211
styrene   88, 89
succinic acid   269
succinyl coenzyme A   270, 271, 281
sucrose   204, 215, 216, 246
sugar beet   215
sugar cane   215
sugars   153, 157, 235, 265
sulphanilic acid   128
sulphanilamide   129
sulphonamides   126, 169, 261
sulphonate esters   106, 113
sulphonation   87–88
sulphones   115
sulphonic acids   115
sulphonyl chlorides   126, 127
sulphoxide   115
sulphur dichloride oxide   107, 158
sulphur dioxide   158
sulphuric acid   66, 68, 85, 86, 87, 108,
      118, 128, 137, 205

tannins   115, 278
(+)-tartaric acid   28, 33
(−)-tartaric acid   28, 33
meso-tartaric acid   28
tartrate ions   163
D-talose   206
tautomerism   147
tautomers   187, 188
tea   115
Teflon   72
template   239
terephthalic acid   96
termination step   63
terpenes   265
γ-terpinene   275
tertiary structure (of proteins)   231
Terylene®   70, 96, 100, 143. 191
testosterone   258, 260
tetrachloromethane   3, 60, 62
tetracycline   108, 261
tetrafluoroethene   100, 101
tetrahedron   2
tetrahydrofolic acid   254
tetrahydrofuran   101, 102, 103
tetramethylsilane   53
tetrapyrroles   265

thermal cracking   74
thiamine   251
thiamine pyrophosphate   251
thioethers   106
thiolesters   153
thioketal   153
thiols   13, 112, 175
thionyl chloride   117, 158
thiophen   85
thiourea   229
threonine   222
D-threose   206
thymidine   237
thymidylic acid   237
thymine   236, 237, 238
thyrotropin   258
thyroxine   257, 258
tin(II)chloride   135
tocopherols   251, 256
Tollens reagent   163, 205, 211, 216
toluene   95, 97
tomatoes   180, 275
tranquilizers   262
transfer RNA   239
transition state   14
trialkylboranes   68
2,4,6-tribromoaniline   128, 129, 135
1,3,5-tribromobenzene   135
2,4,6-tribromophenol   116
tricarboxylic acid cycle   269
trichloromethane   3, 42, 62, 100, 101
2,4,6-trichlorophenol   101, 102
trifluoroethanoic acid   235
triglycerides   196–200, 269
trimethylamine   123
2,4,6-trinitrophenol   112
triplets   53
triterpenes   274
tryptophan   222, 278, 279
Tyrian purple   1
tyrosine   95, 222, 276, 277, 278, 279,
      280

ubiquinone   248
umbelliferone   278
αβ-unsaturated carbonyl
      compounds   176, 180–182, 186
αβ-unsaturated nitriles   182
uracil   236, 237
urea   2, 132, 232
urea formaldehyde plastics   132
urethanes   176
uridine   237
uridylic acid   237

urine   163, 259
uroporphyrinogen III   282
uterus   234, 260
u.v. spectra   177

valency   2
valine   222, 280
Valium® 262
Van der Waals forces   18, 60
vanillin   79
Van't Hoff   2
vasopressin   234, 235, 258, 260
vegetable oils   71
vinegar   1
vinyl chlorides   100, 101, 184
'vinyl' polymers   201
viscose   217
vitamin A   156, 180
11-*cis*-vitamin A   166, 167
trans-vitamin A   166, 167
vitamin B$_1$ 138
  B$_6$ 155
  B$_{12}$ 108, 111, 242, 243, 282
  D   259
  D$_2$ 255
  D$_3$ 255
  E   79, 80, 273
  K$_1$ 182, 257
  K$_2$ 257, 273

Von Berzelius, J. J. F.   2
vulcanization   179

Wacker process   168
Warfarin® 278
watermelons   180
Watson   239
waxes   196
wheat   218
wheatgerm   251
Wilkins   239
Wohler   2
Wolff–Kischner reaction   90, 165
Woodward, R. B.   253

xanthate esters   217, 218
X-ray crystallography   30, 56, 230, 232
  diffraction   3, 55–56, 79, 253
xylene   97
xylose   204
D-xylose   206
xylulose-5-phosphate   246, 247, 248,
  268, 269

yeast   117, 183, 251

Z-isomer   32–33
zinc   70, 165
zinc amalgam   166
zwitterion   224